APPLIED OPTICS

AND

OPTICAL DESIGN

A. E. CONRADY

Edited and completed by
RUDOLF KINGSLAKE

IN TWO PARTS
PART TWO

DOVER PUBLICATIONS, INC.
NEW YORK

Published in Canada by General Publishing Company, Ltd., 30 Lesmill Road, Don Mills, Toronto, Ontario.
Published in the United Kingdom by Constable and Company, Ltd., 3 The Lanchesters, 162–164 Fulham Palace Road, London W6 9ER.

This Dover edition, first published in 1992, is a reissue of the edition first published by Dover in 1957 (Part I) and 1960 (Part II). The Dover edition of Part I was an unabridged and corrected republication of the work originally published by the Oxford University Press, London, in 1929. Part II was originally published by Dover Publications, Inc., in 1960; this part was a posthumous publication edited and completed by Dr. Rudolf Kingslake (then Director of Optical Design, Eastman Kodak Company, Rochester, New York) with the assistance of Dr. Fred H. Perrin of the Eastman Kodak Company Research Laboratories.

Manufactured in the United States of America
Dover Publications, Inc., 31 East 2nd Street, Mineola, N.Y. 11501

Library of Congress Cataloging in Publication Data

Conrady, A. E. (Alexander Eugen)
    Applied optics and optical design / A. E. Conrady ; edited and completed by Rudolf Kingslake.
        p.   cm.
    Includes bibliographical references and index.
    ISBN 0-486-67007-4 (pbk. : v. 1). — ISBN 0-486-67008-2 (pbk. : v. 2)
    1. Optics.   2. Optical instruments.   I. Kingslake, Rudolf.   II. Title.
QC371.C62   1991
535—dc20   91-33198
                                                                        CIP

THIS VOLUME IS DEDICATED
BY HIS DAUGHTER HILDA TO
PROFESSOR CONRADY'S STUDENTS
THROUGHOUT THE WORLD,
WITHOUT WHOM TEACHER AND
WRITER SPEAK IN VAIN

FOREWORD

WHEN Professor Conrady, my father, died in 1944 at the age of 78 years, he left an incomplete manuscript of Part II of his book *Applied Optics and Optical Design.* Years of poor health aggravated by the difficult conditions of World War II in London had made it too difficult for him to finish his task.

The disposition of his papers was left to the judgment of the family, and the optical world is indebted to his youngest surviving daughter, Irene, who painstakingly collected and preserved for the remainder of the war all papers she could find, making duplicates where there were none, and later obtaining the necessary Government permission to send a copy of the work abroad for our study and care.

After considerable thought and the receipt of one or two generous offers to go through his papers, I finally decided that, quite apart from any personal relationship, my husband, Rudolf Kingslake, was really his best qualified disciple to edit this work. Many of Professor Conrady's students have become very capable lens designers in industry, while others have remained largely academic or have branched into fields of optics other than lens design. Professor Conrady attached great importance to the valuable combination of a thorough academic training with years of actual experience in industry, knowing only too well certain inevitable limitations of the purely academic, however gifted, and the much more serious limitations of the untrained self-taught man in scientific industry. Twenty years of experience in optical industry, mostly in the United States, following ten years of University teaching, combined with almost thirty years of lecturing in lens design, all of which was based largely on Professor Conrady's work, have proved invaluable to my husband in the editing and preparing of this work for publication.

I leave it to Dr. Kingslake to be specific in his Preface with regard to the handling of the manuscript. Many hours have been spent in this exacting labour, and it is hoped that this resulting Part II will be as Professor Conrady would have wished it, and that his students all over the world who have waited so long will find a useful and worthy sequel to Part I.

This being a posthumous publication, it was thought fitting to append a brief biography of my father, whose life was rather unusual, together with a list of his scientific papers.

HILDA G. CONRADY KINGSLAKE

EDITOR'S PREFACE

IN attempting to reconstruct what Professor Conrady most probably intended to publish as Part II of his book *Applied Optics and Optical Design*, our principal source was the typewritten notes that he issued to students between 1921 and his retirement in 1931. He had expanded and partly rewritten large sections of these notes, which are reproduced here virtually unchanged; the only alterations are editorial in character and were made necessary by the fact that some sections of the original Part II lecture notes had already been printed in the published edition of Part I. These typewritten notes constitute the entire chapters on optical path differences (XII, XIII, and XVII), on optical tolerances (XIV), on the $(d' - D')$ method of achromatism (XV), on microscope objectives (XVI), and on symmetrical photographic objectives (XIX). Most of the new Chapter XI was left by Conrady, handwritten in its final form ready for publication, bearing the heading 'Chapter XI' and continuing the numerical series of section numbers and figure numbers from the end of Chapter X in Part I. Unfortunately the manuscript ended abruptly near the conclusion of section [87]; but enough data were given to enable that section to be completed.

The remaining source of manuscript was Conrady's personal set of handwritten notes covering a supplementary course of lectures which he gave to senior students during 1925 and 1926. Much of this material had already been included in Chapter VI of the published Part I, and there was some degree of repetition in the 1926 series, but these notes provided almost the only source of material on photographic objectives. After slight editing and rearranging, and with some changes in notation so as to be consistent with the rest of the book, these notes now constitute section [88] on photovisual objectives, sections [115], [116], and [117] on simple achromatic eyepieces, section [119] on anastigmatic landscape lenses, sections [122] and [123] on symmetrical anastigmats, and most of sections [124] and [125] on the Petzval portrait lens.

Conrady's published scientific papers contain much material which he naturally incorporated into the text of this book; an exception is the path-difference derivation of the optical sine theorem, which has been reprinted in section [96] of Part II by kind permission of the Royal Astronomical Society.

Professor Conrady wrote always for the student and for the reader who would teach himself the technique of optical design by his own personal effort. Feeling that such a reader would be disappointed to find only uncompleted numerical examples to illustrate the methods described in the text, I have undertaken to complete some of these examples and to add others where none existed in the original manuscript. I also took the liberty of writing full instructions for, and worked examples of, the design of a simple telephoto lens and of a Cooke triplet objective. The procedures in these instances follow strictly Conrady's notation and methods, of course, and they are intended to make up obvious omissions which Conrady would undoubtedly have filled himself had he been able to do so. The sections that I have written in accordance with this understanding are the following:

section [87]F, last half, on the Gauss condition; section [88], last part, on photo-visual objectives; section [125], last half, on the Petzval portrait lens; the whole of section [126] on telephoto lenses; and the whole of section [127] on the Cooke triplet lens. I was also responsible for section [113], in which the results of the OPD studies in Part II are related to the closely similar results derived by strictly geometrical methods in Chapter VI of Part I.

After a long period of unchange, the methods used in the design of lenses are today suffering a drastic upheaval as a result of the introduction of, first, the electric desk calculator, and now the electronic digital computer. In Conrady's time a designer would be willing to take almost any trouble to avoid the labour of tracing numerous rays by six-figure logarithms; he would arrange his calculations so that he could extract the greatest possible amount of information from every ray he traced, and he would use many theorems and auxiliary formulae to yield information on the properties of a lens in a region beyond that in which the rays had been traced. Conrady's own contributions to this aspect of the science are considerable: the sine theorem, the $(d' - D')$ method of achromatism, the fundamental laws of oblique pencils, the calculation of OPD along a traced ray, the use of optical tolerances based on the Rayleigh limit, and the Seidel aberration contribution formulae, are noteworthy instances. The analytical design methods given in the new Chapter XI indicate a number of ways by which an approximate solution can be reached quickly without tracing any rays at all, and in all his work Conrady would carry such procedures as far as he could, to the point where the need to consider finite thicknesses and apertures made ray-tracing unavoidable. One feels a sense of regret that the modern designer with his high-speed computing machinery may be tempted to ignore these elegant analytical and approximate solutions in favour of blind trial-and-error ray-tracing. However, it is gratifying to observe that in some computing establishments Conrady's formulae are being rediscovered and found to be highly suitable for employment in the writing of elaborate programmes for the automatic or semi-automatic design of the simpler types of lenses.

My wife has been of very great help to me in editing and completing Part II. She also drew the diagrams for both parts of the book, and appropriately provided the Foreword for Part II.

We are both very grateful to Dr. F. H. Perrin of the Kodak Research Laboratories for his characteristically careful work in critically reading the manuscript and making many valuable editorial suggestions.

We must also express the great indebtedness of the optical world to Mr. Hayward Cirker, president of Dover Publications Inc., for reprinting Part I and for undertaking the publication of Part II.

R. K.

Rochester, New York.
August 1958.

CONTENTS

CHAPTER XI

ADDITIONAL SOLUTIONS BY THE THIN-LENS METHOD
(Axial Correction Only)

IT was shown in Chapters II, V, VI, and VII of Part I that approximate solutions [82] for reasonably thin lens systems satisfying certain conditions can be secured with very little trouble by the simple algebraic formulae which neglect the thicknesses entirely, and that these rough solutions can then be readily corrected trigonometrically by the differential methods which were given for that indispensible final operation.

Our present purpose is not to supersede or modify those methods, but to make valuable additions to them and to obtain an extended survey of all the possibilities and limitations of thin systems by a more complete and general discussion of the equations.

The procedure consists in first fixing the focal length f' of the system, which may necessitate a simple preliminary calculation by the thin-lens paraxial equations. As some type of achromatism is nearly always required, the next step calls for the determination of the strength or net curvature of the component lenses which will secure this. For the most usual case of complete achromatism of a combination of two thin lenses the formulae are

$$c_a = \frac{1}{f'} \cdot \frac{1}{\delta N_a} \cdot \frac{1}{V_a - V_b}; \quad c_b = \frac{1}{f'} \cdot \frac{1}{\delta N_b} \cdot \frac{1}{V_b - V_a}.$$

The spherical aberration and the sagittal coma for central passage of the oblique pencils through the thin system are then given for any distance of the object and for any bending of the constituent lenses by the equations

$$LA' = SA^2 \cdot l'^2_k \cdot \sum(\text{Spherical } G\text{-sums of individual lenses})$$

$$Coma'_s = SA^2 \cdot H'_k \cdot \sum(\text{Coma } G\text{-sums of individual lenses})$$

The G-sums of the individual thin lenses can be worked out in two interchangeable forms to be referred to as the 'original' (i.e. as first proved) and the 'alternative' (as subsequently transformed) G-sum; they are

Original Spher. G-sum $= G_1 c^3 - G_2 c^2 c_1 + G_3 c^2 v_1 + G_4 c c_1{}^2 - G_5 c c_1 v_1 + G_6 c v_1{}^2$

Alternative Spher. G-sum $= G_1 c^3 + G_2 c^2 c_2 - G_3 c^2 v'_2 + G_4 c c_2{}^2 - G_5 c c_2 v'_2 + G_6 c v'_2{}^2$

Original Coma G-sum $= \frac{1}{4} G_5 c c_1 - G_7 c v_1 - G_8 c^2$

Alternative Coma G-sum $= \frac{1}{4} G_5 c c_2 - G_7 c v'_2 + G_8 c^2$

The signs of the terms, and especially the changes of sign in the 'original' and 'alternative' sums must of course be most carefully observed. The total or net curvature of a particular lens $c = c_1 - c_2$ is connected with its focal length by $c(N-1) = 1/f'$. The individual curvatures of the left and right surfaces of the lens are $c_1 = 1/r_1$ and $c_2 = 1/r_2$ respectively. The reciprocals of the left- and right-hand external intersection lengths of the lens standing in *air* are $v_1 = 1/l_1$

[82] and $v'_2 = 1/l'_2$; hence by the thin-lens equations $v'_2 = v_1 + 1/f' = v_1 + c(N-1)$. The eight G-values are pure functions of N with the explicit values (Part I, pages 95 and 324): $G_1 = \frac{1}{2}N^2(N-1)$; $G_2 = \frac{1}{2}(2N+1)(N-1)$; $G_3 = \frac{1}{2}(3N+1)(N-1)$; $G_4 = (N+2)(N-1)/2N$; $G_5 = 2(N^2-1)/N$; $G_6 = (3N+2)(N-1)/2N$; $G_7 = (2N+1)(N-1)/2N$; $G_8 = \frac{1}{2}N(N-1)$. A table of their numerical values and their logs was given in Part I, page 513. Since all the G-values have $(N-1)$ as one factor combined with other very simple terms, there are many simple relations between them, and as these are useful in certain transformations, we shall collect the principal relations here:

$$(1) \begin{cases} G_2 - G_4 - \frac{1}{2}G_5(N-1) = 0; \quad G_3 - \frac{1}{2}G_5 - G_6(N-1) = 0; \\ 3G_1 - 2G_3(N-1) - G_4 + G_6(N-1)^2 = 0; \quad \frac{1}{2}G_5 + 2G_7(N-1) - 4G_8 = 0; \\ G_6 - G_4 = (N-1); \quad G_4 - G_5 + G_6 = 0; \quad G_3 - G_2 - G_8 = 0; \\ 4G_7 - 2G_6 = (N-1); \quad 4G_8 - G_3 = \frac{1}{2}(N-1)^2. \end{cases}$$

In Part I we used the TL equations exclusively for the solution of definite problems, and the complete form in which they are stated above was then the most convenient one and fitted in with the trigonometrical correction. In the present chapter one of our chief aims will be to include the object distance expressed by v_1 or v'_2 as one of the variables, and the $l'_k{}^2$ in the outside factor of the G-sums in the equations for LA' would then become objectionable because l'_k necessarily would vary when v_1 or v'_2 are treated as variable. We shall evade this complication by omitting the outside factors in the equation for both LA' and $Coma'_s$ and by discussing simply the value of the G-sums. This is justified firstly by the fact that interest is nearly always limited to zero-value of both aberrations or, at any rate, to low values of the order of the tolerances, for it is obvious that the aberrations can only become zero or small when the G-sums are zero or small. But there are very good additional arguments—from our present point of view—against the $l'_k{}^2$ in the expression for LA': this term arose out of the transfer of the contributions of individual surfaces to the final image position by the law of longitudinal magnification, and it is thus closely associated with the misleading changes in magnitude of longitudinal spherical aberration according to the convergence of pencils. This becomes clear if we convert LA' into the more reliable measure of angular aberration by (10)****: $AA'_p = LA'_p \cdot u'/l'$. For our present equation u' and l' must be given the suffix k, and u'_k must be put equal to SA/l'_k. Hence we can convert our equation for LA' into AA' by the factor $SA/l'_k{}^2$ and find, for any thin system,

(2) $\quad LA' \cdot SA/l'_k{}^2 = AA' = SA^3 \cdot \Sigma(\text{Spher. } G\text{-sums of individual lenses}),$

which is simply proportional to the cube of the aperture. We obtain a still more significant and useful justification of the objection to the factor $l'_k{}^2$ in the original equation for LA' if we compare the latter with its tolerance according to OT(2) (see page 137 of Part I). Evidently the ratio $LA'/Tolerance$ is a direct measure of the real seriousness of the aberration, a value of unity meaning that the aberration is just at the Rayleigh limit, values over unity, that the aberration is decidedly serious, values under unity, that the aberration is comparatively harmless. The equation OT (2) is

$$LA'\text{-tolerance} = 4\lambda/N' \sin^2 U'.$$

For our thin systems in air we must put $N' = 1$, and, as we are restricted to [82] primary aberration, we must put $\sin U'_k = SA/l'_k$. Hence LA'-tolerance $= 4\lambda \cdot l'_k{}^2/SA^2$, and if we now form the ratio that measures the real seriousness of the aberration, we find

(3) $LA'/Tolerance = (SA^4/4\lambda) \cdot \Sigma(\text{Spher. } G\text{-sums of constituent lenses}),$

the factor $l'_k{}^2$ again cancelling out.

When we have obtained a solution in terms of the spherical G-sums only, we can find the corresponding angular aberration by applying simply the factor SA^3, or we can determine the seriousness of the aberration by applying the factor $SA^4/4\lambda$, or we can apply the original factor $SA^2 \cdot l'_k{}^2$ to find the ordinary LA' for direct comparison with a trigonometrical calculation or with a prescribed amount of longitudinal aberration.

In the case of coma we shall also discuss the G-sums only, although no direct difficulty renders this necessary. Any residual value of this sum found for a particular solution can then be turned into $Coma'_s$ by the factor $SA^2 \cdot H'_k$ or into OSC' by the simple factor SA^2. The most convenient test as to the seriousness of the residual aberration will usually be the comparison of the corresponding OSC' with out limit ± 0.0025.

Introducing the easily remembered compound symbols SGS for the spherical G-sum and CGS for the coma G-sum, we have the two general equations

$$SGS = \Sigma(\text{Spherical } G\text{-sums of constituent lenses})$$
$$CGS = \Sigma(\text{Coma } G\text{-sums of constituent lenses}).$$

For the usual binary combination of a crown and a flint lens which we shall first discuss, we can write the sums explicitly. Using the alternative G-sums for the left-hand component with suffix 'a' and the original G-sums for the right-hand component with suffix 'b', they are

(4) $SGS = (G_1{}^a c_a{}^3 + G_1{}^b c_b{}^3) + G_2{}^a c_a{}^2 c_2 - G_2{}^b c_b{}^2 c_3 - (G_3{}^a c_a{}^2 - G_3{}^b c_b{}^2) v'_2$
$\qquad\qquad + G_4{}^a c_a c_2{}^2 + G_4{}^b c_b c_3{}^2 - G_5{}^a c_a c_2 v'_2 - G_5{}^b c_b c_3 v'_2 + (G_6{}^a c_a + G_6{}^b c_b) v'_2{}^2$

$\quad CGS = \tfrac{1}{4} G_5{}^a c_a c_2 + \tfrac{1}{4} G_5{}^b c_b c_3 - v'_2(G_7{}^a c_a + G_7{}^b c_b) + (G_8{}^a c_a{}^2 - G_8{}^b c_b{}^2)$

We shall regard the G-values as constants fixed by the refractive indices of the chosen kinds of glass and the values of c_a and c_b as similarly fixed by the achromatic condition; but we shall treat c_2 and c_3 as independent variables depending on bending and v'_2 as another independent variable depending on the object distance. SGS and CGS may then be regarded as dependent variables; but as we have only two liberties in modifying the design, namely, the two independent bendings or values of c_2 and c_3, we shall have to introduce restrictive assumptions which will lead to a number of different solutions according to which of the potential variables have definite values assigned to them.

For cemented lenses we must have $c_2 = c_3$, and a further collection of terms results, with a reduction of the variables by one and a reduction of the liberties of variation in design to the solitary one of a bending of the complete objective as a whole.

[82]   As the true significance of our algebraic results and their importance in the actual designing of thin systems will be most clearly and fully brought into view by numerical examples, we will at once work out the coefficients in the fundamental equations (4) for the same glasses which were employed in most of the examples in Chapters V, VI, and VII:

(a) $N_d = 1.5407$, $N_f - N_c = 0.00910$;   (b) $N_d = 1.6225$, $N_f - N_c = 0.01729$.

However, as some conclusions of theoretical interest will depend on uncomfortably small differences of relatively large numbers that are calculated separately, we will make sure of four-figure accuracy by calculating certain data more closely than would be necessary, or in fact would have any practical value, in using the TL-method as a rough first approximation in the solution of actual designing problems.   The V-values in the glass lists $[= (N_d - 1)/(N_f - N_c)]$ are only given with one decimal and thus do not correspond very closely to the actual dispersion, especially as the rounding-off to the retained decimal is occasionally faulty. Calculating to two decimals, we find $V_a = 59.42$ (list value 59.4) and $V_b = 36.00$ (list 36.0).   Calculation of c_a and c_b for $f' = 10$ as before, then gives:

$$c_a = [9.6714] = 0.4692 \text{ (Chap. V value, 0.470)}$$
and $$c_b = [9.3926n] = -0.2469 \text{ (Chap. V } -0.247).$$

As a hint of general utility we will here add that the values of c_a and c_b may be checked by putting them into the equation from which the TL, Chr formulae were derived,   namely, $c_a(N_a - 1) + c_b(N_b - 1) = 1/f'$,   giving   in   our   case $0.4692 \times 0.5407 - 0.2469 \times 0.6225 = 0.10004$, as close to the exact value 0.1 as could be expected by four-figure logs.   The values used in earlier chapters give 0.10042 or 0.42 per cent. error, which does not matter *practically* because we found that the actual focal length computed trigonometrically differed from the intended value up to 4 per cent. even for the fairly mild cases to which the examples were restricted.

On putting the values of c_a and c_b and the G-values from the special table into (4), we find the general equations for all possible bendings of the two lenses and all conceivable real or virtual object-distances:

General $SGS = \begin{cases} 0.05397 + 0.24294c_2 - 0.08056c_3 - 0.22328v'_2 + 0.29147c_2{}^2 \\ -0.17163c_3{}^2 - 0.83676c_2v'_2 + 0.49694c_3v'_2 + 0.21995v'_2{}^2 \end{cases}$

General $CGS = 0.06093 + 0.20917c_2 - 0.12422c_3 - 0.13496v'_2.$

APLANATIC OBJECTIVES WITH BROKEN CONTACT

We will first solve the general equations for the special case of a telescope objective having both SGS and CGS equal to zero, thus securing aplanatism in Abbe's sense of the word.   As for a telescope objective we assume $l_1 = \infty$ or $v_1 = 0$, we must calculate $v'_2 = v_1 + c_a(N_a - 1)$ as simply $c_a(N_a - 1) = [9.4044] = 0.25375$ and introduce this numerical value into the general equations, which then become

$SGS = 0.01148 + 0.03062c_2 + 0.04554c_3 + 0.29147c_2{}^2 - 0.17163c_3{}^2$,   required $= 0$

$CGS = 0.02668 + 0.20917c_2 - 0.12422c_3,$ required $= 0$

Solution of the second equation for c_3 gives the equation of condition: [82]

$$c_3 = \left(\frac{0.20917}{0.12422}\right)c_2 + \left(\frac{0.02668}{0.12422}\right) = [0.2263]c_2 + [9.3320],$$

and introducing this into the equation for SGS, we find the simple quadratic in c_2:

$$0 = SGS = -0.19517c_2{}^2 - 0.01684c_2 + 0.01334,$$

the solutions of which are $c_2 = -0.3081$ or $= +0.2219$, the latter being the useless form of two pronounced meniscus lenses. The useful solution, on being introduced into the equation of condition, gives $c_3 = -0.30402$, in close agreement with section [59], page 332.

CEMENTED TELESCOPE OBJECTIVES

In many cases interest is limited to cementable forms. As these require $c_2 = c_3$, a number of the terms in the general solution can be combined and we obtain:

For cementable objectives only:

$$SGS = 0.05397 + 0.16238c_2 - 0.22328v'_2 + 0.11984c_2{}^2 - 0.33982c_2v'_2 + 0.21995v'_2{}^2$$
$$CGS = 0.06093 + 0.08495c_2 - 0.13496v'_2.$$

To find the two possible crown-in-front forms we again have to introduce, as above, $v'_2 = [9.4044]$, which gives for SGS the simple quadratic equation

$$SGS = 0.01148 + 0.07616c_2 + 0.11984c_2{}^2 \qquad \text{for crown in front,}$$

and for coma the simple linear equation

$$CGS = 0.08495c_2 + 0.02668 \qquad \text{for crown in front.}$$

The retention of v'_2 in the general solutions enables us to solve also for flint-in-front telescope objectives, and this is one great advantage of thus retaining v'_2. Evidently our objectives will produce parallel bundles of rays emerging from their right-hand surfaces if we place the object-points in the anterior focal plane, which simply means letting $l_1 = -f'$ or $v_1 = -1/f' = -0.1$ for our numerical example. As calculated, such an objective would act like the collimator-lens of a spectroscope, but it would evidently be equally available as a telescope objective if the eyepiece is placed next to the crown element so that the flint lens is 'in front'. Hence we shall obtain correct solutions for 'reversed Steinheil objectives' by putting $v_1 = -1/f'$ or $v'_2 = -1/f' + c_a(N_a - 1)$ into the general equations, or for our examples $v'_2 = -0.1 + 0.25375 = +0.15375$; we thus find

$$SGS = 0.02485 + 0.11014c_2 + 0.11984c_2{}^2 \qquad \text{for reversed Steinheil}$$
$$CGS = 0.08495c_2 + 0.04018. \qquad \text{for reversed Steinheil}$$

We can now determine the contact curvature c_2 of the four possible telescope objectives by solving the two quadratic equations for SGS. The neatest and

[82] quickest way of obtaining the solution in a form equally suitable for numerical discussion and for the more instructive plotting of the SGS parabola is the following:

The equations are of the form $y = ax^2 + bx + c$ and give as the general solution for $y = 0$

$$x = -\frac{b}{2a} \pm \sqrt{\frac{b^2 - 4ac}{4a^2}},$$

which, when applied to our SGS equations, determine the c_2 of the spherically corrected forms. But we can easily extract more information. As the two zero-solutions differ by the value of the root, but in opposite directions, from the rational part $-b/2a$, the latter obviously locates the axis of the vertical parabola; hence if we put $x = -b/2a$ into the original equation, we shall obtain the minimum (for concave systems the maximum) value of SGS as

$$SGS \text{ at pole} = a(-b/2a)^2 + b(-b/2a) + c = -(b^2 - 4ac)/4a,$$

which can thus be most conveniently calculated. For plotting we then have three values of SGS at equal intervals ($=$ value of the root) of the abscissae so that the dip-method of section [47] can be applied.

By comparison of the SGS equation for crown-in-front with the standard form, we find $a = 0.11984$, $b = 0.07616$, $c = 0.01148$, and the general solution gives:

For zero aberration: $c_2 = -0.3178 \pm 0.0719$

Minimum aberration at pole: $SGS = -0.000620$.

For the reversed Steinheil forms, we have to use $a = 0.11984$, $b = 0.11014$, $c = 0.02485$ and we find

For zero aberration: $c_2 = -0.4595 \pm 0.0617$

Minimum aberration: $SGS = -0.000456$.

As a linear function of c_2, CGS can be interpolated or plotted from two values, and it is convenient to calculate it for the two values of c_2 which give zero spherical aberration. We thus obtain the following results:

	1st sol. $SGS = 0$	Min. SGS	2nd sol. $SGS = 0$
Crown-in-front	$c_2 = -0.3897$	$c_2 = -0.3178$	$c_2 = -0.2459$
corresponding G-sum	$CGS = -0.00643$	$SGS = -0.000620$	$CGS = +0.00579$
Reversed Steinheil	$c_2 = -0.5212$	$c_2 = -0.4595$	$c_2 = -0.3978$
corresponding G-sum	$CGS = -0.00410$	$SGS = -0.000456$	$CGS = +0.00639$

For plotting, a horizontal scale of 1 inch $= 0.1$ of curvature will be suitable and a vertical scale of 1 inch $= 0.001$ of SGS and of 1 inch $= 0.01$ of CGS. A graph like Fig. 103 results, and as the ordinary and the reversed Steinheil forms are referred to the same scale of c_2, each point of this scale means an objective of definite form, no matter whether it is to be used as a crown-in-front or as a flint-in-front lens. Hence the obvious fact that the two parabolas intersect each other very close to the zero axis at about $c_2 = -0.393$ means that an objective of that [82] contact curvature will perform equally well and almost perfectly when used as a telescope object-glass *either way round*.   Such 'reversible' objectives have often been advocated on account of their 'foolproof' merit.   The left solution for the Steinheil type and the right solution for the ordinary type show that reversibility is not a usual property of a well-corrected object-glass, for if the respective other parabola were extended to the ordinate of these outside solutions, it evidently would indicate an appalling amount of spherical under-correction if these lenses were used the wrong way round.   As we derived the reversed Steinheil form from the general solution for crown-in-front objectives simply by varying the value of

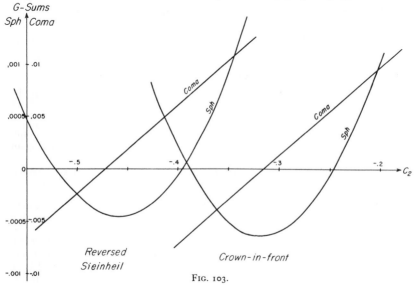

FIG. 103.

v'_2, it is evident that the behaviour of a given objective on either reversal or on a mere change of object distance can be deduced from our general solutions.   The study of thin objectives from this point of view will prove of considerable interest and importance.

THE HERSCHEL CONDITION

WE shall realize the nature of the problem and of all the possibilities most clearly [83] and quickly by beginning with a few definite cases and the graph resulting from these.   We will select the two crown-in-front solutions with $c_2 = -0.3897$ and -0.2459 respectively and the aplanatic form with $c_2 = -0.3081$ and $c_3 = -0.30402$. By putting the numerical value of c_2 for the two cemented objectives into the equations for SGS and CGS of cementable objectives and the values of c_2 and c_3 of the aplanatic form into the absolutely general equations for SGS and

[83] CGS, we obtain simple quadratic equations in v'_2 for the SGS and simple linear equations for the CGS which express the aberrations of each of the selected forms for any value of v'_2. The equations are:

For cemented O.G.
with $c_2 = -0.3897$
$\begin{cases} SGS = 0.21995v'_2{}^2 - 0.09085v'_2 + 0.00890 \\ CGS = -0.13496v'_2 + 0.02782 \end{cases}$

For cemented O.G.
with $c_2 = -0.2459$
$\begin{cases} SGS = 0.21995v'_2{}^2 - 0.13972v'_2 + 0.02129 \\ CGS = -0.13496v'_2 + 0.04004 \end{cases}$

For the aplanatic
object-glass
$\begin{cases} SGS = 0.21995v'_2{}^2 - 0.11655v'_2 + 0.01541 \\ CGS = -0.13496v'_2 + 0.03425 \end{cases}$

These equations can be solved in the same manner and by the same convenient scheme which was applied above to the solution for the bendings of cemented lenses, with the following results:

	$SGS = 0$	Min. SGS	$SGS = 0$
Cemented O.G. with $c_2 = -0.3897$	$v'_2 = 0.1597$	$v'_2 = 0.2065$	$v'_2 = 0.2533$
Corresponding CGS or SGS	$CGS = 0.00627$	$SGS = -0.000481$	$CGS = -0.00637$
Cemented O.G. with $c_2 = -0.2459$	$v'_2 = 0.2537$	$v'_2 = 0.3176$	$v'_2 = 0.3815$
Corresponding CGS or SGS	$CGS = 0.00580$	$SGS = -0.000898$	$CGS = -0.01145$
Aplanatic Object-glass	$v'_2 = 0.2534$	$v'_2 = 0.2649$	$v'_2 = 0.2765$
Corresponding CGS or SGS	$CGS = -0.00005$	$SGS = -0.000030$	$CGS = -0.00307$

The graph representing these results becomes more directly useful if v_1, the reciprocal of the actual object-distance, is substituted for the v'_2. As we have for all possible objectives obtainable from our general solution (4) $v'_2 = v_1 + c_a(N_a - 1)$ or $v_1 = v'_2 - c_a(N_a - 1)$, we have only to subtract $c_a(N_a - 1)$, or in our example 0.2537, from the v'_2 values in order to obtain the corresponding v_1. The graph then takes the form of Fig. 104, the scales employed for Fig. 103 being again suitable.†

In discussing this graph, we must bear in mind that spherically corrected achromatic lenses are used almost exclusively for the production of real inverted images of real objects, which means v_1 values between 0 and $-1/f'$ or, in our case, between 0 and -0.1. Positive values of v_1 correspond to virtual objects and are of no interest in the present discussion; hence our attention must be concentrated on the range between -0.1 and zero.

We see at once that the usual cemented form with a nearly equi-convex crown

† In plotting Fig. 104, it should be noted that the aplanatic graph can be accurately drawn, although the three calculated points are too close together to admit of reasonable precision, by noting that all three parabolas have the same curvature or dip by reason of identical coefficients of $v'_2{}^2$, and all three coma-lines have the same slope by reason of identical coefficients of v'_2. Hence, after laying down the point of minimum SGS at $v'_2 = 0.2649$ corresponding to $v_1 = 0.0112$, the parabola can be accurately completed by using on the ordinate of this point the dip and interval of one of the other two parabolas. Also, as the aplanatic objective was corrected for coma at $v'_2 = 0.2537$ or $v_1 = 0$, this gives the crossing point of the coma-line, which can then be drawn parallel to the other two.

component—in our case $c_2 = -0.2459$—is very unfavourable for objects at short [83] distances because the parabola crosses the zero-axis at $v_1 = 0$ at a very steep angle, indicating a rapid rise of the spherical aberration for approaching objects.

The aplanatic form gives a parabola which almost coincides with the zero-axis for some distance near its pole, and although the closest approach lies on the wrong side of $v_1 = 0$, it is evident that the spherical aberration will be small until the object approaches within 4 or 5 times the focal length. The aplanatic form is therefore decidedly favourable, because even laboratory telescopes for reading

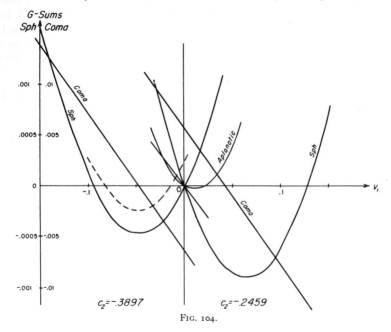

FIG. 104.

thermometers, barometers, gas-tubes, &c., are rarely used at shorter distances than those just estimated as being admissible for any ordinary aplanatic objective.

The parabola for the unusual cemented form $(c_2 = -0.3897)$, which has a comparatively flat surface at its left side, naturally confirms our previous discovery that it is also nearly free from spherical aberration for $v_1 = -0.1$, that is, for objects at its anterior focus. But we now learn also how the aberration changes at all intermediate distances; for these it is always negative or over-corrected, but with a maximum numerical amount of only $SGS = 0.000481$. We will apply our equation (3) for the seriousness of the spherical aberration to find out at what aperture of a lens of 10-inch focal length this residual would reach the Rayleigh limit, or $LA'/Tolerance = 1$. With $\lambda = 0.00002$ for inch measure, the equation then gives

$$1 = SA^4/0.00008 \times SGS \quad \text{or} \quad SA^4 = 0.00008/0.000481 = 0.166;$$

[83] whence by Barlow's tables $SA^2 = 0.407$ or $SA = 0.638$ inch. Hence a 10-inch lens of 1.27 inch aperture, or working at $f/7.9$, would give respectable real images of real objects at all possible distances, and moreover, it would do so if placed either way round in its cell, thus being completely foolproof! At the point of greatest spherical aberration, the coma would be zero—also a favourable property—whilst at the worst point, $v_1 = 0$, the coma would correspond to $CGS = -0.00642$, equivalent, by $OSC' = CGS \cdot SA^2$, to $OSC' = -0.00642 \times 0.407 = -0.0026$. This is only just beyond our empirical limit. A possibility of further improving this remarkable cemented lens substantially is indicated by the dotted parabola in Fig. 104: By using a denser barium crown, the over-correction at the pole could easily be halved, leading to only half the previous numerical value of SGS, but now at the three points $v_1 = 0$, $v_1 = -0.05$, and $v_1 = -0.1$; it would then be spherically corrected within half the Rayleigh limit throughout the whole range of distances for which real objects can yield real images. The method of discussion just exemplified should be carefully studied, as the successful use of the Herschel condition depends largely upon it.

We can now put the Herschel condition into a general algebraic form. As the variation of SGS with the reciprocal of the object-distance plots as a vertical parabola, which we shall call the Herschel parabola, it is evident that the variation will have a minimum value if the pole of the parabola is laid upon that value of v_1 which is midway between the extreme values to be included in the use of the object-glass; and as the pole is mathematically defined by $\partial SGS/\partial v_1 = 0$, this equation, when satisfied by the mean value of v_1, is the proper form of the Herschel condition. Moreover, as $v'_2 = v_1 + c_a(N_a - 1)$, in which the second term is constant for a given system, we have $dv'_2 = dv_1$. We can therefore write equally well $\partial SGS/\partial v'_2$ as expressing the condition, provided that we make this partial differential coefficient vanish for that value of v'_2 which corresponds to the mean v_1 according to $v'_2 = v_1 + c_a(N_a - 1)$. Differentiating the first of (4) with reference to v'_2 only, we thus obtain the condition:

$$\partial SGS/\partial v'_2 = -(G_3{}^a c_a{}^2 - G_3{}^b c_b{}^2) - G_5{}^a c_a c_2 - G_5{}^b c_b c_3 + 2v'_2(G_6{}^a c_a + G_6{}^b c_b) = 0$$

for mean v'_2. This equation can be used to solve for the correct form of a Herschel objective in exactly the same way as the very similar coma condition is employed in solution for aplanatic objectives.

To obtain the best possible compromise for the whole range of v'_2, we must however aim at arranging the dotted position of the Herschel parabola in Fig. 104 in such a manner that the finished objective may have an over-correction for mean v'_2 equal to its under-correction for the two extreme values. With reference to the pole of the parabola as origin, the equation for SGS becomes reduced to its quadratic term in $v'_2{}^2$; hence—as in the formula for the 'dip' in [48.5] (1)—we can calculate the full range of SGS within the limits 'max v'_2' and 'min v'_2' as

$$\text{Full range of } SGS = (G_6{}^a c_a + G_6{}^b c_b) \cdot \left(\frac{\text{max } v'_2 - \text{min } v'_2}{2}\right)^2.$$

One half of this, with reversed sign, will have to be the value of SGS for the mean v'_2. We therefore arrive at the following complete method of solution:

Herschel Condition: In the first equation of (4), insert the numerical values of [83] the G-coefficients; select the v'_2 limits within which good correction is desired. By differentiation form

$$\partial SGS/\partial v'_2 = -(G_3{}^a c_a{}^2 - G_3{}^b c_b{}^2) - G_5{}^a c_a c_2 - G_5{}^b c_b c_3 + 2v'_2(G_6{}^a c_a + G_6{}^b c_b) = 0$$

for mean v'_2, then put in the numerical value of mean v'_2 and solve for c_3, thus obtaining an equation of condition to be used for the elimination of c_3 from the SGS equation.   Determine also in numerical form, except for c_2 and c_3, the general SGS equation:

$$SGS = A + Bc_2 + Cc_3 + Dc_2{}^2 + Ec_3{}^2 = -\tfrac{1}{8}(G_6{}^a c_a + G_6{}^b c_b)(\max v'_2 - \min v'_2)^2,$$

$A, B, \ldots E$ simply standing for the resulting numerical factors; eliminate c_3 by the Herschel condition; then solve the final simple quadratic in c_2.   Only one of the two solutions is of practical value, just as in solutions for aplanatism.

A numerical example will render the method quite clear.   Sacrificing the excessive range of our foolproof cemented solution, we will seek the best possible solution for an objective with a slight air-gap covering the still very generous range from $v_1 = 0$, or objects at infinity, to $v_1 = -0.05$, corresponding to objects at twice our adopted focal length of 10 and therefore to natural-size inverted images.   Even a laboratory or optical-bench object-glass is not likely to be used outside this range, and in the reversed position such an object-glass would be suitable for a reading microscope giving primary images magnified from -1 times to infinity.   As for our numerical examples, we have $v'_2 = v_1 + 0.2537$; we must introduce max $v'_2 = 0.2537$, mean $v'_2 = 0.2287$, and min $v'_2 = 0.2037$.   Differentiating the general numerical equation for SGS with reference to v'_2, we obtain the Herschel condition:

$$\partial SGS/\partial v'_2 = 0.4399 v'_2 - 0.83676 c_2 + 0.49694 c_3 - 0.22328 = 0.$$

Introducing mean $v'_2 = 0.2287$, we get

$$0 = -0.83676 c_2 + 0.49694 c_3 - 0.12268,$$

which leads to the equation of condition, $c_3 = 1.68382 c_2 + 0.24687$. We next introduce mean $v'_2 = 0.2287$ into the general SGS equation, which thus gives:

$$SGS = 0.01441 + 0.05157 c_2 + 0.03309 c_3 + 0.29147 c_2{}^2 - 0.17163 c_3{}^2$$
$$= -\tfrac{1}{8} \times 0.21995 \times (0.05)^2.$$

Using the equation of condition to eliminate c_3 and working out the right-hand side also, we get

$$-0.19512 c_2{}^2 - 0.03542 c_2 + 0.01212 = -0.00007$$

with the solution

$$c_2 = -0.09076 \pm 0.26592 = -0.35668 \quad \text{or} \quad +0.17516 \text{ (useless)}.$$

Putting the good solution into the equation of condition then determines $c_3 = -0.35371$, after which the complete prescription can be worked out by c_a and c_b

[83] in the usual way, giving $c_1 = 0.11252$ and $c_4 = -0.10681$. These values show that the resulting object-glass is nearly equi-convex externally, with a slight air-gap. As the highest numerical amount of SGS within the contemplated range of object distances is 0.00007 by the right-hand side of the SGS equation, we can determine the admissible aperture as before by:

Admissible $SA^4 = 4\lambda/SGS$; which, for inch measure, or for an object-glass of 10-inch focal length, gives, with $\lambda = 0.00002$,

Admissible $SA^4 = 8/7$; therefore admissible $SA = 1.034$ inch.

As far as spherical aberration is concerned, the system might therefore be made over 2 inches in aperture, when it would work at the high ratio of $f/4.8$. But before accepting this aperture as safe, we must also test for coma. Putting $c_2 = -0.35668$ and $c_3 = -0.35371$, also either max $v'_2 = 0.2537$ or min $v'_2 = 0.2037$ into the general CGS equation, we find for max v'_2, $CGS = -0.00398$, and for min v'_2, $CGS = 0.00277$. We see at once that if these values were turned into OSC' by applying the factor SA^2, they would with the value just found give figures in excess of our 0.0025 limit [Equation OT (6) on page 395 of Part I]. Hence coma is going to decide the really safe aperture, and this is the usual situation. Taking the highest numerical value, namely 0.00398, found for $v_1 = 0$, we have the condition

$$0.0025 \geqslant CGS \cdot SA^2$$
or,   Admissible $SA^2 = 0.0025/CGS = 0.627$.

Therefore $SA = 0.792$ and we must be satisfied with an aperture of 1.584 inches or an object-glass working at $f/6.3$. At that aperture, the spherical aberration will reach only 0.35 of the Rayleigh limit and will therefore be quite harmless.

When the trigonometrical correction of the TL solution is made for a Herschel objective, it will usually be sufficient to accept the form of the first lens as definitive and to bend the second lens only so as to attain the proper amount of overcorrection at mean v'_2. In that case the method employed in [49] for the Dennis Taylor objective will be suitable. A trigonometrical method for the direct inclusion of the Herschel condition in the verification will be deduced subsequently.

There is a simple and important relation between the Herschel condition and the zero-coma condition which becomes manifest if we compare the equation of the former with four times the equation of the latter:

$$\partial SGS/\partial v'_2 = 2v'_2(G_6{}^a c_a + G_6{}^b c_b) - G_5{}^a c_a c_2 - G_5{}^b c_b c_3 - (G_3{}^a c_a{}^2 - G_3{}^b c_b{}^2)$$
$$4 \cdot CGS = -4v'_2(G_7{}^a c_a + G_7{}^b c_b) + G_5{}^a c_a c_2 + G_5{}^b c_b c_3 + 4(G_8{}^a c_a{}^2 - G_8{}^b c_b{}^2).$$

The terms depending on c_2 and c_3—on bending—are identical but of opposite signs, and the other terms come near fulfilment of the same relation because, as a glance at the special tables shows, we have *nearly* $G_6 = 2G_7$, and $G_3 = 4G_8$. We obtain an interesting result on adding the two equations and simplifying the sum. By the equations collected under (1), we have accurately

$$4G_7 - 2G_6 = N - 1; \qquad 4G_8 - G_3 = \tfrac{1}{2}(N-1)^2.$$

Hence we can write the sum of the two equations: [83]

$$\partial SGS/\partial v'_2 + 4CGS = -v'_2[c_a(N_a-1)+c_b(N_b-1)] + \tfrac{1}{2}c_a{}^2(N_a-1)^2 - \tfrac{1}{2}c_b{}^2(N_b-1)^2.$$

Now we have by TL (2): $c(N-1) = 1/f'$, and also $1/f'_a + 1/f'_b = 1/f'$ of system. With these the last equation becomes

$$\partial SGS/\partial v'_2 + 4CGS = -v'_2(1/f') + \tfrac{1}{2}[(1/f'_a)^2 - (1/f'_b)^2]$$
$$= -v'_2/f' + \tfrac{1}{2}(1/f'_a - 1/f'_b)/f'$$
$$= (-v'_2 + 1/2f'_a - 1/2f_b')/f'.$$

If we now introduce $v'_2 = v_1 + c_a(N_a-1) = v_1 + 1/f'_a$, we find

$$\partial SGS/\partial v'_2 + 4CGS = -(v_1 + 1/f'_a - 1/2f'_a + 1/2f'_b)/f' = -(v_1 + 1/2f')/f'.$$

For an objective satisfying the Herschel condition for the object distance corresponding to the v_1 in the last equation, we have $\partial SGS/\partial v'_2 = 0$; hence the expression on the right gives the value of $4CGS$ at the axis of the Herschel parabola, and we can state our result in the form:

On axis of Herschel parabola: $CGS = -\tfrac{1}{4}(v_1 + 1/2f')/f'.$

This becomes zero for $v_1 = -1/2f'$, or when the Herschel condition is fulfilled for natural-size inverted images. Hence $M' = -1$ represents the only case in which the Herschel condition and the zero-coma condition can be simultaneously fulfilled. It is important to remember this, for it follows evidently that with $M' = -1$ at the mean object distance, we shall be able to attain the largest useful aperture and the longest range in v_1-values or both.

In the most usual case of fulfilment of the Herschel condition for $v_1 = 0$, we shall have $CGS = -1/8f'^2$ or on multiplying by SA^2 to turn CGS into OSC':

$$OSC' = -\tfrac{1}{8}(SA/f')^2.$$

For the limit value of $OSC' = 0.0025$, this gives $(SA/f')^2 = 0.02$ or $SA/f' = 0.1414$, corresponding to an aperture of $f/3.5$. It is therefore possible to satisfy the Herschel condition for a short range in v_1 near its zero point at even the highest usual aperture ratios. This is, however, not the best way at large apertures; when there is a conflict between several desiderata—in the present case between freedom from coma and small variation of spherical aberration near a selected v_1-value—it is always best to compromise. We saw in the preliminary studies that the aplanatic form for $v_1 = 0$ had zero coma and zero spherical aberration at that value, but that the spherical aberration increased steeply towards negative values of v_1, thus putting a possibly too narrow limit on the admissible range in object distance. Evidently a correction at $v_1 = 0$ intermediate between fulfilment of the zero-coma and the Herschel conditions would be best and should be sought for to the extent that the individual case demands.

The important points to be remembered when a range of object distances must be covered with good correction of coma and spherical aberration throughout are:

(1) The Herschel condition gives a most emphatic verdict in favour of the *unusual* form of the cemented telescope objective and this form will very often prove good enough if rather dense barium crown is used;

[83]   (2) The form of object-glass which is strictly aplanatic for mean v_1 is decidedly favourable although as a rule it is not quite the best;

(3) Owing to the agreement between the Herschel and the zero-coma conditions for $M' = -1$, the longest range of v_1 and therefore of magnification can be covered if the mean v_1 is chosen equal to $-1/2f'$;

(4) When the mean v_1 cannot be placed at this best position, a compromise between the Herschel and the zero-coma conditions will usually be most satisfactory;

(5) Because SGS plots as a vertical parabola against v_1, the best compromise for spherical correction calls for overcorrection at midrange and undercorrection of nearly the same amount at the ends of the range.

It should be noted that our study of the Herschel condition has supplied a confirmation of the important conclusion arrived at in Chapter IX concerning the impossibility of perfect systems as there defined; for we have learnt that, whatever form may be given to the two component lenses of an ordinary thin object-glass, the coma plots as an inclined straight line and the spherical aberration as a vertical parabola against v_1; hence the coma can only be corrected for one object-distance and the spherical aberration for two object-distances.   Corresponding conclusions will subsequently be added, which are not restricted to thin systems but cover centred systems of any construction.

It was Sir John Herschel (1792–1871) who first suggested (*Phil. Trans.* **111**, 1821, pp. 222–266) that the second liberty of varying the design, which is afforded by two-lens objectives with a small air-space, should be utilized to establish spherical correction for two different object distances.   He was apparently the pioneer in using reciprocals of radii, intersection- and focal lengths, and in thus simplifying the algebra of optics.   With the characteristic thoroughness of a pioneer, he extended the reciprocal relation to the spherical aberration itself, measuring it in our notation by $1/L'_k - 1/l'_k$, which in sufficient primary approximation is equal to our $LA'/l'_k{}^2$.   The equation with which he worked was thus, in our notation,

$$LA'/l'_k{}^2 = SA^2 \cdot SGS,$$

and he thus got rid automatically of the awkward factor $l'_k{}^2$ on the right.   His equivalent of our SGS was expressed in terms of v_1 for the object distance, as we might do by replacing v'_2 by $v_1 + c_a(N_a - 1)$.   After ordering the equation according to powers of v_1, he discussed

$$LA'/l'_k{}^2 = SA^2[A + Bv_1 + Cv_1{}^2]$$

in which A, a quadratic function of c_2 and c_3, is evidently the aberration constant for $v_1 = 0$; B, a linear function of c_2 and c_3, is the rate of change of that constant near $v_1 = 0$; and C is identical with our factor $v'_2{}^2$.

He then concluded that the aberration would be zero for all object distances if $A = B = C = 0$, but he noticed that $C = 0$ was incompatible with the achromatic condition.   He overcame this obstacle by pointing out that for telescopes only very small values of v_1 are of interest so that $Cv_1{}^2$ may be treated as negligible.   He therefore solved only for $A = 0$ and $B = 0$, which is easily seen to be the equivalent to our solution for the pole of the Herschel parabola at $v_1 = 0$.   He did not discuss the residual aberration at finite values of v_1 at all and thus missed its parabolic law [83] of change.   He spoke very slightingly of d'Alembert's suggestions that spherical aberration should be corrected for two colours, or alternatively that the extra-axial aberrations should be taken into account.   He pointed out quite correctly that these defects were very small for the low values of relative aperture that were then common, but he failed to realize that the same objection applied in equal degree to his own suggestion.   Nevertheless it is a great pity that his suggestion was not accepted at once and widely, for we have learnt that fulfilling the Herschel condition is almost equivalent to fulfilling the sine condition up to decidedly high values of the relative aperture.

SOME GENERAL SOLUTIONS

A. Analysis of the General Quadratic Equation

OUR general equation for SGS is a quadratic.   If we reduce the number of vari- [84] ables in it to two by either eliminating the others or by assigning definite values to them, the equation can be plotted as a conic section.   We have already treated SGS itself as one variable, retaining either c_2 or v'_2 as the other variable, and we have found the resulting parabolas, which represent the law of change of the spherical aberration when a lens is bent or when the object distance is changed, to be extremely helpful.   We shall now regard SGS as prescribed, usually as at or near zero, in which case we can retain two of the three variables on the right.   As a rule the equation then becomes a general quadratic, which may plot as an ellipse, a hyperbola, or a parabola, or perchance the equation may have no real solution. We shall therefore begin by developing a simple and efficient method for interpreting and plotting a general quadratic equation in two variables.   Adopting the standard form

$$Ax^2 + 2Bxy + Cy^2 + 2Dx + 2Ey + F = 0,$$

in which x and y are the two variables and $A, B, \ldots F$ are constants, we can exclude the case when $A = B = C = 0$ as the equation would then be linear and could at once be plotted as an inclined straight line.   Similarly $A = B = 0$ would be an obvious parabola with its axis horizontal, and $B = C = 0$ a parabola with its axis vertical.   It would not be difficult to single out a few more cases of easy interpretation, but these are dealt with as readily by the general method.   It may be stated that the terms containing the variables to the first power have been given a numerical factor 2 to reduce the numerical factors in the solution to a minimum.

For the general discussion, we rearrange the terms so as to obtain a quadratic equation in y:

$$Cy^2 + (2Bx + 2E)y + (Ax^2 + 2Dx + F) = 0$$

and solve this equation for y in terms of x and the constants.

(1) If $C = 0$, the first term would disappear and we should have

$$y = -(Ax^2 + 2Dx + F)/2(Bx + E),$$

which evidently would give one real value of y for every value of x throughout the range $-\infty$ to $+\infty$, and y would change sign by the usual suddenness via infinity

[84] when the denominator changed sign at $x = -E/B$ by passing through zero. Hence the curve represented by the equation extends into infinity at the right and left as well as up and down and must be a hyperbola with a vertical asymptote at $x = -E/B$. It can be discussed in detail and plotted by the general method to be given later.

(2) If C has a finite value, the equation in y is quadratic and by a slight simplification gives the solution

$$y = [-(Bx+E) \pm \sqrt{(B^2-AC)x^2+2x(BE-CD)+(E^2-CF)}]/C$$

which shows that there will be two values of y for each value of x, but that some values may be imaginary. Differentiation of the solution for y with reference to x gives

$$dy/dx = -(B/C) \pm [(B^2-AC)x+(BE-CD)]/[C \times \text{Root in solution for } y].$$

This differential coefficient determines the direction of the two tangents of the curve at any value of x. Taken together, the solutions for y and for dy/dx determine the type of curve by the magnitude and sign of the factor (B^2-AC) in the root expression.

(a) If $(B^2-AC) = 0$, we shall have

$$y = [-(Bx+E) \pm \sqrt{2x(BE-CD)+(E^2-CF)}]/C$$
$$dy/dx = -(B/C) \pm (BE-CD)/[C \times \text{above root}].$$

The root expression will be zero for $x = -(E^2-CF)/2(BE-CD)$. It will be negative, and y imaginary, for larger values of x if $(BE-CD)$ is negative, or for smaller values of x if $(BE-CD)$ is positive. The curve represented by our equation in either case extends only to one side of the ordinate at the critical value of x, but it reaches infinity on that side because the root expression will steadily rise in value positively as the numerical value of x grows. This suggests a parabola and this guess is confirmed by dy/dx, for as its $\pm$ part has the root in the denominator, this part becomes smaller and smaller for increasing numerical values of x; hence the two arms of the curve become more and more nearly parallel as infinity is approached, and this is an exclusive property of the parabola. It also follows that the axis of this parabola makes with the x-axis an angle whose tangent is $-B/C$.

We can now plot the parabola with very little trouble. For the critical value of x, dy/dx becomes infinite because the root in the denominator is zero, hence the critical ordinate at $x = -(E^2-CF)/2(BE-CD)$ is a tangent of the parabola and the contact point can be found by putting that value of x into $y = -(Bx+E)/C$. We then change x by a suitable amount in the direction of real solutions and determine the corresponding values of y by the full equation for y. We thus secure three known points of the parabola P_{-1}, P and P_{+1} in Fig. 105, and as the tangent at P is parallel to the chord $P_{-1}P_{+1}$, a line joining P to the bisecting point of $P_{-1}P_{+1}$ is a diameter of the parabola and is parallel to its axis. Hence the three known points satisfy the conditions required by the dip-method of plotting parabolas when referred to an auxiliary axis of abscissae drawn at right angles to the diameter through P.

There is one special case, namely, when $(BE-CD)$ is also zero; it is easily seen [84] that the parabola then degenerates into two parallel straight lines which become imaginary if (E^2-CF) should prove negative.

(b) If (B^2-AC) is positive, we have to discuss the complete equations for y and for dy/dx. In the root

$$\sqrt{(B^2-AC)x^2+2x(BE-CD)+(E^2-CF)}$$

the now necessarily positive first term will overpower any possible negative value of the second and third terms when x reaches sufficiently large negative or positive values, and there will be real solutions for y for all large numerical values of x to

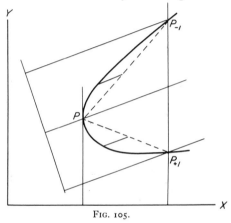

FIG. 105.

infinity.   Hence the curve extends to infinity to left as well as to right and must be a hyperbola.   This conclusion is confirmed by the equation for dy/dx, for in the $\pm$ part of

$$dy/dx = -(B/C)\pm[(B^2-AC)x+(BE-CD)]/$$
$$C\sqrt{(B^2-AC)x^2+2x(BE-CD)+(E^2-CF)},$$

the first terms in both numerator and denominator will predominate in value more and more as x grows numerically, and hence this part will tend towards the limit

$$\pm\frac{(B^2-AC)x}{C\sqrt{(B^2-AC)x^2}} = \pm\frac{(B^2-AC)x}{Cx\sqrt{B^2-AC}} = \pm\frac{\sqrt{B^2-AC}}{C}.$$

It follows that, at remote distances from the origin, the arms of the curve have the two fixed directions

$$\frac{-B+\sqrt{B^2-AC}}{C} \quad \text{and} \quad \frac{-B-\sqrt{B^2-AC}}{C},$$

which evidently are the direction-constants of the asymptotes of the hyperbola and

[84] are worth calculating in the practical applications as a check on the general solution to be given.

(c) If $(B^2 - AC)$ is negative, then $(B^2 - AC)x^2$ under the root will lead to imaginary solutions for all sufficiently large positive or negative values of x; therefore the curve cannot reach infinity to either right or left and must be an ellipse if it exists. To avoid wasting time on the complete final solution, it is advisable to ascertain in this case whether there is a real solution. As the expression under the root

$$\sqrt{(B^2 - AC)x^2 + 2x(BE - CD) + (E^2 - CF)}$$

is necessarily negative for all sufficiently large numerical values of x, it can only reach the positive values required for a real solution by passing through zero; therefore we can determine whether a real solution exists by equating the expression under the root to zero and solving for x, finding

$$x = [-(BE - CD) \pm \sqrt{(BE - CD)^2 - (E^2 - CF)(B^2 - AC)}]/(B^2 - AC).$$

If the expression under this new root proves to be positive, there will be a real solution for an ellipse of finite size. If it proves to be zero, the ellipse shrinks to a point at $x = -(BE - CD)/(B^2 - AC)$, which in all cases marks the centre of the ellipse. If the expression under the last root proves to be negative, there is no real solution and further work would be wasted, except when in our optical applications, the negative value is so small that a small change in F would neutralize it. Then there *may* be a good solution at or near $x = -(BE - CD)/(B^2 - AC)$ and the associated y found by $y = -(Bx + E)/C$, owing to the effects of the necessarily finite lens thicknesses and of higher aberrations.

It is quite possible to complete the solution along the lines adopted for the preliminary discussion, but the following general method is preferable when it has been ascertained that the general equation represents a hyperbola or a real ellipes.

(3) Returning to the equation

$$Ax^2 + 2Bxy + Cy^2 + 2Dx + 2Ey + F = 0,$$

in which x and y are referred to the axes AX and AY in Fig. 106, we can remove the linear terms in x and y from the equation by shifting the origin to A' with coordinates 'a' and 'b' as referred to the original axes, keeping $A'X'$ parallel to AX and $A'Y'$ parallel to AY.

The figure shows that we shall have to introduce $x = a + x'$, $y = b + y'$ into the general equation to effect the transformation. The result of this substitution is

$$Ax'^2 + 2Bx'y' + Cy'^2 + 2x'(Aa + Bb + D) + 2y'(Ba + Cb + E)$$
$$+ Aa^2 + 2Bab + Cb^2 + 2Da + 2Eb + F = 0.$$

The terms in x' and y' will disappear if

$$Aa + Bb + D = 0 \quad \text{and also} \quad Ba + Cb + E = 0,$$

and solving these two equations for the unknown 'a' and 'b' gives

$$a = (CD - BE)/(B^2 - AC) \quad \text{and} \quad b = (AE - BD)/(B^2 - AC).$$

The long absolute term of the transformed equation can be rearranged in the [84]
form

$$a(Aa+Bb+D)+b(Ba+Cb+E)+Da+Eb+F$$

and as the two factors in parenthesis are zero for the values of 'a' and 'b' just
adopted, the transformed equation now is

$$Ax'^2+2Bx'y'+Cy'^2+F' = 0, \quad \text{in which} \quad F' = Da+Eb+F,$$

but is subject to the condition that the above values of 'a' and 'b' are used.

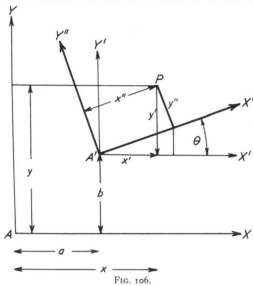

FIG. 106.

As the terms of the transformed equation are not changed in value if we replace
x' by $-x'$ and also y' by $-y'$, it follows that every chord passing through A' is
bisected at A'; in other words, in Fig. 106 the A' defined by the selected values of
'a' and 'b' marks the *centre* of the ellipse or hyperbola.

Finally we get rid of the term in $x'y'$ by turning the system of coordinates through
an appropriate angle θ into its final position $A'X''$ and $A'Y''$. As we are here
employing pure analytical geometry, we must count θ positive in the counter-
clockwise direction from $A'X'$ to $A'X''$. The transformation formulae are easily
taken from Fig. 106 by the aid of the auxiliary lines added to the diagram. They
are $y' = y'' \cos \theta + x'' \sin \theta$ and $x' = x'' \cos \theta - y'' \sin \theta$, and their introduction into
the equation in x' and y' gives:

$$x''^2[A \cos^2 \theta + 2B \sin \theta \cos \theta + C \sin^2 \theta]$$
$$+ x''y''[2(C-A) \sin \theta \cos \theta + 2B(\cos^2 \theta - \sin^2 \theta)]$$
$$+ y''^2[A \sin^2 \theta - 2B \sin \theta \cos \theta + C \cos^2 \theta] + F' = 0.$$

[84] The coefficient of $x''y''$ will disappear if

$$(C-A) \sin 2\theta + 2B \cos 2\theta = 0 \quad \text{or if} \quad \tan 2\theta = 2B/(A-C),$$

and we adopt the value of θ following from the last equation to bring the equation of our curve into its simplest form. Without any sacrifice as regards the validity of our solution, we can stipulate that 2θ shall always be taken out as an angle of less than 90°, with the sign of $2B/(A-C)$. For, since the alternative values of 2θ as determined by its tangent are $(2\theta + q\pi)$, q being any whole number, the selection of $q = 0$ simply means that we take our chance as to which of the two axes of the curve will fall into the X''-direction.

As stated above, the coefficients of x''^2 and y''^2 can be greatly simplified. Calling them A' and C', we see at once that, because $\sin^2 \theta + \cos^2 \theta = 1$, we have $A' + C' = A + C$, and if we form their difference we obtain

$$A' - C' = (A-C)(\cos^2 \theta - \sin^2 \theta) + 2B \sin 2\theta = (A-C) \cos 2\theta + 2B \sin 2\theta.$$

By the solution for $\tan 2\theta$ we can substitute either $2B = (A-C) \tan 2\theta$ or $(A-C) = 2B \cotan 2\theta$, in which we may replace $\tan$ by $\sin/\cos$ or $\cotan$ by $\cos/\sin$. The first substitution gives

$$A' - C' = (A-C)(\cos 2\theta + \sin^2 2\theta/\cos 2\theta) = (A-C)/\cos 2\theta$$

and the alternative substitution gives

$$A' - C' = 2B(\cos^2 2\theta/\sin 2\theta + \sin 2\theta) = 2B/\sin 2\theta.$$

We therefore obtain A' and C' as half sum and half difference from

$$A' + C' = A + C$$

and $\quad A' - C' = \quad$ either $\quad (A-C)/\cos 2\theta \quad$ or $\quad = 2B/\sin 2\theta.$

The practical computer's rule for choosing between the alternative solutions for $(A' - C')$ is to take the first if 2θ is numerically less than 45° and the second if 2θ exceeds 45°, simply because the differences for interpolation will thus be smallest.

The equation for our curve is now

$$A'x''^2 + C'y''^2 + F' = 0$$

and it assumes the usual standard form of the equation of an ellipse by dividing throughout by $-F'$:

$$\frac{x''^2}{-F'/A'} + \frac{y''^2}{-F'/C'} = 1.$$

It shows that, with reference to the final axes $A'X''$ and $A'Y''$,

the semi-axis in the X'' direction is $\sqrt{-F'/A'}$

the semi-axis in the Y'' direction is $\sqrt{-F'/C'}$.

We can adhere to this interpretation even when the equation really represents a hyperbola by adopting the perfectly legitimate view that a hyperbola is an ellipse in which one of the axes is imaginary.   Alternatively we can, in the case of a hyperbola, first determine the signs of $-F'/A'$ and $-F'/C'$ and then give to the two terms on the left of the standard equation opposite signs leading to real values of both semi-axes.   We shall adopt the first alternative.   The two poles of the actual hyperbola branches then lie at the ends of the *real* semi-axis.

In optical applications, we shall not meet the tilted parabola of Fig. 105 but only hyperbolas and ellipses.

(4) *Summary of method for ellipse and hyperbola:*

(a) Order the given equation in conformity with the standard form

$$Ax^2 + 2Bxy + Cy^2 + 2Dx + 2Ey + F = 0$$

and tabulate the numerical values of the constants A to F.

(b) Form $(B^2 - AC)$.   If this proves to be positive, then the equation represents a hyperbola.   The direction constant of the asymptotes $(-B \pm \sqrt{B^2 - AC})/C$ may be calculated as a check on the final solution.   If $(B^2 - AC)$ is found to be negative, the equation represents a real or an imaginary ellipse.   To avoid wasting time over the latter, ascertain that $(BE - CD)^2$ is greater than $(E^2 - CF)(B^2 - AC)$ before proceeding to the general solution.

(c) Calculate the coordinates of the centre of the conic section:

$$a = (CD - BE)/(B^2 - AC) \quad \text{and} \quad b = (AE - BD)/(B^2 - AC)$$

and the new absolute term $F' = aD + bE + F$.

(d) Find the direction of one axis of the ellipse or hyperbola by calculating as an angle below 90° and with the sign of the right-hand side:

$$\tan 2\theta = 2B/(A - C).$$

Half of this angle, i.e., θ, determines the direction of the final X''-axis, the angle to be applied counter-clockwise from the original X-axis if positive, clockwise if negative.   On squared paper, the axis is best drawn by the natural tangent of θ, *not* by a protractor.

(e) Calculate $A' + C' = A + C$

and $A' - C' = $ either $(A - C)/\cos 2\theta$ or $= 2B/\sin 2\theta$,

using the sine or the cosine formula according to whether 2θ is greater or less than 45°.   Then

$$\text{Semiaxis in } X''\text{-direction} = \sqrt{-F'/A'}$$
$$\text{Semiaxis in } Y''\text{-direction} = \sqrt{-F'/C'}.$$

If the preliminary test in (b) has been used, both will be real in the case of an ellipse.   One will be imaginary in the case of a hyperbola, but must be evaluated as $\sqrt{-1}$ times the real value of the root.   It will be seen that the calculations are quite simple, but vigilant care is required as regards *signs*.

Since we shall meet both the ellipse and the hyperbola in the practical examples to be given, we shall, to save repetition, leave the best methods of drawing the

[84] graphs for incorporation in these examples. In most cases it is advisable to work out the solution numerically, as a solution in general algebraic terms becomes hopelessly complicated, but a partial exception can occasionally be highly instructive.

B. ALL SPHERICALLY CORRECTED CEMENTED OBJECTIVES ON ONE GRAPH

[85] THE general equation for the SGS of cemented objectives was given in section [82]; it is a general quadratic in the two variables on the right, c_2 and v'_2. As it is sometimes desirable to solve for a prescribed amount of SGS or to admit an uncorrected residue within the tolerance, we shall turn the original equation into one of zero-value by combining the residual SGS_0 with the absolute term. With a view to plotting c_2 in the horizontal or X-direction and v'_2 in the vertical or Y-direction, we order the equation accordingly and compare it with our adopted standard form:

$$0.11984c_2{}^2 - 0.33982c_2v'_2 + 0.21995v'_2{}^2 + 0.16238c_2 - 0.22328v'_2$$
$$Ax^2 \quad + \quad 2Bxy \quad + \quad Cy^2 \quad + \quad 2Dx \quad + \quad 2Ey$$
$$+ (0.05397 - SGS_0) = 0$$
$$+ \quad F \quad = 0.$$

Remembering signs and the factor 2 in some of the standard coefficients, we record prominently

$$A = 0.11984; \quad B = -0.16991; \quad C = 0.21995; \quad D = 0.08119;$$
$$E = -0.11164; \quad F = 0.05397 - SGS_0$$

and we can now work quite mechanically through the schedule of the summary. Since $B^2 - AC = 0.028869 - 0.026359 = +0.002510$ is positive, the equation will plot as a hyperbola.

The check-formula for the slope of the asymptotes gives

$$\text{tan of angle of asymptotes} \quad -\frac{B}{C} \pm \frac{\sqrt{B^2 - AC}}{C} = 0.77249 \pm 0.22778$$

or 0.54472 and 1.00027 as the tangents of the angles between the X-axis and the two asymptotes of the hyperbola; the asymptotes therefore have a slope upwards towards the right of 28° 35′ and 45° 1′ respectively.

The centre of the hyperbola is next located at

$$a = (CD - BE)/(B^2 - AC) = -0.001111/0.002510 = -0.4426 \text{ in } X\text{-direction}$$
$$b = (AE - BD)/(B^2 - AC) = \quad 0.000416/0.002510 = \quad 0.1658 \text{ in } Y\text{-direction,}$$

and the new absolute term of the transformed equation is

$$F' = aD + bE + F = -0.035935 - 0.018510 + 0.05397 - SGS_0$$
$$= -0.000475 - SGS_0.$$

We next have for the final turning of the coordinates

$$\tan 2\theta = 2B/(A - C) = -0.33982/-0.10011 = 3.3945,$$

whence $\qquad 2\theta = 73\text{-}35\text{-}8$; $\theta = 26\text{-}47\text{-}34$; $\tan\theta = 0.7479$ [85]

for drawing graph. The value of θ is proved correct by comparing it with the mean of the asymptote angles:

$$\tfrac{1}{2}(28\text{-}35 + 45\text{-}1) = 36\text{-}48.$$

A' and C' are then found by

$$A' + C' = A + C \qquad = \quad 0.33979$$
$$\text{and} \quad A' - C' = 2B/\sin 2\theta = -0.35424 \quad \text{giving by}$$

half algebraic sum $\quad A' = -0.00722$
half algebraic difference $\quad C' = \quad 0.34702$

With these we then find the two semiaxes of the hyperbola, and calculating for $SGS_0 = 0$ or perfect spherical correction, we have

X''-semiaxis $= \sqrt{-F'/A'} = \sqrt{0.000475/-0.00722} = 0.2565\sqrt{-1}$ imaginary,

Y''-semiaxis $= \sqrt{-F'/C'} = \sqrt{0.000475/\ 0.34702} = 0.0370$ real.

The poles of the hyperbola branches therefore lie in the obtuse angles of the asymptotes on the Y''-axis and at a distance 0.0370 from the centre defined by 'a' and 'b'. The branches will run across the diagram from south-west to northeast (in terms of the orientation of a geographical map!), and on account of the small difference in the direction of the asymptotes, they will have only a slight curvature.

The graph will be more convenient and instructive, and the calculation will be facilitated, if the vertical coordinate reads directly the reciprocal of the object distance v_1 instead of the practically uninteresting v'_2. Since $v'_2 = v_1 + c_a(N_a - 1)$, which in our case becomes $v_1 + 0.2537$, or a *constant* difference, only a final vertical shift of the origin is required for the transformation. This affects only the ordinate b of the centre of our hyperbola, which with reference to v'_2 was calculated above as $+0.1658$. This will be expressed in terms of v_1 by adding $v_1 - v'_2 = -c_a(N_a - 1)$, in our case -0.2537, and the desirable change will therefore be effected by plotting the centre of the curve at

$$b_{v_1} = b_{v'_2} + (v_1 - v'_2) = +0.1658 - 0.2537 = -0.0879,$$

all the other calculated data remaining unchanged, Fig. 107.

It is particularly important in the cases with which we are now dealing, when *calculated* angles and lengths in various directions are to be plotted, that the squared paper be tested for *uniform* spacing of its lines in *both* directions. (Irregularities of spacing of 0.01 inch are quite common, and as paper shrinks at very different rates in the directions of length and width, the vertical scale is apt to be decidedly different from the horizontal scale!) It must also be tested for the correctness of the right angle between the horizontal and vertical lines, which may be wrong by as much as a whole degree! As, chiefly by reason of shrinkage, the spacing may

[85] disagree very obviously with a correct scale, although it may be uniform, it is advisable to use a strip of the actual graph paper for the measurements.

There is rarely any advantage in plotting on a large scale; for our present case a scale of one inch for every 0.1 of c_2 and of v_1 will clearly reveal everything that is of interest. Adopting that scale, we begin by laying down the original X- and Y-axes at a convenient crossing of main divisions of the paper and marking $c_2 = -0.1, -0.2$ &c. and $v_1 = 0, \pm 0.1, \pm 0.2$ &c. We then locate the centre of our curve at $c_2 = a = -0.4426$ and at $v_1 = b_{v_1} = -0.0879$. We next mark the X''-axis by its tangent, which is 0.7479, marking say 5 inches to the right of the

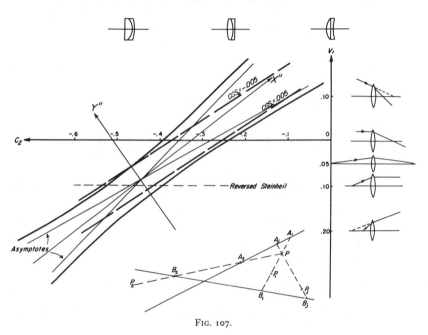

FIG. 107.

centre, then $5 \times 0.7479 = 3.74$ inches vertically and drawing the axis through the point thus found and the centre. We then add the Y''-axis at right angles with the X''-axis and on it, as the *real* axis of the hyperbola, locate the poles of the two branches at the calculated distance, which is 0.0370 from the centre. We now lay down the asymptotes, either by the tangent method described for the X''-axis with the calculated values 0.5447 and 1.0003, or by the numerical values of the semiaxes, marking the X''-semiaxis = 0.2565 in the X''-direction, then the Y''-semiaxis = 0.0370 at right angles to the X''-axis. Lines from the centre through the points thus found are the asymptotes. By using both methods, a final check on the accuracy of the data (or sometimes of the squared paper!) is secured.

Having now given the asymptotes and one point—the pole—of the hyperbola, [85] we can plot the latter in a few minutes and with high accuracy by a property of the hyperbola illustrated separately in the lower part of Fig. 107. The theorem is:

'On a straight line crossing a hyperbola and its asymptotes, the two segments between asymptote and adjacent hyperbola arm are equal.'

If P and P_1 in the figure are points of a hyperbola with the asymptotes, as drawn, then on the straight line $A_1PP_1B_1$ we must have $A_1P = P_1B_1$. The theorem holds equally for points on the two separated branches, say for P and P_5. If P_5 is a point on the hyperbola, then $P_5B_5 = A_5P$. We derive our method of plotting by using this theorem to locate hyperbola points. Assuming P to be the only known point, lay any straight line A_jPB_j through it; measure, or set dividers to, A_jP; cut off an equal distance from B_j to P_j; then P_j is also a hyperbola point. When, as in our present optical case, the hyperbola branches are very slightly curved, it is best to work from each pole towards the *other* branch to avoid intersections at highly acute angles.

The hyperbola of our graph marks all possible cemented objectives of the chosen glass and focal length for which $SGS = 0$, or those which are spherically corrected; these can therefore be read off at once for every value of v_1 or the object distance. Thus the two branches cut the original X-axis laid through $v_1 = 0$ at $c_2 = -0.390$ and -0.246, in agreement with the earlier solution for crown-in-front telescope objectives. At $v_1 = -0.1$ we read off $c_2 = -0.521$ and -0.398 for reversed Steinheil objectives, also in agreement with the earlier direct solution. But the present graph also gives direct readings for any other value of v_1 and is therefore remarkably useful. Beginners will find it very helpful in clearly realizing the full significance of the graph to add a few sketches of lens forms over the X-axis and of conjugate distances along the Y-axis: For $c_2 = -0.47$ the first surface is plane, for $c_2 = -0.235$ the crown is equiconvex, for $c_2 = 0$ the contact surface is plane; $v_1 = 0$ means that parallel light from the left is concentrated at the posterior focus, $v_1 = -0.1$ means that light from the anterior focus emerges as a parallel bundle on the right, $v_1 = +0.1$ means that arriving light aiming at the posterior focus is brought to focus at $\frac{1}{2}f'$, and so forth.

The graph also gives full information concerning the behaviour of an object glass of any given shape with reference to the Herschel condition; for as any ordinate corresponds to a given value of c_2 and therefore to a fixed lens form, we can read off the two values of v_1 for which any shape of object glass will be spherically corrected. We then know from the previous study that that lens will be suitable for use throughout that range and a little beyond if the aperture is fixed at a value admissible by the tolerance.

As the parabolic law of the change in SGS resulting from bending or from a change in object-distance always holds, and as the curvature of the respective parabolas is determined by the coefficients of $c_2{}^2$ and of $v'_2{}^2$ in the starting equation for SGS, we can easily draw the bending parabolas (Fig. 103) for any given value of v_1 by reading Fig. 107 for the two values of c_2 which give spherical correction. These values therefore locate the points in Fig. 103 at which the bending parabola

[85] cuts the zero axis. Calling these special values of c_2, c'_2 and c''_2 respectively, we then calculate for the pole of the parabola

$$\text{Min. } SGS = -\tfrac{1}{4}A(c''_2 - c'_2)^2,$$

A being the coefficient of $c_2{}^2$ in the original SGS equation. Plotting this value of c_2 midway between c'_2 and c''_2 we have the necessary three points at equal intervals of the abscissae for the use of the dip method.

In the case of the Herschel condition, Fig. 107 shows for a given shape of objective the two values v'_1 and v''_1 for which SGS equals zero, and a simple calculation of

$$\text{Min. } SGS = -\tfrac{1}{4}C(v''_1 - v'_1)^2$$

gives the midway pole of the Herschel parabola, Fig. 104, which can then be drawn.

These parabolas, or merely the calculated values of Min. SGS, facilitate discussions of tolerances, and they can also be used to find differential coefficients graphically by the method described in section [48]. They also demonstrate that objectives whose c_2- and v_1-values locate them between the two branches of the hyperbola in Fig. 107 are spherically overcorrected—the more seriously so the farther they lie from the nearest branch—and that objectives which fall outside the hyperbola branches are undercorrected.

The slope of the X''-axis in Fig. 107 justifies the rough rule given in one of the last paragraphs of section [48] for the variation of the 'middle bending' in the case of short conjugates. Evidently the desirable middle bending is one near the pole of the bending parabola, and therefore close to the X''-axis in Fig. 107. Now we determined the slope of the X''-axis by calculating that a change of 0.7479 in the v_1 direction corresponded to a change of 1.000 in the c_2 direction, or by $\delta c_2/\delta v_1 = 1/0.7479 = 1.34$. Hence a change in v_1, the $1/l$ of section [48], calls for 1.34 times that change in bending; the convenient figure 1.5 was given in the rough rule as more suitable for quick mental calculation and amply close for practical purposes. The sufficiency of this rule for all ordinary pairs of glasses depends upon the remarkable fact that the coefficients of A, B, and C in the general SGS equation, on which the slopes of the X''-axis and of the asymptotes depend, vary little for all the usual combinations and are almost pure functions of the focal length alone. To prove this, we extract the values of A, B, and C for cemented lenses from section [82], equation (4), remembering that for cemented lenses $c_2 = c_3$, and find

$$A = G_4{}^a c_a + G_4{}^b c_b; \quad B = -\tfrac{1}{2}(G_5{}^a c_a + G_5{}^b c_b); \quad C = G_6{}^a c_a + G_6{}^b c_b.$$

Introduction of the explicit G-values then gives

$$A = \frac{N_a+2}{2N_a}(N_a-1)c_a + \frac{N_b+2}{2N_b}(N_b-1)c_b = \frac{N_a+2}{2N_a}\cdot\frac{1}{f'_a} + \frac{N_b+2}{2N_b}\cdot\frac{1}{f'_b}$$

$$B = -\left(\frac{N_a+1}{N_a}(N_a-1)c_a + \frac{N_b+1}{N_b}(N_b-1)c_b\right) = -\left(\frac{N_a+1}{N_a}\cdot\frac{1}{f'_a} + \frac{N_b+1}{N_b}\cdot\frac{1}{f'_b}\right)$$

$$C = \frac{3N_a+2}{2N_a}(N_a-1)c_a + \frac{3N_b+2}{2N_b}(N_b-1)c_b = \frac{3N_a+2}{2N_a}\cdot\frac{1}{f'_a} + \frac{3N_b+2}{2N_b}\cdot\frac{1}{f'_b}.$$

The surviving functions of N vary little in numerical value for the usual range of indices from 1.5 to 1.6:

$(N+2)/2N$ gives 1.167 for $N = 1.5$ and 1.125 for $N = 1.6$

$(N+1)/N$ gives 1.667 for $N = 1.5$ and 1.625 for $N = 1.6$

$(3N+2)/2N$ gives 2.167 for $N = 1.5$ and 2.125 for $N = 1.6$,

hence we shall not make a serious error if we use average values for $N = 1.55$ in both terms of each equation, thus securing $1/f'_a + 1/f'_b = 1/f'$ as a common factor. This trick gives approximately:

$$A = 1.144/f'; \quad B = -1.645/f'; \quad C = 2.144/f',$$

which, for our $f' = 10$, will be noticed to give a very fair approach to the actual values.   They will be found useful for estimates of the dip of the 'bending' and Herschel parabolas and as an additional rough check in calculations by the SGS methods.

The coefficients D, E, and F of the SGS equation are less fixed in value, and a change of glass consequently leads to very appreciable shifts of the centre of the hyperbola and especially of the curve itself.   These changes in the hyperbola are of a very interesting type.   We will study them first for our example without a change of glass.   The two semiaxes of the hyperbola are proportional to the root of F', and we found $F' = -0.000394 - SGS_0$, SGS_0 standing for any definite amount of spherical under- or over-correction for which we may desire to solve. If we desired spherical under-correction, SGS_0 would have to be positive and F' would rise in numerical value.   The semiaxes would grow and the hyperbola branches would become more and more widely separated.   But if we solved for a negative SGS_0 or spherical over-correction, then F' would diminish in numerical value and the hyperbola branches would move closer to the asymptotes.   For $SGS_0 = -0.000475$, an amount which by our earlier discussions would be well inside the Rayleigh limit up to considerable apertures for our 10-inch lenses, we should reach $F' = 0$, or the zero-value of both semiaxes of the hyperbola.   That is the case when the hyperbola degenerates into two intersecting straight lines, the asymptotes.   If we now ask for still higher spherical over-correction, F' will change from the previous negative to new positive values, with the result—as A' and C' always retain the same value—that the X''-semiaxis becomes real and the previously real Y''-semiaxis becomes imaginary.   This means that for sufficiently high, but really quite moderate, spherical over-correction, the hyperbola branches break through the asymptote barrier and become acutely curved lines fitting the acute angles bisected by the X''-axis, moving farther and farther away from the centre of the hyperbola as the over-correction is increased.

Changes in the kinds of glass will produce the same sequence of modifications even for a permanent zero-value of SGS_0.   If, while retaining substantially the original flint glass and the V-value of the crown glass, we lower the index of the latter, say by using ordinary hard crown instead of the light barium crown, F' will grow in numerical value and the hyperbola branches will move farther away from the asymptotes.   On the contrary, if we raise the index of the crown glass above the

[85] original 1.54, the hyperbola branches will approach the asymptotes more closely. At about $N = 1.59$ or 1.60 we shall pass through the interesting case when the asymptotes themselves become the graph representing the SGS equation, and for still higher values of the crown-glass index, the hyperbola branches will migrate into the acute angles of the asymptotes.

We can finally make an important addition to our study of cemented objectives by including the coma, for which we found the general solution (for cemented objectives!):

$$CGS = 0.08495c_2 - 0.13496v'_2 + 0.06093.$$

Turned into a zero-equation for any prescribed value CGS_0 of the coma-sum, this becomes

$$0.08495c_2 - 0.13496v'_2 + (0.06093 - CGS_0) = 0,$$

and being a linear equation in c_2 and v'_2 it will plot as an inclined straight line and will be determined for any given value of CGS_0 by two points of this line. All the 'coma lines' for successive values of CGS_0 will be parallel and will be separated by intervals proportional to CGS_0. We will now add two such coma lines to Fig. 107, choosing $CGS_0 = +0.005$ and -0.005, respectively, because this amount will by $OSC' = SA^2 \cdot CGS$ correspond to our OSC'-limit of ± 0.0025 for the reasonable semi-aperture of 0.7071 inch for our 10-inch lenses, and will thus give a clear indication of the tolerance. To obtain the necessary plotting points on the $v_1 = -0.1$ and $v_1 = +0.2$ levels of Fig. 107, we must calculate the c_2 of the last equation with $v'_2 = v_1 + 1/f'_a = -0.1 + 0.2537 = +0.1537$ and $+0.2 + 0.2537 = +0.4537$ respectively and find:

for $CGS_0 = -0.005$ the points $v_1 = -0.1$, $c_2 = -0.5319$;

and $v_1 = +0.2$, $c_2 = -0.0553$

for $CGS_0 = +0.005$ the points $v_1 = -0.1$, $c_2 = -0.4142$;

and $v_1 = +0.2$, $c_2 = +0.0624$.

Heavy broken lines marked with the respective values of CGS_0 have been drawn in Fig. 107 through the points thus defined. Bearing in mind that the line for zero coma would lie exactly midway between the two lines that were actually drawn, we easily see:

(1) As the coma lines have a slope intermediate between that of the two asymptotes, there can be no intersection of the zero-coma line with either branch of the hyperbola; hence, with our chosen glasses, an aplanatic cemented objective is totally impossible at any value of v_1.

(2) The actual line for $CGS_0 = -0.005$ crosses the left-hand branch of the hyperbola for low values (algebraically!) of v_1, indicating that this branch will yield the nearest approach to aplanatism for negative values of v_1 below about $v_1 = -0.05$, which represents equal conjugate distances.

(3) The actual line for $CGS_0 = +0.005$ indicates in the same way that for values of v_1 greater than about -0.05, the objectives from the right-hand branch of the hyperbola approach most closely to aplanatism.

(4) As the relation of the coma lines with reference to the asymptotes varies only

slightly with changes of glass, it is easily seen that true aplanatism will only become [85] possible when the hyperbola branches have made their interesting break through the asymptotes into the acute angles of the latter, either by a sufficient allowance of spherical over-correction with the present kinds of glass, or by changing to a crown of higher index.

C. All Aplanatic Objectives on One Graph

We learnt in Chapters VI and VII that exact aplanatism can be achieved with both crown-in-front and flint-in-front telescope objectives if a slight air-gap is allowed between the component lenses. We can now solve this problem in its utmost generality. In section [82] we obtained the equations:

$$\text{General } SGS = \begin{cases} 0.05397 + 0.24294c_2 - 0.08056c_3 - 0.22328v'_2 + 0.29147c_2{}^2 \\ -0.17163c_3{}^2 - 0.83676c_2v'_2 + 0.49694v'_2c_3 + 0.21995v'_2{}^2 \end{cases}$$

$$\text{General } CGS = 0.06093 + 0.20917c_2 - 0.12422c_3 - 0.13496v'_2,$$

which in this form contain three variables on the right and are therefore not suitable for plotting as a plane curve. But if we decide that only coma-free objectives are to be considered, then the second equation with $CGS = 0$ supplies the necessary condition by which one of the three variables can be eliminated from the SGS equation. We shall therefore eliminate c_3. If $CGS = 0$ is transposed into a solution for c_3 it gives:

$$0.12422c_3 = 0.20917c_2 - 0.13496v'_2 + 0.06093,$$

or $$c_3 = [0.2263]c_2 + [0.0360n]v'_2 + [9.6906].$$

We use this equation to replace c_3 by c_2 and v'_2 in the general SGS equation, the safest procedure being to write down first the terms not containing c_3 and then add the equivalents of the terms containing c_3. This procedure gives:

SGS	$=$	$0.29147c_2{}^2 - 0.83676c_2v'_2 + 0.21995v'_2{}^2 + 0.24294c_2 - 0.22328v'_2 + 0.05397$
$-0.08056c_3 =$		$-0.13564c_2 + 0.08752v'_2 - 0.03951$
$0.49694c_3v'_2 =$		$0.83676c_2v'_2 - 0.53990v'_2{}^2 + 0.24375v'_2$
$-0.17163c_3{}^2 =$		$-0.48663c_2{}^2 + 0.62798c_2v'_2 - 0.20258v'_2{}^2 - 0.28351c_2 + 0.18293v'_2 - 0.04129$

$$\text{Sum} = SGS = -0.19516c_2{}^2 + 0.62800c_2v'_2 - 0.52253v'_2{}^2 - 0.17621c_2 + 0.29092v'_2 - 0.02683$$

To provide for the possibility of solving for a coma-free objective with a prescribed amount SGS_0 of spherical over- or under-correction, we form a zero-equation by throwing SGS_0 to the right and are then ready to discuss the general quadratic in c_2 and v'_2 by the method of the summary, with $A = -0.19516$; $B = 0.31400$; $C = -0.52253$; $D = -0.08810$; $E = 0.14546$; $F = (-0.02683 - SGS_0)$. Now $B^2 - AC = 0.09860 - 0.10198 = -0.00338$ is negative, hence the equation represents an ellipse, if it represents any real curve. Furthermore, the test advised in the summary proves that $(BE - CD)^2 > (E^2 - CF)(B^2 - AC)$, and therefore the ellipse is real. It should be noted that, of the three terms in this test, the first is required for the calculation of 'a' in the final solution and the third is already known, hence computing $(E^2 - CF)$ is the only added labour!

[85]   The standard method now gives in quick succession:

$a = -0.1066$; $b = 0.2144$; $F' = 0.01375 - SGS_0$; $\theta = 31\text{-}14\text{-}2$, $\tan\theta = 0.6064$;
$A' = -0.00474$; Semiaxis in X''-direction $= 1.703$ } for $SGS_0 = 0$
$C' = -0.71294$; Semiaxis in Y''-direction $= 0.1389$ }

In plotting the curve, it will again be convenient to render the graph direct-reading in terms of v_1 instead of the uninteresting v'_2, and as 'b' is the only direct ordinate in the calculated data, we can effect this desirable change by plotting

$$b_{v_1} = \text{calculated } b_{v'_2} - 1/f'_a = 0.2144 - 0.2537 = -0.0393 \text{ increase.}$$

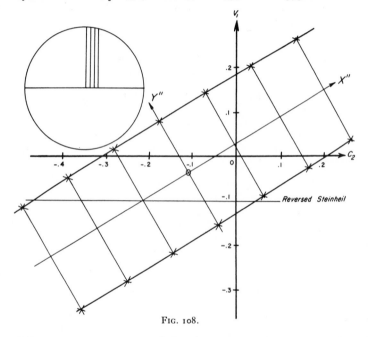

FIG. 108.

The scale of 1 inch per 0.1 of c_2 and v_1 will again be suitable. It would be a senseless waste of paper to try to draw the whole ellipse, which would be 34.06 inches long on the adopted scale, because interest is limited to conjugate distances covering at most $v_1 = +0.1$ to $v_1 = -0.2$ or -0.3; only the central part need be represented, Fig. 108.

We first lay down the centre of the ellipse at $c_2 = a = -0.1066$ and $v_1 = b_{v_1} = -0.0393$. We then draw the X''-axis through this centre so that its tangent is $+0.6064$. To draw the interesting part of the long ellipse, we strike a circle in a convenient blank corner of the paper with the short semiaxis as radius. The length of this semiaxis is 0.1389, which is represented by 1.389 inches in the plot. We

then draw a few ordinates at equal intervals.   Since an ellipse may be regarded as   [85]
an oblique elongated projection of a circle having the short axis as diameter, the
ellipse will have the same ordinates as our auxiliary circle at intervals of the
X''-axis increased in the proportion of long semiaxis to short semiaxis.   In our
case, the intervals are $0.1 \times 1.703/0.1389 = 1.23$ inches.   We therefore mark such
intervals with a strip of the actual graph paper as scale to both sides of the centre
along the X''-axis and find the ellipse points marked by crosses in Fig. 108.   Then
we can draw the curves, in the present case, on account of the very slight curvature,
as a succession of ruled straight lines between the successive points.

As the present graph implies a *condition*, it gives directly only the c_2 of the crown
lens which will yield an aplanatic objective for any value of v_1 or of the object
distance.   The form of the flint lens is then determined by putting the values of c_2
and v_1 into the zero-coma condition

$$c_3 = [0.2263]c_2 + [0.0360n]v'_2 + [9.6906],$$

which requires further that the selected v_1 is first turned into the corresponding
v'_2 by the equation $v'_2 = v_1 + 1/f'_a$, which becomes $v_1 + 0.2537$ in our case.

Thus for $v_1 = 0$, or a crown-in-front telescope objective, the left-hand arm of
our ellipse reads, as nearly as the scale of the graph allows, $c_2 = -0.3081$, thereby
fixing the form of the crown component.   With $v'_2 = 0 + 0.2537$ and $c_2 =
-0.3081$, the zero-coma condition then gives $c_3 = -0.3039$ in almost absolute
agreement with the value -0.3402 found by direct algebraic solution in section
[82].   At $v_1 = -0.1$ we should find in the same way the two possible solutions for
reversed Steinheil aplanatic telescope objectives.   At $v_1 = -0.05$ we should
obtain the solution for an aplanatic objective with equal conjugate distances of
opposite sign, or for $M' = -1$, as frequently is the condition in periscopes.

The right-hand arm of the ellipse throughout the usual range of v_1 gives the
venturesome alternative forms with meniscus crown components, which are rarely
worth working out.

It is theoretically interesting to note that, because our graph yields an ellipse of
limited though considerable length, it is impossible to find an aplanatic objective for
high numerical values of v_1 (short object distance).   In practice this limitation
never becomes felt as the range of v_1 actually covered is far greater than real
designing problems ever call for.

We will briefly discuss the case when the general equation is to be solved for
coma-free objectives with a prescribed finite value of SGS_0 instead of complete
spherical correction.   As SGS_0 only enters into the absolute term F and thence
into the final absolute term $F' = 0.01375 - SGS_0$, the ellipse will retain the same
centre, the same angle θ of the X''-axis, and the same *ratio* of long to short axis, but
as the two semiaxes are proportional to $\sqrt{F'}$, the *size* of the ellipse will change.
For increasingly negative values of SGS_0, or for spherically over-corrected coma-
free objectives, F' and with it the size of the ellipse will steadily grow, hence
solutions for any amount of spherical over-correction are obtainable.   But for
increasingly positive values of SGS_0, F' and with it the size of the ellipse will
shrink until, for $SGS_0 = 0.01375$, F' becomes zero and the ellipse shrinks to a
single point at its centre.   Still larger values of SGS_0 would reverse the sign of F'

[85] and lead to imaginary semiaxes or no solution. Hence the centre of our ellipse defines the most spherically under-corrected coma-free objective as having $c_2 = -0.1066$ and $v_1 = -0.0393$, c_3 being found by the zero-coma condition. The impossibility of making coma-free objectives that are also highly under-corrected, although interesting theoretically, is of no practical importance. As we found earlier, for reasonable apertures of our 10-inch objectives the Rayleigh limit of spherical aberration corresponds to SGS values of the order of 0.001 or less, so the impossible values of SGS above 0.01375 would represent under-correction of such extravagant magnitude as to be practically out of the question.

Similar problems arise in connection with the coma correction. If we required objectives with a prescribed finite value of CGS instead of zero, the coma condition would become

$$0.12422c_3 = 0.20917c_2 - 0.13496v'_2 + (0.06093 - CGS_0),$$

and it is easily seen that the use of this equation in the elimination of c_3 from the SGS equation would alter the coefficients D, E, and F, but not A, B, and C. Hence the ellipse would retain the direction and the ratio of its axes but its centre would shift and its size would change, with further possibilities of eventual imaginary solutions. But again the limit for CGS was found to be of the order of 0.005 whilst the fixed numerical coefficient which it alters is 0.061; therefore the changes in the ellipse resulting from any reasonably probable value of the desired CGS will in practice be fairly small.

It may be added that by substituting the Herschel condition for the zero-coma condition in the above solution, we can obtain a very similar absolutely general solution for all possible Herschel objectives. This solution is, however, of less interest than that for aplanatic objectives, partly because fulfilling the Herschel condition is not often important and partly because, as we learnt in the discussion of the Herschel condition, when it does acquire importance, the best solution usually depends on a clever compromise between spherical correction, coma correction, and the Herschel condition itself. When carried out, the solution for spherically corrected objectives satisfying the Herschel condition gives:

Equation of Condition: $c_3 = [0.2263]c_2 + [9.9471n]v'_2 + [9.6525]$, and with this the equation of the Herschel ellipse:

$$0 = -0.19516c_2^2 + 0.51168c_2v'_2 - 0.35452v'_2{}^2 - 0.15236c_2 + 0.20781v'_2$$
$$- (0.01686 - SGS_0).$$

The solution is:

$$a = -0.1136; \quad b_{v_1} = -0.0419; \quad \tan\theta = 0.7359;$$
$$X''\text{-semiaxis} = 1.414; \quad \text{and} \quad Y''\text{-semiaxis} = 0.1593.$$

The Herschel ellipse is therefore broader and shorter than the aplanatic ellipse, has a slightly shifted centre, and lies at a decidedly steeper angle. When plotted on Fig. 108, it is found to cross the aplanatic ellipse at $v_1 = -0.05$, as it must do by reason of the identity of the two conditions for equal conjugates or $M' = -1$. The other two crossing points of the two ellipses at high numerical values of v_1 indicate only that the crown lens has the same form; the flint lenses are very different for the Herschel and the aplanatic cases respectively on account of the difference [85] in the two equations of condition.

D. ALL SPHERICALLY CORRECTED OBJECTIVES FOR GIVEN CONJUGATES

We obtain another highly instructive graph if we reduce the general SGS equation to two independent variables by assigning selected definite values to SGS_0 and to v'_2 and plotting the resulting equation as a conic section against c_2 and c_3 as abscissae and ordinates respectively.

The equation to be plotted will be

$$0.29147c_2{}^2 - 0.17163c_3{}^2 + c_2(0.24294 - 0.83676v'_2) + c_3(-0.08056 + 0.49694v'_2)$$
$$+ (0.05397 - SGS_0 - 0.22328v'_2 + 0.21995v'_2{}^2) = 0$$

and by comparison with our standard quadratic equation gives

$$A = 0.29147; \quad B = 0; \quad C = -0.17163; \quad D = 0.12147 - 0.41838v'_2;$$
$$E = -0.04028 + 0.24847v'_2; \quad F = 0.05397 - SGS_0 - 0.22328v'_2 + 0.21995v'_2{}^2.$$

As $B^2 - AC$ is obviously positive, the graph will be a hyperbola, and as $B = 0$, the direction constants of the asymptotes will be $\pm \sqrt{-A/C} = \pm 1.3032$. The axes of the hyperbola are vertical and horizontal, and only a parallel shift of the co-ordinates is required to obtain the standard form of the equation as referred to the centre of the hyperbola. A' becomes A and C' becomes C, and the equations for a, b, and F' reduce to

$$a = -D/A; \quad b = -E/C; \quad F' = -D^2/A - E^2/C + F.$$

These can be easily worked out with retention of v'_2 and SGS_0 in algebraic symbols to give;

$$a = -0.41674 + 1.4354v'_2; \quad b = -0.23469 + 1.4477v'_2;$$
$$F' = 0.01279 - SGS_0 + 0.00883v'_2 - 0.02088v'_2{}^2.$$

From the value of F' we find the semiaxes, also in general terms:

In c_2-direction:

Semiaxis $= \sqrt{(-F'/A)} = \sqrt{-0.04388 + 3.431SGS_0 - 0.03030v'_2 + 0.07164v'_2{}^2}$

In c_3-direction:

Semiaxis $= \sqrt{(-F'/C)} = \sqrt{0.07452 - 5.826SGS_0 + 0.05145v'_2 - 0.12166v'_2{}^2}$.

We can include the coma by putting the general CGS equation into the form

$$(0.06093 - CGS_0 - 0.13496v'_2) + 0.20917c_2 - 0.12422c_3 = 0,$$

and we see that this plots as a straight line at a fixed slope, but shifts parallel with itself according to the values of CGS_0 and/or v'_2 put into the individual solutions.

We shall now draw the actual graphs for spherically corrected crown-in-front telescope objectives by introducing $SGS_0 = 0$ and $v'_2 = 1/f'_a = 0.2537$ into the

[85] general solution. The results are: $a = -0.05258$; $b = +0.13259$; $F' = 0.01368 - SGS_0$; semiaxis in c_2 direction $= 0.2167$ (imaginary); and semiaxis in c_3 direction $= 0.2824$ (real).

Plotted on the principles laid down in connection with Fig. 107, these data yield the graph in Fig. 109. Any point of either hyperbola branch represents a spherically corrected objective with c_2 equal to its abscissa and c_3 equal to its ordinate. The graph necessarily includes the two cementable objectives which were solved for earlier. For these objectives $c_2 = c_3$ and they lie on a straight line drawn through the origin at an angle of $45°$. Such a line will, on a properly drawn graph, cut the

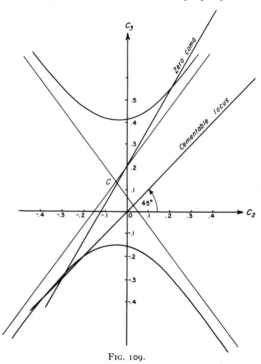

FIG. 109.

lower hyperbola branch in two points closely agreeing with the earlier solution. The zero-coma condition gives the linear equation

$$c_2 = [9.7737]c_3 - 0.12756.$$

For $c_3 = +0.5$, $c_2 = +0.1694$; for $c_3 = -0.2$, $c_2 = -0.2463$.

The zero-coma line drawn through these points cuts the two hyperbola branches in two points agreeing with the earlier direct solution for aplanatic objectives, the lower branch yielding the useful solution, the upper branch, the freakish solution of two meniscus lenses.   As the coma line shifts parallel with itself in strict proportion   [85] to the amount of coma solved for, objectives taken from the graph will have the more coma the farther they lie from the zero-coma line.   Therefore the left-hand part of the upper branch and the right-hand part of the lower branch will yield undesirable combinations, and, as these parts also lie far from the 'cementable' line, such combinations will be characterized by a large difference between c_2 and c_3.

As the semiaxes of the hyperbola are proportional to $\sqrt{F'}$ and as we found $F' = 0.01368 - SGS_0$, it is evident that the semiaxes will grow if we solve for spherically over-corrected objectives and will shrink for under-corrected objectives. It follows that objectives whose c_2- and c_3-values locate them in the hollows of the hyperbola branches are over-corrected and those located in the much larger area between the branches are under-corrected.   For the huge under-correction represented by $SGS_0 = 0.01368$, the asymptotes themselves become the graph of the equation; for still greater under-correction, the real and imaginary axes would change places and the hyperbola branches would migrate into the obtuse angles of the asymptotes.   In the latter case the zero-coma line—which remains immovable —would no longer cut the hyperbola branches; our present graph thus confirms the conclusion drawn from Fig. 108 that coma-free objectives are only possible up to a certain definite limit of spherical under-correction.

The general solutions for the values of a, b, and F' with v'_2 at any positive or negative magnitude enable us to draw graphs like Fig. 109 for any conjugate distances.   If there were a practicable method of constructing three-dimensional models, the continuity of all the resulting hyperbolas would yield a surface of the second order if a v'_2 axis at right angles with the c_2- and c_3-axes of Fig. 109 were added as the third dimension.   For spherically corrected objectives, the resulting model would take the form of a 'hyperboloid of one sheet', very elongated in the v'_2-direction and steeply tilted with reference to both the c_3- and v'_2-axes.   This is merely mentioned because such a model would display the logical connection between the three separate plottings shown in Figs. 107, 108, and 109.   Fig. 109 is of course a section at right angles to the v'_2-axis and at $v'_2 = 0.2537$.   Fig. 107 is the projection upon the c_2-v'_2 plane of the section of the lower part of the hyperboloid by a plane containing the v'_2-axis and at $45°$ with the c_2-v'_2 plane.   Fig. 108 is the projection upon the c_2-v'_2 plane of the section of the complete hyperboloid by the inclined plane represented by $CGS = 0$.

It is not difficult to conclude that all conceivable cemented lenses—undercorrected, corrected, and over-corrected—would yield a model in the form of a hyperbolic paraboloid if SGS is represented in the Z-direction, and that all conceivable coma-free objectives would in the same way yield an elliptic paraboloid.

The chief value of the general solutions dealt with in the present section lies in the remarkable broadness with which they present all the possibilities and limitations of thin systems of two components.   In the working out of definite designing problems, the specialized methods of Chapters V and VII of Part I will usually be preferable on account of their more direct relations with the trigonometrical correcting process which must always follow the approximate analytical solution.

TRIPLE CEMENTED OBJECT-GLASSES

[86] IN section [64] D, reasons were stated why cemented combinations are so widely preferred in all optical systems of small and medium size, up to 2 or 3 inches aperture or so.   It was also shown that the usual two-lens cemented combination causes difficulties when aplanatism is required because there are not sufficient liberties for varying the design, the choice of dense barium crown glass being only a partial remedy calling for compromises on account of the limited choice of available types.   We can secure a freely and continuously available added liberty for varying the design by cementing a third lens to the original pair.   The most usual and also the most useful forms of these triple cemented objectives consist of either a flint lens cemented between two crown lenses or—what is nearly always preferable— of a crown lens cemented between two flint lenses of the same kind of glass.   If properly designed, the latter type admits of apertures as large as $f/3$ or more, and this adds greatly to the value of the type.

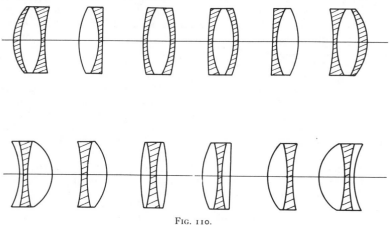

Fig. 110.

We can imagine all possible forms of both types to be produced by beginning with an ordinary binary combination of the two selected kinds of glass and then dividing either the crown lens or the flint lens into two parts by a spherical surface. These two parts will retain the exact combined power of the original lens and will produce the same kind of achromatism, and they will continue to do this when bent into any possible shape, in accordance with the laws of thin lenses deduced in Chapter I of Part I.   Moreover, the three lenses now constituting the complete system may be arranged in any order without disturbing either the power or the achromatism of the thin system.   By thus dividing the flint lens, we can produce the successive shapes shown in the upper row of Fig. 110, and by splitting the crown lens, we obtain the succession in the lower row, with of course all intermediate and still more extreme flanking possibilities.   It should be noted that the process may include a flint lens split into a convex and a correspondingly stronger concave part or a crown lens split into a weak concave and a very strong convex part. Such [86] divisions are not likely to prove favourable, but our perfectly general solution will include them automatically. The solution will in the same way'include the simple binary combination—either way round—and these have therefore been included in the diagrams. It would obviously be a very laborious task to explore all these possibilities empirically by successive trigonometric trials, especially as bendings of each possible division would also require attention; the general thin-lens solution of the problem is therefore decidedly valuable as it will reduce the final trigono-metric work to a mere differential correction.

In the general solution, it will be convenient to distinguish the two kinds of optical glass by the suffix o for the outside lenses and i for the inside lens. We can then calculate the respective total curvatures by the usual TL,Chr solution:

$$c_i = 1/f' \cdot \delta N'_i \cdot (V_i - V_o) \qquad c_o = 1/f' \cdot \delta N'_o (V_o - V_i)$$

with inclusion of the terms in R if a prescribed chromatic over- or under-correction is to be included.

To determine the spherical aberration and the coma, we must then work out the G-sums by Seidel Aberrations III. Very much depends in the present rather intricate case on the skilful choice of the variables by which the bendings and the splitting up of the outside total curvature c_o are introduced into the equations. We shall obtain a remarkably simple and convenient solution by noting that in these triple lenses there is one special case of complete symmetry, namely, when the inside lens is exactly equiconvex or equiconcave and when the outside lenses are also exactly equal in strength, and when this combination works on equal conju-gates of $-2f'$ and $+2f'$ to the left and the right respectively.

By expressing all possible modifications in terms of the resulting departure from this completely symmetrical case, both the algebra of the proof and the calculations of concrete cases are greatly simplified. The central or inside lens always has the fixed total curvature c_i and can be modified only by bending. For the working out of the G-sums, we will call the curvatures of the four refracting surfaces from left to right c_1, c_2, c_3, and c_4 and the total curvatures of the three components by the usual names c_a, c_b, and c_c. In the completely symmetrical case, we shall evidently have $c_a = \frac{1}{2}c_o$ and $c_c = \frac{1}{2}c_o$ and always $c_b = c_i$. If we disturb the symmetry by dividing c_o unequally between the two outside lenses, one will gain as much as the other loses because the sum must always equal c_o. We call x the amount by which c_a differs from the value $\frac{1}{2}c_o$; hence the general values will be

$$c_a = \tfrac{1}{2}c_o + x \quad \text{and} \quad c_c = \tfrac{1}{2}c_o - x.$$

The curvatures of the inside lens in the symmetrical case will be $c_2 = \frac{1}{2}c_i$ and $c_3 = -\frac{1}{2}c_i$. Bending will change both by the same amount, which we call y; hence the general values of the curvature of the cemented surfaces will be

$$c_2 = \tfrac{1}{2}c_i + y \quad \text{and} \quad c_3 = -\tfrac{1}{2}c_i + y.$$

But as $c_1 - c_2 = c_a$ and $c_3 - c_4 = c_c$, the external curvatures now follow by $c_1 = c_a + c_2$ and $c_4 = c_3 - c_c$ as:

$$c_1 = \tfrac{1}{2}(c_o + c_i) + (x + y) \quad \text{and} \quad c_4 = -\tfrac{1}{2}(c_o + c_i) + (x + y),$$

[86] and we have thus determined all the curvature values.

In the completely symmetrical case, $v_1 = -1/2f'$ or, since $1/f' = c_o(N_o-1) + c_i(N_i-1)$, then $v_1 = -\frac{1}{2}[c_o(N_o-1) + c_i(N_i-1)]$; if we denote the departure from this special value of v_1 by z, we have to employ in the working out of the G-sums

$$v_1 = -\tfrac{1}{2}[c_o(N_o-1) + c_i(N_i-1)] + z.$$

If we now decide to use the alternative G-sums for component 'a' and the original G-sums for components 'b' and 'c', we shall require $v'_2 = v_1 + c_a(N_o-1)$ and $v'_3 = v'_2 + c_i(N_i-1)$. Then we find, with the general v_1 already determined,

$$v'_2 = -\tfrac{1}{2}c_i(N_i-1) + x(N_o-1) + z \quad \text{and} \quad v'_3 = \tfrac{1}{2}c_i(N_i-1) + x(N_o-1) + z.$$

Introducing the c- and v-values into the G-sums for spherical aberration as explicitly restated in section [82], we now find

$SGS =$

First lens
$$\left\{ \begin{array}{l} G_1{}^o(\tfrac{1}{2}c_o+x)^3 + G_2{}^o(\tfrac{1}{2}c_o+x)^2(\tfrac{1}{2}c_i+y) \\ - G_3{}^o(\tfrac{1}{2}c_o+x)^2[-\tfrac{1}{2}c_i(N_i-1) + x(N_o-1) + z] + G_4{}^o(\tfrac{1}{2}c_o+x)(\tfrac{1}{2}c_i+y)^2 \\ - G_5{}^o(\tfrac{1}{2}c_o+x)(\tfrac{1}{2}c_i+y)[-\tfrac{1}{2}c_i(N_i-1) + x(N_o-1) + z] \\ + G_6{}^o(\tfrac{1}{2}c_o+x)[-\tfrac{1}{2}c_i(N_i-1) + x(N_o-1) + z]^2 \end{array} \right.$$

Second lens
$$\left\{ \begin{array}{l} + G_1{}^i c_i{}^3 - G_2{}^i c_i{}^2(\tfrac{1}{2}c_i+y) + G_3{}^i c_i{}^2[-\tfrac{1}{2}c_i(N_i-1) + x(N_o-1) + z] \\ + G_4{}^i c_i(\tfrac{1}{2}c_i+y)^2 - G_5{}^i c_i(\tfrac{1}{2}c_i+y)[-\tfrac{1}{2}c_i(N_i-1) + x(N_o-1) + z] \\ + G_6{}^i c_i[-\tfrac{1}{2}c_i(N_i-1) + x(N_o-1) + z]^2 \end{array} \right.$$

Third lens
$$\left\{ \begin{array}{l} + G_1{}^o(\tfrac{1}{2}c_o-x)^3 - G_2{}^o(\tfrac{1}{2}c_o-x)^2(-\tfrac{1}{2}c_i+y) \\ + G_3{}^o(\tfrac{1}{2}c_o-x)^2[\tfrac{1}{2}c_i(N_i-1) + x(N_o-1) + z] \\ + G_4{}^o(\tfrac{1}{2}c_o-x)(-\tfrac{1}{2}c_o+y)^2 \\ - G_5{}^o(\tfrac{1}{2}c_o-x)(-\tfrac{1}{2}c_i+y)[\tfrac{1}{2}c_i(N_i-1) + x(N_o-1) + z] \\ + G_6{}^o(\tfrac{1}{2}c_o-x)[\tfrac{1}{2}c_i(N_i-1) + x(N_o-1) + z]^2. \end{array} \right.$$

Multiplying out and ordering the result as an equation in x, y, and z, it is found that all the cubic terms, such as x^3, xy^2, which are contributed by the first lens are cancelled by similar terms of opposite sign from the third lens. Moreover, all the linear terms in x, y, or z similarly disappear, either obviously, in the case of those from the first and third lenses, or, in the case of the second lens, because the coefficients vanish by reason of the relations collected in equation [82] (1). Hence the result is a pure quadratic equation in x, y, and z. The contribution of the second lens to the absolute term can be reduced to simply $\tfrac{1}{4}G_1{}^i c_i{}^3$ by a further application of [82] (1), and the equation then becomes

$$\begin{aligned} SGS = \; & x^2[3G_1{}^o c_o + G_2{}^o c_i - 2G_3{}^o c_o(N_o-1) + G_3{}^o c_i(N_i-1) - G_5{}^o c_i(N_o-1) \\ & \qquad + G_6{}^o c_o(N_o-1)^2 - 2G_6{}^o c_i(N_o-1)(N_i-1) + G_6{}^i c_i(N_o-1)^2] \\ & + xy[2G_2{}^o c_o + 2G_4{}^o c_i - G_5{}^o c_o(N_o-1) + G_5{}^o c_i(N_i-1) - G_5{}^i c_i(N_o-1)] \\ & + y^2(G_4{}^o c_o + G_4{}^i c_i) \\ & + xz[-2G_3{}^o c_o - G_5{}^o c_i + 2G_6{}^o c_o(N_o-1) - 2G_6{}^o c_i(N_i-1) + 2G_6{}^i c_i(N_o-1)] \\ & + yz(-G_5{}^o c_o - G_5{}^i c_i) + z^2(G_6{}^o c_o + G_6{}^i c_i) \\ & + \tfrac{1}{4}[G_1{}^o c_o{}^3 + G_1{}^i c_i{}^3 + G_2{}^o c_o{}^2 c_i + G_3{}^o c_o{}^2 c_i(N_i-1) + G_4{}^o c_o c_i{}^2 \\ & \qquad\qquad + G_5{}^o c_o c_i{}^2(N_i-1) + G_6{}^o c_o c_i{}^2(N_i-1)^2]. \end{aligned}$$

A final application of equation [82] (1) and use of the explicit G-values in other [86] parts of the more complicated coefficients then gives the final computing equation included in the summary.

The coma sum is much simpler. Put into the sequence 'first-third-second lens', the standard equation gives

$$
\begin{aligned}
CGS = \ & \tfrac{1}{4}G_5^o(\tfrac{1}{2}c_o+x)(\tfrac{1}{2}c_i+y) && -G_7^o(\tfrac{1}{2}c_o+x)[-\tfrac{1}{2}c_i(N_i-1)+x(N_o-1)+z] \\
& && +G_8^o(\tfrac{1}{2}c_o+x)^2 \\
& +\tfrac{1}{4}G_5^o(\tfrac{1}{2}c_o-x)(-\tfrac{1}{2}c_i+y) && -G_7^o(\tfrac{1}{2}c_o-x)[\tfrac{1}{2}c_i(N_i-1)+x(N_o-1)+z] \\
& && -G_8^o(\tfrac{1}{2}c_o-x)^2 \\
& +\tfrac{1}{4}G_5^i(\tfrac{1}{2}c_i+y)c_i && -G_7^ic_i[-\tfrac{1}{2}c_i(N_i-1)+x(N_o-1)+z] \\
& && -G_8^ic_i^2.
\end{aligned}
$$

On multiplying out and ordering, all the quadratic terms as well as the complete absolute term cancel out—in some cases by applying equation (1)—and the surviving terms are

$$
\begin{aligned}
CGS = \ & x[\tfrac{1}{4}G_5^o(c_i+c_o)+G_7^oc_i(N_i-1)-G_7^ic_i(N_o-1)] \\
& +y(\tfrac{1}{4}G_5^oc_o+\tfrac{1}{4}G_5^ic_i)-z(G_7^oc_o+G_7^ic_i).
\end{aligned}
$$

The use of the explicit G-values in the coefficient of x then gives the final computing equation of the summary.

TL SOLUTION FOR TRIPLE CEMENTED OBJECT-GLASSES

The glass data being distinguished by the suffix o for the outside lenses and the suffix i for the inside lens, we solve for achromatism by

$$c_i = 1/f'\cdot\delta N'_i(V_i-V_o) \quad \text{and} \quad c_o = 1/f'\cdot\delta N'_o(V_o-V_i)$$

or with the inclusion of the R-terms for chromatic over- or under-correction. Since x is the departure from the equal division of the outside glass, y the departure from the symmetrical form of the inside lens, and z the departure from equal conjugate distances, we then calculate the G-sums by

$$
\begin{aligned}
SGS = \ & x^2\left[G_4^oc_o-c_i\frac{N_o-1}{2}-c_i\frac{(N_o-1)^2}{N_i}+2c_i\frac{N_i(N_o-1)}{N_o}\right]+y^2(G_4^oc_o+G_4^ic_i) \\
& +z^2(G_6^oc_o+G_6^ic_i)+xy\left[2G_4^o(c_o+c_i)-2c_i(N_o-N_i)\cdot\frac{N_o-1}{N_o}\cdot\frac{N_i-1}{N_i}\right] \\
& +xz\left[-G_5^o(c_o+c_i)+2c_i(N_o-N_i)\frac{N_o-1}{N_o}\cdot\frac{N_i-1}{N_i}\right]+yz(-G_5^oc_o-G_5^ic_i) \\
& +\tfrac{1}{4}[G_1^oc_o^3+G_1^ic_i^3+G_2^oc_o^2c_i+G_3^oc_o^2c_i(N_i-1)+G_4^oc_oc_i^2 \\
& \qquad\qquad +G_5^oc_oc_i^2(N_i-1)+G_6^oc_oc_i^2(N_i-1)^2] \\
CGS = \ & x\left[\tfrac{1}{4}G_5^o(c_o+c_i)-\tfrac{1}{2}c_i(N_o-N_i)\frac{N_o-1}{N_o}\cdot\frac{N_i-1}{N_i}\right] \\
& +y(\tfrac{1}{4}G_5^oc_o+\tfrac{1}{4}G_5^ic_i)-z(G_7^oc_o+G_7^ic_i).
\end{aligned}
$$

[86] The prescription for the lens will be

$$c_1 = \tfrac{1}{2}(c_o + c_i) + x + y; \quad c_2 = \tfrac{1}{2}c_i + y; \quad c_3 = -\tfrac{1}{2}c_i + y; \quad c_4 = -\tfrac{1}{2}(c_o + c_i) + x + y.$$

If z is solved for, it will correspond to $v_1 = z - 1/2f'$. If a prescribed v_1 is to be realized, we must use $z = v_1 + 1/2f'$ in working out the equations.

When calculating the coefficients, it will be found helpful to prepare and to collect all the entering quantities, N_o, N_i, $(N_o - N_i)$, $(N_o - 1)$, $(N_i - 1)$, c_o, c_i, and $(c_o + c_i)$, together with their logs, and also the logs of the G's for glass o from (1) to (7), and for glass i for (1), (4), (5), (6), and (7). On account of the curiously mixed form of some of the terms, close attention to signs and symbols is indispensable. For our examples, we shall employ the same two kinds of glass as before and solve for crown inside, according to the upper row in Fig. 110. We shall then have $c_o = -0.2469$ and $c_i = +0.4692$ and we shall find, for focal length 10, also as before

$$SGS = 0.11909x^2 + 0.29862xy + 0.11994y^2 + 0.21995z^2 - 0.43700xz - 0.33989yz + 0.00053,$$
$$CGS = 0.10924x + 0.08496y - 0.13499z.$$

Comparison of this solution with the previous general solution for a binary cemented objective will show the identity of a number of coefficients, which is accounted for by the fact, already noted, that our new solution covers also the binary forms by introducing $x = \mp \tfrac{1}{2}c_o$. This causes either c_a (upper sign) or c_c (lower sign) to become zero, thus producing simple crown-in-front and flint-in-front cemented objectives respectively.

All Possible Triple Cemented Aplanatic Objectives

By using the zero-coma condition to eliminate one of the three variables on the right of the SGS equation, we obtain a single quadratic equation in two variables that can be plotted as a conic section to represent all possible aplanatic objectives. Selecting y for elimination, the zero-coma condition gives

$$0.08496y = -0.10924x + 0.13499z,$$

or

$$y = [0.1092n]x + [0.2011]z = -1.2858x + 1.5889z,$$

and substitution of this value of y turns the SGS equation into

$$-0.06657x^2 - 0.01557xz - 0.01730z^2 + 0.00053 - SGS_o = 0,$$

which on comparison with the standard quadratic equation (changing the y of the latter to z!) gives

$$A = -0.06657; \quad B = -0.00779; \quad C = -0.01730; \quad D = E = 0;$$
$$F = 0.00053 - SGS_0.$$

As $B^2 - AC = 0.000061 - 0.001152$ is negative, the equation plots as an ellipse, which is proved to be real by the test provided for this purpose.   As there are no [86] terms in x and z, the origin of the coordinates is the centre of the ellipse, and we have only to determine the slope of the axes.   By $\tan 2\theta = 2B/(A-C)$ we find $\theta = 8° 46' 26''$ and for the drawing $\tan \theta = 0.1543$.   Then $A' + C' = A + C = -0.08387$, $A' - C' = (A-C)/\cos 2\theta = -0.05167$, whence $A' = -0.06777$ and $C' = -0.01610$.   These then give

$$\text{Semiaxis in } X''\text{-direction} = 0.0884$$

$$\text{Semiaxis in } Z''\text{-direction} = 0.1814.$$

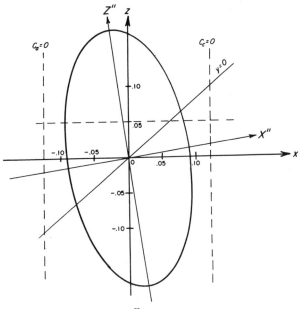

FIG. 111.

When plotted, these data give a small, nearly upright ellipse, showing in the first place that aplanatic triple lenses of the chosen type and kinds of glass are only possible for comparatively low values of z and therefore also—by $v_1 = z - 1/2f'$—of v_1.   But, as the usual applications of highly corrected lenses for the formation of real images of real objects only cover the range from $z = -0.05$ to $z = +0.05$, the restricted range of the ellipse is not likely to affect practical problems.

Every point of the ellipse in Fig. 111 gives directly the x and z of an aplanatic triple lens, and the zero-coma equation of condition then gives the corresponding y.

Thus we obtain the two possible aplanatic telescope objectives at $z = v_1 + 1/2f' = 0 + 0.05 = 0.05$ by reading the corresponding abscissae as $x = -0.091$ or $x = +0.079$.   The equation of condition $y = [0.1092n]x + [0.2011]$ then gives

[86] for $x = -0.091$, $y = +0.1964$, and for $x = +0.079$, $y = -0.0221$. The prescription for the two objectives is then found by

$$c_1 = \tfrac{1}{2}(c_o + c_i) + x + y, \quad c_2 = \tfrac{1}{2}c_i + y, \quad c_3 = -\tfrac{1}{2}c_i + y, \quad c_4 = -\tfrac{1}{2}(c_o + c_i) + x + y,$$

or

First Form, $x = -0.091$		Second Form, $x = 0.079$	
$c_1 = 0.2166$	$r_1 = 4.617$	$c_1 = 0.1680$	$r_1 = 5.952$
$c_2 = 0.4310$	$r_2 = 2.320$	$c_2 = 0.2125$	$r_2 = 4.705$
$c_3 = -0.0382$	$r_3 = -\,26.18$	$c_3 = -0.2567$	$r_3 = -3.895$
$c_4 = -0.0057$	$r_4 = -175.4$	$c_4 = -0.0543$	$r_4 = -18.42$

As triple lenses are normally only resorted to when the double form will not yield sufficient aperture, the choice will fall upon the second form because its shortest radius is 3.895 against 2.320 in the first form. It may be taken as a rough rule that a clear aperture equal to the shortest radius is likely to prove practicable; by that criterion, the second form will admit of more than 1.5 times the aperture of the first form and will moreover be cheaper at any given aperture.

Another solution of general interest is the one for oppositely equal conjugate distances or for $M' = -1$, which on the graph means $z = 0$ by definition of z. The corresponding value of x is read as ± 0.089. For $x = +0.089$ the equation of condition gives $y = -0.1144$ and then, by the process already described, the prescription:

$c_1 = +0.0858$	$r_1 = 11.66$
$c_2 = +0.1202$	$r_2 = 8.319$
$c_3 = -0.3490$	$r_3 = -2.865$
$c_4 = -0.1366$	$r_4 = -7.321$

The alternative solution $x = -0.089$ gives in this case simply the same prescription with reversed signs and sequence of the radii, so that the stated solution is really unique. The distribution of curvatures is very unsymmetrical in spite of the symmetrical position of the conjugate points, and the shortness of r_3 would restrict the attainable aperture in a very objectionable way. This triple lens would in fact be only moderately superior to a skilfully chosen binary combination of dense barium crown and ordinary flint on the principles of equations [48.5] (4) and [64] (d). We must therefore enquire how a more symmetrical form might be attained. With the glass actually used for our triple objectives, the strictly symmetrical form would have $x = y = z = 0$ and therefore, by the general SGS equation, $SGS = +0.000532$. This amount of spherical under-correction was shown in section [83] to represent the Rayleigh limit at an aperture of about $f/8$. We may therefore estimate that skilful juggling with tolerances together with the predominant over-correcting tendency of the higher aberrations would enable us to attain a very respectable aperture with the strictly symmetrical form. The surprisingly large departure from symmetry which the TL solution proves to be necessary for the removal of the small positive SGS residual of the symmetrical form is accounted for by the fact that its internal component represents a minimum-aberration lens whose spherical correction is but slightly altered by small bendings, just as in the similar case of [48.5] (4). The best procedure in the present case consists of a change of glass. We must try to increase the over-correcting effect of the two internal contact surfaces, and this can be achieved, as in all cases of this kind, by diminishing the V-difference $V_i - V_o$ whilst approximately retaining the original indices, or alternatively by increasing the N-difference $N_o - N_i$ and retaining the original V-values. In this way a practically perfect result can be secured with a perfectly symmetrical form by a few systematic trigonometrical trials.

In the case of the second form of the aplanatic triple telescope objective, the departure from symmetry of the internal lens is not so great as to curtail the attainable aperture to any serious extent; this form is therefore quite suitable for the ordinary trigonometrical correcting process. But if we desire, chiefly by reason of workshop economy, to explore the possibilities of a strictly equiconvex internal lens, we can do so by a small addition to the graph. We shall require $y = 0$ and must also have coma freedom or $y = 0 = [0.1092n]x + [0.2011]z$. The special condition thus found for the relation of x and z can be plotted on the graph in Fig. 111 as a straight line passing through the origin and rising towards the right at the rate of [0.1092] vertically for [0.2011] horizontally. It is marked '$y = 0$' in Fig. 111. Any combination of x and z read off this line will yield a coma-free objective with an equiconvex internal lens, but only the one from the intersection of the $y = 0$ line with the aplanatic ellipse will be spherically corrected from the strict TL point of view, those falling inside the ellipse being under-corrected, those falling outside, over-corrected. For a coma-free telescope objective, the $y = 0$ line reads $x = +0.062$ at $z = +0.05$, and as this lies only a little inside the ellipse, the spherical under-correction will be small. We can determine it by putting $x = 0.062$, $y = 0$, $z = 0.05$ into the general SGS equation and find $SGS = +0.000184$ or only about 1/3 of that of the symmetrical lens for equal conjugates. It would be venturesome to try to predict whether this solution would turn out spherically over- or under-corrected when tested trigonometrically with suitable thicknesses; a good result is probable.

The general SGS equation may be plotted in several other ways. By putting in the proper numerical value of z, for instance, $z = +0.05$ for telescope objectives, and also a definite value of SGS_0, we obtain a general quadratic in the two variables x and y. This quadratic plots as a hyperbola and gives in a single graph all conceivable forms of the triple lens which yield the prescribed spherical correction at the selected object distance. By putting in $y = 0$ and proceeding in the usual way, we obtain a hyperbola in x and z which gives all possible forms with equiconvex internal lens. In either case, a zero-coma line can be added to indicate the coma-free forms. But as interest is nearly always limited to the latter, the ellipse of Fig. 111 is by far the most useful graph.

Finally, it should be remembered that the general solution for triple lenses can be turned into one for binary cemented lenses by putting $x = \mp \frac{1}{2}c_0$, the double sign indicating crown or flint in front respectively.

EXTENSIONS OF THE TRIGONOMETRICAL CORRECTING
METHODS

[87] THE methods given in Part I for the final correction of systems of the telescope-objective type will in all ordinary cases give the best possible solution with ease and certainty. Nevertheless, there are exceptional problems when unusual or additional conditions are introduced, and these can be solved more efficiently by the supplementary methods now to be given. One such case was included in Part I, section [64] C, where complete aplanatic correction was secured by separating the two components of an ordinary objective instead of making the curvature of their adjacent surfaces slightly different, as is customary. This procedure led to the introduction of the remarkably effective method of trigonometric interpolation. Now this is a method of linear interpolation which would break down seriously if it were applied to large alterations in the form of a system. Therefore it was pointed out at once in the first application of the method in [64] C that it is important to try to approximate progressively the intended state of correction in choosing the required successive trial-forms. In the simple case referred to, this proved to be possible by comparatively simple and more or less empirical guesses; our chief aim in the present section will be to find systematic methods for securing this highly important progressive approach to the final form of any required system. From the purely mathematical point of view, there would be an attractive alternative method of extending the range of validity of the interpolation method, namely, to take into account not only the first, the linear, terms of Taylor's series, but also the second, the quadratic, and possibly even the third, the cubic, terms. In the simplest case of seeking fulfilment of only one condition—spherical correction—we successfully applied this method in section [48.5] (3) and proved its effectiveness. But at present we shall be chiefly concerned with the simultaneous fulfilment of three conditions by three corresponding changes in the system, namely, bending the two separate components and varying the air-space between them. With reference to [64] C, the equation for each one of the three conditions, of the form

$$f_{(x+\delta x),\ (y+\delta y),\ (z+\delta z)} = f_{x,y,z} + \delta x\, \frac{\partial f}{\partial x} + \delta y\, \frac{\partial f}{\partial y} + \delta z\, \frac{\partial f}{\partial z}$$

would require to be extended by the quadratic terms

$$+ \tfrac{1}{2}\delta x^2\, \frac{\partial^2 f}{\partial x^2} + \tfrac{1}{2}\delta y^2\, \frac{\partial^2 f}{\partial y^2} + \tfrac{1}{2}\delta z^2\, \frac{\partial^2 f}{\partial z^2} + \delta x \delta y\, \frac{\partial^2 f}{\partial x\, \partial y} + \delta x \delta z\, \frac{\partial^2 f}{\partial x\, \partial z} + \delta y \delta z\, \frac{\partial^2 f}{\partial y\, \partial z},$$

and instead of having to add to the starting form, which determines $f_{x,\,y,\,z}$, three modifications by which the three first differential coefficients can be determined, we should require nine independent modifications of the starting form to secure the value of the six second differential coefficients in addition to the first three. As a result, the final solution for the values of δx, δy, and δz which would give the desired correction would not call for just a simple solution of three linear equations as before; it would call for a formidable solution of three simultaneous general quadratic equations in three unknowns. It is clear that we may safely ignore this

alternative. If, as may happen in difficult cases, the first trial-forms depart too [87]
far from the initial form to promise a close solution by linear interpolation, then it
will be cheaper to use the knowledge gained for a new solution that will demand
much smaller changes.

In addition to the case treated in section [64] C, the design of a telescope objective may be rendered more difficult and may call for the method of trigonometric
interpolation as the line of least resistance, when a third condition is added to the
usual two for spherical correction and freedom from coma. Besides a reduction
of the secondary spectrum, which is merely a question of selecting suitable glass,
the added condition might be either the reduction or the complete correction of
the *zonal* spherical aberration, or else fulfilment of the 'Gauss condition', that is,
simultaneous spherical correction for two different colours.

We shall study the zonal spherical aberration quite closely in following chapters
on account of its great importance in the design of microscope objectives. In
telescope objectives it hardly ever reaches serious magnitude when compared with
the tolerance under the Rayleigh limit. The cheap cemented type of telescope
objective with a nearly equiconvex crown lens is a particularly unfavourable one as
regards zonal spherical aberration. Nevertheless, the exceptionally bold example
of 2 inches aperture and only 8 inches focal length which was extensively discussed
in Part I was found to have zonal aberration of only two-thirds of the Rayleigh
limit, which would be quite harmless in the presence of the huge secondary
spectrum, the heavy offence against the sine condition, and, as we may anticipate,
also a very serious non-fulfilment of the Gauss condition. In the better forms of
the ordinary telescope objective, in which both spherical aberration and coma are
corrected by the introduction of suitable bendings plus a slight air-gap, the
zonal aberration amounts to a very small fraction of that in a comparable cheap
cemented form and may safely be regarded as completely negligible. We must,
moreover, bear in mind that, on account of the want of homogeneity in all optical
glass and of imperfect annealing, telescope objectives larger than 2 or 3 inches in
aperture hardly ever bear out the calculation as regards spherical correction; the
latter has to be secured by 'figuring' one of the refracting surfaces, and any calculated zonal aberration will automatically be included in the correction by figuring
without causing any additional cost.

Figuring has no sensible effect—in telescope objectives—on any of the other
aberrations, for we learnt in Part I that achromatism depends on the correct
proportioning of the total curvatures of the crown and flint lenses and is not
affected by *minute* changes such as might result from figuring; and as the differently
coloured components of any single entering white ray never become widely
separated in their progress through a telescope objective, any conceivable amount
of figuring will affect all the colours to practically the same extent. This means
that any *difference* in spherical aberration will also be unaffected and any offence
against the Gauss condition disclosed in the original calculation will appear in the
finished figured objective by exactly the same amount. Finally, we learnt that the
offence against the sine condition can only be usefully varied by bendings of the
objective that are relatively very large, so that mere figuring will leave the *OSC'*
quite unchanged. It follows that the almost certain necessity of more or less

[87] figuring of the finished telescope objective fully justifies a comparatively rough correction of the general spherical aberration and complete neglect of the zonal aberration; on the other hand, achromatism, freedom from coma, and fulfilment of the Gauss condition can only be determined by exact calculation followed by correspondingly exact execution of the calculated data; these aberrations are quite incapable of sensible improvement by any practicable amount of subsequent figuring. Many existing large objectives are permanently defective because this important distinction was insufficiently appreciated.

Since these considerations leave the Gauss condition as the only aberrational condition which might have to be added to the usual correction of spherical aberration and coma, we shall first try to secure some preliminary knowledge of the probable magnitude of this defect. Unlike zonal aberration, the offence against the Gauss condition does not vary much for the various usual forms of two-lens objective provided that the usual kinds of glass are used and that there is no sensible air-space between the components. We shall therefore secure a useful estimate by the results for the much-discussed two-inch objective. According to the little table in section [45] (page 188), this objective, when spherically corrected for the brightest light, has $LA'_r = +0.007$ for C-light and $LA'_v = -0.008$ for F-light. We shall measure the offence against the Gauss condition by $LA'_r - LA'_v$, which in the present case amounts to $+0.015$. As the spherical tolerance for this objective was found to be 0.005, the offence against the Gauss condition is decidedly serious, the spherical aberration of either colour being one and one-half times the Rayleigh limit. We must, however, modify this result so as to arrive at an estimate for the usual apertures of astronomical object-glasses. For photographic telescopes $f/10$ is quite common. At $f/10$ our objective would have 0.4 of its originally calculated aperture, and as by section [82] the 'seriousness' of spherical aberration grows with the fourth power of the aperture, our objective at $f/10$ would have oppositely equal amounts of spherical aberration for C and F amounting to $1\frac{1}{2} \times 0.4^4 = 1\frac{1}{2} \times 0.0256 = 0.0378$ or about 0.04 times the Rayleigh limit. At $f/10$ and at the calculated focal length of 8 inches, the offence against the Gauss condition would therefore be quite negligible. It will, however, grow proportionately when the objective is made to a larger scale. Hence it will reach the full Rayleigh limit when our now $f/10$ objective is made to 25 times the original scale so that its focal length is 200 inches and its aperture is 20 inches. Very few $f/10$ objectives as large as this exist.

A similar calculation for the more usual $f/15$ ratio of large object-glasses easily leads to the result that the Gauss condition would then be offended against to the extent of the single Rayleigh limit for C and F at 1000 inches of focal length, corresponding to 67 inches clear aperture. Such a giant does not exist and is not likely to be produced 'in our time'.

It will be realized that these figures throw a heavy doubt upon the supposed advantage of satisfying the Gauss condition in telescope objectives of the usual type. The fact that a moderate number of existing large objectives have been designed so as to satisfy the Gauss condition and that some of them have given very good results does not invalidate our conclusion; it is simply a case of *post hoc non propter hoc*. The results in these cases are good because the best type of achromatism has been realized and because the figuring has been highly successful, but with regard [87] to the Gauss condition, we shall subsequently see that in probably every case the results are good not on account of its fulfilment but *in spite of this handicap*.

The case becomes different for objectives of the photovisual type with greatly diminished secondary spectrum because the closer union of all colours makes uniform spherical correction more valuable, whilst the severe curvatures of these objectives greatly aggravate the possible magnitude of the offence against the Gauss condition.   For that reason, the correction of photovisual objectives will be our chief subject in this section.

As in all our practical computing work, we seek a method which will give a *trigonometrically exact* solution with the least total expenditure of time and trouble. For our present problems, the solution depends upon the finding of simple and convenient, though not theoretically exact, formulae for the requisite changes in the bendings of the two components of an object-glass and of the small air-space which will approximately fulfil the conditions.

A. The simplicity of the formulae depends largely upon the rules for approximate calculations with small quantities, which are extensively employed in all branches of applied science for the estimation of small corrections.   These rules are based upon Taylor's theorem reduced to its first-order terms, the most useful ones being the following:

(1) If δ stands for a small proper fraction, n for any number, positive or negative, whole or fractional, then nearly:

$$(1 \pm \delta)^n = 1 \pm n\delta; \quad 1/(1 \pm \delta)^n = (1 \pm \delta)^{-n} = 1 \mp n\delta.$$

Example: $\delta = 0.015$ and $n = 3$ give $1.015^3 = 1.045$ (true result, 1.04568); $0.985^3 = 0.955$ (true result, 0.95567).   The rule can be indefinitely extended, provided the δ are sufficiently small:

$$(1 \pm \delta_1)^{n_1} \cdot (1 \pm \delta_2)^{n_2} \ldots /(1 \pm \delta_k)^{n_k}(1 \pm \delta_{k+1})^{n_{k+1}} \ldots$$
$$= 1 \pm n_1\delta_1 \pm n_2\delta_2 \ldots \mp \delta_k n_k \mp \delta_{k+1}n_{k+1} \ldots$$

The error in all cases is of the order of δ^2.   The advantages are, from an algebraic point of view, that we get rid of powers and fractions and secure simple linear equations, and from the numerical computer's point of view, that we secure a good approximation by simple mental arithmetic or slide rule.   When δ is very small, say below 0.001, a more precise result is obtained than could be obtained by logs of any usually accessible number of decimal places.

(2) Algebraic equations usually do not yield directly factors or divisors of the form $(1 + \delta)$; but if p and q are two terms (possibly of quite complicated type) which can be shown to be nearly equal, then we can make the transformation

$$p/q = (q + p - q)/q = 1 + (p - q)/q$$

in which $(p-q)/q$ is now of the δ-order of smallness so that the rules for small corrections can be applied to it.   The transformation into $p - q$ may take various forms according to individual cases.   Supposing p to be the quantity which it is desirable to retain, we should put

$$p \cdot q = p(p + q - p) = p^2\left(1 + \frac{q-p}{p}\right).$$

[87] An interesting alternative that is frequently useful is to note that

$$p = \tfrac{1}{2}(p+q) + \tfrac{1}{2}(p-q) \quad \text{and} \quad q = \tfrac{1}{2}(p+q) - \tfrac{1}{2}(p-q),$$

whence
$$p \cdot q = \left(\frac{p+q}{2}\right)^2 - \left(\frac{p-q}{2}\right)^2 \simeq \left(\frac{p+q}{2}\right)^2$$

to a first approximation. The product of two nearly equal quantities is therefore sensibly equal to the square of their arithmetic mean, or the geometric mean of two nearly equal quantities may be replaced by the arithmetic mean.

B. The correcting formulae for our present purpose assume that we have trigonometrical results for a sufficiently corrected starting form of the projected system with components in axial contact or as close to contact as a small difference in curvature may allow; formulae are required to determine with sufficient approximation the changes in the aberrations resulting from the introduction of a small air-space or a small bending of the second component or both. These changes

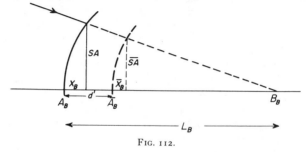

FIG. 112.

depend largely upon the contraction of the cone of rays in the air-space, and we therefore begin by determining this.

Distinguishing the two components by suffixes A and B respectively, we shall know by the initial trigonometrical calculation the data of the rays arriving at the first surface of the second component. With reference to Fig. 112, in which the starting position of the first surface of component B is shown in full lines and its shifted position in broken lines, we know $A_B B_B = L_B$. Let the original curvature of the surface be $1/r = c_B$, the longitudinal shift be d', and let the curvature be changed to $c_B + \delta c_B$. Dropping perpendiculars from the respective points of incidence upon the optical axis, we define at the original surface the semi-aperture SA and the depth of curvature X_B, and, introducing a bar to distinguish data of the shifted component, we similarly define $\overline{SA}$ and $\bar{X}_B$. Similar right-angled triangles with SA and $\overline{SA}$ as vertical sides then give the contraction of the cone of rays measured as

$$(\text{1}) \qquad \frac{\overline{SA}}{SA} = \frac{L_B - d' - \bar{X}_B}{L_B - X_B} = \frac{L_B - X_B - d' - (\bar{X}_B - X_B)}{L_B - X_B} = 1 - \frac{d' + (\bar{X}_B - X_B)}{L_B - X_B},$$

which is thus put into the standard form suitable for the application of the rules for

small corrections. For the X we use the parabolic formula (8p) of section [17]: [87]
$x = \frac{1}{2}y^2/r = \frac{1}{2}c \cdot y^2$ or, with our present replacement of y by SA:

$$X_B = \frac{1}{2}c_B \cdot SA^2;$$

$$\bar{X}_B = \frac{1}{2}(c_B + \delta c_B) \cdot \overline{SA^2} = \frac{1}{2}(c_B + \delta c_B)\left(1 - 2\frac{d' + (\bar{X}_B - X_B)}{L_B - X_B}\right)SA^2.$$

The last comes from squaring equation (1), using the rules for small corrections.
In multiplying out the last formula, we treat the product of the small δc_B and the
small correcting term in the second bracket as a second-order term to be neglected.
Re-introducing $X_B = \frac{1}{2}c_B \cdot SA^2$, we find

$$\bar{X}_B = X_B + \frac{1}{2}\delta c_B SA^2 - 2X_B\frac{d' + (\bar{X}_B - X_B)}{L_B - X_B}.$$

Solved for $\bar{X}_B - X_B$, this equation gives

$$(\bar{X}_B - X_B)\left(1 + 2\frac{X_B}{L_B - X_B}\right) = \frac{1}{2}\delta c_B SA^2 - 2d'\frac{X_B}{L_B - X_B},$$

or by bringing the second term on the left to a common denominator,

$$(\bar{X}_B - X_B)\frac{L_B + X_B}{L_B - X_B} = \frac{1}{2}\delta c_B \cdot SA^2 - 2d'\frac{X_B}{L_B - X_B}.$$

Division throughout by $L_B + X_B$ then gives

$$\frac{\bar{X}_B - X_B}{L_B - X_B} = \frac{1}{2}\delta c_B\frac{SA^2}{L_B + X_B} - 2d'\frac{X_B}{(L_B - X_B)(L_B + X_B)},$$

and introduction of this into (1) produces

$$\frac{\overline{SA}}{SA} = 1 - \frac{d'}{L_B - X_B} - \frac{1}{2}\delta c_B\frac{SA^2}{L_B + X_B} + 2d'\frac{X_B}{(L_B - X_B)(L_B + X_B)}$$

$$= 1 - \frac{d'}{(L_B - X_B)}\left(1 - \frac{2X_B}{L_B + X_B}\right) - \frac{1}{2}\delta c_B\frac{SA^2}{L_B + X_B},$$

which by an obvious simplification gives the final result

(2)
$$\frac{\overline{SA}}{SA} = 1 - \frac{d'}{L_B + X_B} - \frac{1}{2}\delta c_B\frac{SA^2}{L_B + X_B}.$$

C. We next seek a simple formula for the approximate estimation of the change
in the spherical correction which results from the introduction of an air-space or a
small change of curvature or both. From the initial calculation of a system with
components close together, we learn that the front component at its own focus
produces an aberration which we shall call LA'_A and that the complete close
system gives a residual aberration called LA'_k at its final focus. The air-space d'
and bending by δc_B are to be such as to lead to complete spherical correction at the
final focus of the modified system. It is convenient to establish this balance

[87] entirely in the air-space; therefore we transfer the final LA'_k to the air-space by the law of longitudinal magnification. The paraxial law $\overline{m} = u'^2_k/u'^2_A$ is rarely near enough in the cases under consideration on account of the enormous spherical aberration LA'_A produced by the first component. We obtain a more reasonable transfer-factor by noting that in the exact equation (10*) of section [26] the transfer factor from surface to surface is $N \cdot l' \cdot L'/N' \cdot l \cdot L$, thus involving both paraxial and marginal data. As with the comparatively small convergence angles in telescope lenses the intersection lengths are almost inversely proportional to the angles U or U' or their sines, we therefore adopt as more reasonable the transfer-factor $u'_k \cdot \sin U'_k/u'_A \cdot \sin U'_A$ and can then claim that in the close position the aberration LA_B *compensated* by the second component was

(3)
$$LA_B = LA'_A - LA'_k \cdot u'_k \sin U'_k/u'_A \cdot \sin U'_A.$$

We now determine, or, rather, estimate, the change in LA_B resulting from the introduction of an airspace d' *only*. Fig. 112 shows that, owing to the contraction of the cone, the second component will now produce or compensate for less spherical aberration. It would be useless to try to work out a theoretically exact equation for the magnitude of the change. But under the special conditions which invariably prevail in highly corrected telescope objectives, we obtain remarkably close estimates by the simple assumption that the spherical aberration *constant* ('a' in $LA' = a \cdot SA^2$) does not change seriously for the small shift of the second component by d'. On that assumption the aberration compensated by the shifted component will be changed simply by the square-of-the-aperture law and will be

$$\overline{LA}_B = LA_B(\overline{SA}/SA)^2 = LA_B\left(1 - 2\frac{d'}{L_B + X_B}\right)$$

by (2), as there is no simultaneous bending.

To estimate the effect of a small bending of the second component by δc_B, we take advantage of the fact that in the systems under consideration a moderate uniform bending of the whole system changes the final aberration very little, because the systems are nearly free from coma and the forms dealt with are therefore always located near the pole of the aberration-parabolas, Fig. 103, for the complete system. Hence we may substitute a bending of the first component by $-\delta c_B$ for the really intended bending of the second component by $+\delta c_B$ without seriously affecting the final result. Such a bending of the first component will, by the first term of Taylor's theorem, change

$$LA'_A \quad \text{to} \quad LA'_A - \delta c_B \frac{\partial LA'_A}{\partial c_A}.$$

We now obtain our complete correcting equation by first introducing into the above equation for $\overline{LA}_B$ the value of LA_B given by (3), finding, by first order terms only,

$$\overline{LA}_B = LA'_A - LA'_k \cdot u'_k \cdot \sin U'_k/u'_A \cdot \sin U'_A - 2LA'_A \frac{d'}{L_B + X_B}.$$

As we require complete spherical correction at the final focus, this $\overline{LA_B}$ must be [87] equal to the LA'_A modified by the transferred bending, whence the condition

$$LA'_A - \delta c_B \frac{\partial LA'_A}{\partial c_A} = LA'_A - LA'_k \cdot u'_k \cdot \sin U'_k / u'_A \cdot \sin U'_A - 2LA'_A \frac{d'}{L_B + X_B}$$

which immediately yields our final correcting equation

$$(4) \qquad 2LA'_A \cdot \frac{d'}{L_B + X_B} - \delta c_B \cdot \frac{\partial LA'_A}{\partial c_A} = -LA'_k \cdot u'_k \cdot \sin U'_k / u'_A \cdot \sin U'_A.$$

D. We must finally estimate the change in OSC' resulting from the modification of the initial form of the system with the components in close contact. As a rule the modified system will require a uniform bending throughout by δc, an additional bending of the second component only by δc_B, and an air-space d'. As the initial system may be considered to be thin, we obtain the change in OSC' resulting from a uniform bending at once by the extremely useful approximate formula 'Sine Theorem IV' on page 393 of Part I:

$$\delta OSC' = 0.86\delta c \cdot SA^2 / f'.$$

Hence the OSC'_k found for the initial system will be changed by the uniform bending to

$$OSC' = OSC'_k + 0.86\delta c \cdot SA^2 / f'.$$

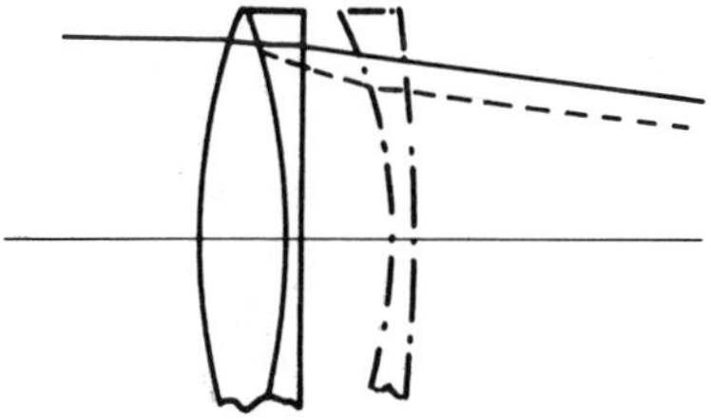

FIG. 113.

We must now estimate the additional change in OSC' which results when the bent initial system has its components separated by d' and when, additionally, the second component is further bent by δc_B. In Fig. 113 the track of a ray through the bent initial system is shown in continuous lines and the altered state on separating and bending the second component in broken lines. We shall call the final data of a paraxial and a marginal ray in the case of the close position u', l', and U', L' respectively. As in telescopes and similar systems, we always have a close approach to aplanatism, it is permissible to assume the exit pupil to be in contact with the last refracting surface, simply because any likely departure from that position would be of moderate magnitude and, by the second fundamental law of oblique pencils, would have only an evanescent effect upon the coma of a nearly aplanatic system. Hence for the close position

$$OSC' = 1 - \frac{l' \cdot u'}{L' \cdot \sin U'}$$

[87] and we must now note that $l'u'$ is accurately the nominal or fictitious incidence-height of the paraxial ray at the last surface whilst $L' \cdot \sin U'$ is strictly the length of a perpendicular dropped from the pole of the last surface upon the emerging marginal ray. However, as the final U' is always fairly small for the systems under present consideration, $L' \sin U'$ can differ only by a very small percentage from the marginal incidence-height. This identification of the meaning of $l'u'$ and $L' \sin U'$ supplies the key to our solution, for the dotted part of Fig. 113 shows that the separation and bending of the second component will introduce the contraction of the cones of rays determined by equation (2) above. Since we may regard the second component as thin, this contraction will persist with the same percentage-amount at the last surface, hence the

$l'u'$ of the close system will be changed to $l'u' \cdot (\overline{SA}/SA)_{\mathrm{paraxial}}$

$L' \sin U'$ of the close system will be changed to $L' \sin U'(\overline{SA}/SA)_{\mathrm{marginal}}$

and we shall therefore have, for the completely changed system

$$\text{Final } OSC' = \mathrm{I} - l'u'(\overline{SA}/SA)_{\mathrm{paraxial}}/L' \sin U'(\overline{SA}/SA)_{\mathrm{marginal}}.$$

Equation (2) gives the marginal contraction directly in our adopted symbols, and as for strict paraxial rays we must, on the right-hand side of the equation, treat SA as being equal to zero, similarly $X_B = \mathrm{o}$, whilst L_B has to be replaced by l_B, we thus have

$$(\overline{SA}/SA)_{\mathrm{paraxial}} = \mathrm{I} - d'/l_B.$$

By introducing these values into the last equation and carrying out the division of the correcting terms by the rules for small departures from unit-value, we obtain

$$\text{Final } OSC' = \mathrm{I} - \frac{l'u'}{L' \sin U'} \left(\mathrm{I} - \frac{d'}{l_B} + \frac{d'}{L_B + X_B} + \tfrac{1}{2}\delta c_B \cdot \frac{SA^2}{L_B + X_B} \right)$$

$$= \mathrm{I} - \frac{l'u'}{L' \sin U'} + \frac{l'u'}{L' \sin U'} \left[d'\left(\frac{\mathrm{I}}{l_B} - \frac{\mathrm{I}}{L_B + X_B} \right) - \tfrac{1}{2}\delta c_B \cdot \frac{SA^2}{L_B + X_B} \right].$$

The first two terms on the right represent the OSC' of the bent but close initial system. They can now be replaced by its value in terms of OSC'_k and of the general δc determined at the beginning of this subsection D. Finally the factor $l'u'/L' \sin U'$ of the long final term equals $\mathrm{I} - OSC'$ of the bent but close system, and as OSC' in telescope systems is always very small, rarely reaching even o.o1, the factor is so close to unity that it may be omitted. In the final bracket, we bring the factor containing d' to a common denominator and obtain our solution:

$$(5) \quad \text{Final } OSC' = OSC'_k + \mathrm{o.86}\delta c \, \frac{SA^2}{f'} + d' \, \frac{L_B + X_B - l_B}{l_B(L_B + X_B)} - \tfrac{1}{2}\delta c_B \cdot \frac{SA^2}{L_B + X_B}.$$

This equation usually gives remarkably close estimates, *but only on one highly important condition*, the ignoring of which would lead to grossly wrong results when a considerable uniform bending has to be applied. This restriction arises from the fact that our remarkably simple equation was secured by *first* applying the general bending by δc. This had to be done because 'Sine Theorem IV' [87] would give quite useless results when applied to a system with separated components and heavy spherical aberration in the air-space. The consequence of this sequence in our operations is that the L_B, l_B, and X_B in the equation are those for the *bent* system and not those for the original system. The use of the latter data might in unfavourable cases make the factor of d' wrong 50 per cent. or even more!

E. We will first apply the correcting formulae to the useful alternative method of attaining aplanatism by only three different radii and a small air-space as introduced in section [64] C. In these cases it will always be sufficient to use as the initial form a cementable combination determined either by the 'middle-bending rules' of Chapter V or by the simple solution for a coma-free cementable lens given at the end of section [59] (page 333); a complete ray-tracing through the cemented form and an additional calculation of the rays from the first component into the air-space then supplies all the data. As δc_B in these cases is permanently zero, equation (4) simplifies to

$$2LA'_A\left(\frac{d'}{L_B+X_B}\right) = -LA'_k u'_k \sin U'_k / u'_A \sin U'_A,$$

and as all items excepting d' are known from the initial ray-tracing, the equation can be transposed into the solution

$$(6) \qquad d' = -\tfrac{1}{2}(L_B+X_B)\cdot\frac{LA'_k}{LA'_A}\cdot u'_k \sin U'_k / u'_A \sin U'_A.$$

Data with suffix k refer to the final focus of the initial form, those with suffix A, to the rays emerging from the initial first component into the air of the air-space. In the present case—with absolutely no initial air-space—L_B is merely another name for L'_A and X_B is the depth of curvature of the initial contact-surface as calculated by one of the equations (8) in Chapter I.

Having thus determined the air-space, we use equation (5) of this subsection to fix the value of the general bending δc which will make the final OSC' vanish. As δc_B is zero, this gives

$$OSC'_k+0.86\delta c\left(\frac{SA^2}{f'}\right)+d'\,\frac{L_B+X_B-l_B}{l_B(L_B+X_B)} = 0,$$

and as δc is the only unknown quantity, we find

$$(7) \qquad \delta c = -1.16\left[OSC'_k+d'\,\frac{L_B+X_B-l_B}{l_B(L_B+X_B)}\right]\cdot f'/SA^2.$$

It is permissible in this case to use in the d'-factor the $L'_A = L_B$, $l'_A = l_B$, and X of the contact-surface for the initial form of the first component because δc will be quite small when the initial form has been determined by one of the methods advised. OSC'_k is the value found at the final focus of the cemented initial form while f' and SA respectively are the focal length and calculated semi-aperture of the latter. For the example in [64] C the initial contact-form gave $LA'_k =$

[87] -0.1019; $\log u'_k = 9.11224$; $\log \sin U'_k = 9.10784$; the tracing of the rays from the first component into the air-space gave $L'_A = L_B = 2.9400$; $l'_A = l_B = 3.8845$; $LA'_A = 0.9445$; $\log u'_A = 9.49518$; $\log \sin U'_A = 9.55960$; and we find for the depth of the contact surface $X_B = -0.2483$. Introduced into equation (6) these data give $d' = 0.02125$ and then by (7) $\delta c = 0.01203$. The final solution in [64] C called for $d' = 0.0214$ and $\delta c = 0.011$; our simple correcting formulae have therefore led to an almost exact solution, and if we had used them, we should, at most, have had to apply a simple linear interpolation to the results of the initial form and of the first modification in order to reach perfect correction.

The correcting formulae, or, rather, a slightly different earlier form of them, were also tried for the flint-in-front form of an objective of the same kinds of glass and with the same bold aperture of 2.5 at focal length 10. The middle-bending rule (footnote in section [48]) gave for the initial form $c_1 = 0.223$, $c_2 = 0.470$, or the prescription: $r_1 = 4.484$; $d'_1 = 0.3$; $r_2 = 2.128$; $d'_2 = 0.5$; r_3 by usual solution $= 1232$ instead of the infinite value given by TL formulae. The final trigonometric results for this were $LA'_k = -0.0936$, $OSC'_k = -0.00083$. The correcting formulae gave $d' = 0.0214$, $\delta c = -0.01351$ or the new prescription $r_1 = 4.774$; $d'_1 = 0.3$; $r_2 = 2.191$; $d'_2 = 0.0214$; $r_3 = 2.191$; $d'_3 = 0.5$; $r_4 = -119.0$ by the usual solution for achromatism, and this yielded $LA'_k = +0.0213$, $OSC'_k = +0.00051$, a decided improvement, but with the correction overdone for both aberrations. Linear interpolations with reference to LA'_k only then led to the conclusion that spherical correction should result with $d' = 0.0214 \times 0.0936/(0.0936 + 0.0213) = 0.0174$, which proved to be nearly correct, for with the resulting new $r_4 = -102.94$ it gave $LA'_k = -0.0018$; $OSC'_k = -0.00009$, the former about one-third of the Rayleigh limit, the second negligible. Clearly a repetition of the simple interpolation would give a result well within the possible inaccuracy of a five-figure calculation and would thus again avoid the complication of a final solution and test by the method of trigonometric interpolation.

The principal object of applying the method of solution to a flint-in-front objective was, however, to determine the zonal spherical aberration resulting from the air-space. Adding the usual zonal ray at 0.7071 of the full aperture, the system with 0.0174 air-space gave at the final focus

$$l' = 9.5159 \qquad L'_z = 9.5231 \qquad L'_m = 9.5177.$$

In the absence of higher aberrations L'_z should have been the mean of l' and L'_m or 9.5168; it was found to be 9.5231. Hence we may claim the difference as the net zonal aberration or $LZA' = 9.5168 - 9.5231 = -0.0063$. The otherwise quite similar crown-in-front objective gave $LZA' = -0.0023$, which was harmless as, for inch measure, the tolerance is ± 0.0075. But for the flint-in-front objective the LZA' is uncomfortably close to the full tolerance; hence we may conclude that the method of securing aplanatism by only three radii plus a small air-space should be restricted to the crown-in-front type, which is also less expensive on account of longer radii and more robust because the harder and durable crown-lens faces outward.

In judging the results for a parallel air-space, it should be borne in mind that they are for the very unusual $f/4$ ratio, with angles of incidence in the air-space up to

more than 48°.   For more usual ratios like $f/6$ or less, the direct result by our   [87]
simple correcting formulae would frequently prove to be sufficiently close to the
desired correction.

F. Aplanatic objectives fulfilling the Gauss condition.   The obvious way to
detect the variation of the spherical correction for two different colours consists in
tracing a paraxial and a marginal ray in each colour.   A direct analytical solution by
the methods applicable to two thin components at a variable separation is quite
possible; nevertheless, it is not compatible with our custom of seeking a trigonometrically exact solution with the least expenditure of time and trouble, for the
analytical solution would not only be very complicated and laborious, but would
also give a very unsatisfactory approximation.   Hence we shall again start with a
close-contact system.   However, as the variation in the spherical correction for
different colours is usually small even in the close form, and as simultaneous zero-
value with a proper final form by no means implies equal spherical aberration when
the aberration is imperfectly corrected, it is necessary to begin with an initial form
of close-contact objectives which gives only very small residuals of spherical
aberration, not much larger than approximately the difference of the LA'_k of the
two separate colours.   The usual TL analytical solution for aplanatic objectives
given in section [59] will rarely fail to give a sufficiently close starting form.   The
four specified rays must then be traced through this, and it will be both easiest and
best to find the last radius by putting the *paraxial* intersection-lengths into Chr. (1)
because it leads to a direct exact solution for coincidence of the paraxial foci.   If
this solution yields a final LA' for the less refrangible colour which is algebraically
greater than the final LA' for the more refrangible colour, then an air-space between
the two components can be found which will secure fulfilment of the Gauss condition.

In order to make it easier to follow the method, and also to justify the criticism
of ordinary binary object-glasses satisfying the Gauss condition, we will interweave
a simple numerical example, choosing the solution in section [42] (2) for an
ordinary object-glass of hard crown and dense flint with ordinary photographic
achromatism (union of D and G') and of 8 units focal length, for which we found
$c_a = 0.5646$, $c_b = -0.2664$.   For the close initial form, ample experience suggested the middle-bending rule or $c_1 = c_a/3 = 0.1882$ for the first surface, whence
for the second surface, $c_2 = c_1 - c_a = -0.3764$, also as likely to give a close approach to spherical correction $c_3 = -0.371$.   It was decided to calculate for $f/6$,
whence initial $Y = y = 0.6667$, for which the thickness of the crown lens was
fixed at 0.2 and that of the flint lens at 0.15.   The starting prescription thus arrived
at is

$r_1 = $ 5.313

$\qquad\qquad\qquad d'_1 = 0.200 \qquad Nd = 1.5155 \qquad Ng' = 1.52630$

$r_2 = -2.657$

$\qquad\qquad\qquad d'_2 = 0$ (air)

$r_3 = -2.695$

$\qquad\qquad\qquad d'_3 = 0.150 \qquad Nd = 1.6225 \qquad Ng' = 1.64539$

$r_4 = $?

[87]   A paraxial and a marginal ray were traced in each colour (by five-figure logs with angles to half seconds) and gave for the rays arriving at the last surface

$$L_r = 26.8290 \qquad l_r = 26.3108 \qquad L_v = 28.4213 \qquad l_v = 27.7542;$$

the paraxial data put into Chr. (1) page 192 then gave $r_4 = -9.607$ and the completion of the ray-tracing gave the final results

$$L'_r = 7.9076 \qquad l'_r = 7.9074 \qquad L'_v = 7.9195 \qquad l'_v = 7.9074$$
$$LA'_r = -0.0002 \qquad\qquad LA'_v = -0.0121$$

with exact paraxial achromatism.   As $LA'_r - LA'_v = +0.0119$ is positive, it will be possible to find an air-space which will cause the Gauss condition to be fulfilled. The sine condition gave $OSC'_r = -0.00070$ and $OSC'_v = -0.00046$, showing that there is also a slight chromatic variation in the coma.   As proof of the small importance of the Gauss condition in the ordinary aplanatic object-glass, even at the very bold $f/6$ ratio, we note that, if the correction is completed without an air-space, slight changes in the difference of curvature at the contact and a general bending of the whole objective would split the differences in both LA' and OSC' so as to produce, as nearly as is of practical interest,

$$\text{Final } LA'_r = +0.0059; \qquad LA'_v = -0.0059;$$
$$\text{Final } OSC'_r = -0.00012; \qquad OSC'_v = +0.00012.$$

The coma correction would be amply sufficient for all colours, and as OSC' is independent of linear scale, also at all apertures of the $f/6$ objective.   For 8 inches focal length, the spherical tolerance at $f/6$ is ± 0.0115 or twice the LA' found for D and G', and the conversions for other f-ratios and focal lengths already exemplified will show that the small variation in spherical aberration would never become serious in the proportions of simple object-glasses that are used in practice.

When the chromatic variation of the spherical aberration has been determined for the close initial form, we shall require a suitable simple formula for the estimation of the air-space which will correct that variation, in order to avoid a series of tiresome trials.   We obtain this formula in the following manner.

With a crown-in-front objective, after the air-space has been introduced, the r- and v-rays from any point on the first surface of the air-space will be dispersed so that the v-ray is more strongly convergent than the r-ray; hence the rays become separated in the air-space, with the result that the r-ray enters the flint lens at a higher SA than the v-ray and thus suffers more over-correction at the flint, as required.   A similar argument could be used for the flint-in-front form.

It will be convenient at this point to transfer the final aberrations LA'_r and LA'_v back into the air-space by the marginal longitudinal magnification formula $(\sin U'_k / \sin U'_A)^2$.   We shall call these transferred values $T(LA'_r)$ and $T(LA'_v)$ respectively.   If we call LA'_{A_r} and LA'_{A_v} the aberrations produced in the air-space by the first component only, then in the close position the second component contributes an amount of aberration equal to $LA'_{A_r} - T(LA'_r)$ in red, and similarly in violet.   On separating the components by an air-space d', the r- and v-rays will become separated, and instead of entering the flint at the same $\overline{SA}$ they will enter respectively at $\overline{SA}_r$ and $\overline{SA}_v$. The separation, plus a small bending of the second [87] component to be separately determined, is to be such that the second component will accurately absorb both the red and the violet aberrations presented to it by the first component. Hence, by means of a small supplementary bending, we shall have to attain the two simultaneous conditions:

$$[LA'_{A_r} - T(LA'_r)]\left(\frac{\overline{SA}_r}{\overline{SA}}\right)^2 = LA'_{A_r}$$

for red, and similarly for violet. Dividing the first by the second, to eliminate the mean $\overline{SA}$, we obtain:

$$\frac{LA'_{A_r} - T(LA'_r)}{LA'_{A_v} - T(LA'_v)}\left(\frac{\overline{SA}_r}{\overline{SA}_v}\right)^2 = \frac{LA'_{A_r}}{LA'_{A_v}}.$$

We now introduce the contraction formula (2) on page 567, neglecting the δc_B term, with reference to a common SA at the second surface in the close position, X_B being the same for both colours, giving

$$\frac{SA_r}{SA} = 1 - \frac{d'}{(L_B + X_B)_r}$$

and on dividing this by the corresponding formula for the violet ray, we get:

$$\frac{\overline{SA}_r}{\overline{SA}_v} = \left(1 - \frac{d'}{(L_B + X_B)_r}\right)\bigg/\left(1 - \frac{d'}{(L_B + X_B)_v}\right)$$

$$= 1 - \frac{d'}{(L_B + X_B)_r} + \frac{d'}{(L_B + X_B)_v} \quad \text{approx.}$$

$$= 1 + d' \cdot \frac{L_{B_r} - L_{B_v}}{(L_{B_r} + X_B)(L_{B_v} + X_B)}$$

Hence

$$\left(\frac{\overline{SA}_r}{\overline{SA}_v}\right)^2 = 1 + 2d' \cdot \frac{L_{B_r} - L_{B_v}}{(L_{B_r} + X_B)(L_{B_v} + X_B)}.$$

This is now ready to be put into the equation stated ten lines above, which then becomes

$$\frac{LA'_{A_r}}{LA'_{A_v}}\left(1 - \frac{T(LA'_r)}{LA'_{A_r}} + \frac{T(LA'_v)}{LA'_{A_v}}\right)\left(1 + 2d'\frac{L_{B_r} - L_{B_v}}{(L_{B_r} + X_B)(L_{B_v} + X_B)}\right) = \frac{LA'_{A_r}}{LA'_A}.$$

Multiplying out and retaining first-order terms only, this gives:

$$\frac{LA'_{A_r}}{LA'_{A_v}}\left(2d'\frac{L_{B_r} - L_{B_v}}{(L_{B_r} + X_B)(L_{B_v} + X_B)} - \frac{T(LA'_r)}{LA'_{A_r}} + \frac{T(LA'_v)}{LA'_{A_v}}\right) = 0$$

whence

(8) $$d' = \tfrac{1}{2}\left(\frac{T(LA'_r)}{LA'_A} - \frac{T(LA'_v)}{LA'_A}\right)\left(\frac{(L_{B_r} + X_B)(L_{B_v} + X_B)}{L_{B_r} - L_{B_v}}\right).$$

[87]   In our numerical example, the transferred final aberrations of the close form are found to be $T(LA'_r) = -0.00002$ and $T(LA'_v) = -0.00191$, and the aberrations of the front component are $LA'_{A_r} = +0.30447$ and $LA'_{A_v} = +0.30448$. By closing the calculation in the air-space, we find: $L_{B_r} = 3.11666$, $L_{B_v} = 3.04625$, and $X_B = -0.08419$. On putting these data into equation (8), we determine the desired air-space d' to be 0.3956.

This separation would lead to the appearance of considerable LA'_k and OSC'_k, and a bending δc_B must be applied to the flint alone to remove the LA' residual, together with a general bending δc of the whole lens to correct the OSC'. To determine the flint bending, we transpose equation (4) (page 569), giving

$$(9) \qquad \delta c_B \cdot \frac{\partial LA'_A}{\partial c_A} = LA'_k \cdot \frac{u'_k \sin U'_k}{u'_A \sin U'_A} + 2LA'_A \cdot \frac{d'}{L_B + X_B}.$$

The new data necessary for using this equation are found to have the following values in our numerical example: $\partial LA'_A / \partial c_A = -1.08$ (from a small trial bending of element A alone); $LA'_k = -0.00014$ for the close form; and $u'_k \sin U'_k / u'_A \sin U'_A = 0.17528$. These give at once the required bending of the second lens as $\delta c_B = -0.07434$.

We must now apply a uniform bending, δc, to the whole lens to correct for OSC'. This bending may be found by using equation (5) (page 570). From that equation the value of OSC'_k for the close form is found to be -0.0007; $SA = 0.6667$; $f' = 8.1$; and we desire the final OSC' to be zero. However, as was pointed out in connection with the derivation of equation (5), the other data in the formula, namely, L_B, l_B, and X_B, are to be those of the original close form *after* the bending δc has been applied to it! We are therefore obliged to make a series of tentative trials before the correct bending δc can be determined. In our example, we first take the data from the original close form, namely, $L_B + X_B = 3.03248$ and $l_B = 3.42114$, and remembering that for this form $\delta c = 0$, we calculate the final OSC' to be -0.01007. This being far too large to be tolerated, we make an arbitrary bending of $\delta c = 0.10$ to only the crown lens of the original close form, giving the new values $L_A + X_B = 3.13183$ and $l_B = 3.19286$. Putting these values, with $\delta c = 0.10$, into equation (5), we calculate the final OSC' to be -0.00071. This would be almost good enough, but a simple linear extrapolation shows that the value of δc required for perfect correction is 0.1076.

We therefore reach the conclusion that by bending the crown lens of our close form by $\delta c_A = 0.1076$, the flint lens by $\delta c_B = 0.1076 - 0.07434 = 0.003326$, and introducing an air-space of 0.3956, we should expect to obtain an aplanatic objective in which the spherical aberration would be the same for both D and G' light.

The full specification of our analytical solution thus becomes:

$$r_1 = 3.3805$$
$$d'_1 = 0.2$$
$$r_2 = -3.7207$$
$$d'_2 = 0.3956$$
$$r_3 = -2.9603$$
$$d'_3 = 0.15$$
$$r_4 = 101.25 \text{ (by Chr. 1)}$$

A careful ray-trace through this system reveals that the spherical aberration is [87] $LA'_D = +0.01046$ instead of the desired zero value, and a small additional bending of the flint is therefore required to eliminate it.   Since $\partial LA'_A/\partial c_A = -1.08$, this would have to be multiplied by $u'_A \sin U'_A/u'_k \sin U'_k = 8.44$ when transferred to the final image, hence $\partial LA'_k/\partial c_B = \partial LA'_k/\partial c_A = 9.12$ approximately.   Since we need $LA'_k = -0.01046$, a bending of $\delta c_B = -0.001137$ is indicated.   Actually this turns out to be insufficient, but this trial bending gives the corrected value of $\partial LA'_k/\partial c_B = 6.57$.   A repetition of the procedure, resulting in a further small bending of the flint lens, leads to the final formula:

$$r_1 = \quad 3.3805$$
$$d'_1 = 0.2 \qquad Nd = 1.5155 \qquad Ng' = 1.52630$$
$$r_2 = -3.7207$$
$$d'_2 = 0.3956 \text{ (air)}$$
$$r_3 = -2.9468$$
$$d'_3 = 0.15 \qquad Nd = 1.6225 \qquad Ng' = 1.64539$$
$$r_4 = 120.43$$
(by Chr. 1).

The results of a careful ray-trace through this system are:

	D	G'
$l' =$	8.48275	8.48275
$L' =$	8.48255	8.48086
zonal $L' =$	8.48562	8.48520
$LA' =$	+0.00020	+0.00189
$LZA' =$	−0.00287	−0.00245

giving

$$LA'_d - LA'_{g'} = -0.00169 \qquad OSC'_d = -0.00014.$$

The sphero-chromatism has thus been reduced to about one-seventh of its original magnitude and its sign has been changed, showing that a repeated design with an

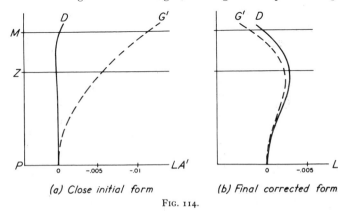

(a) Close initial form   (b) Final corrected form

FIG. 114.

[87] air-space of about 0.350 would be likely to be quite correct. However, as the Rayleigh limit for this lens, in inch measure, is about 0.018, it is clear that all the corrections are amply good enough. The worst remaining defect is a significant residual zonal aberration, in both colours, of the unusual over-corrected form, which is the penalty we must pay for equalizing the two marginal aberrations.

The properties of the original and the final designs are shown graphically in Fig. 114.

PHOTOVISUAL OBJECTIVES

[88] THE thin-lens solution of this problem is given in section [41] of Part I. If the three lenses, having the total curvatures fixed by the TL solution, were all cemented together, the resulting combination could only be bent as a whole, giving *one* liberty for the correction of aberrations other than the chromatic, and it would be possible to produce an objective with spherical correction for only one colour.

Such a procedure, however, would not do justice to the possibilities of such an objective; to reap the full benefit of the removal of the secondary spectrum over the largest possible field, the coma should also be corrected (equivalent to the fulfilment of the sine condition), and it is also highly desirable that the spherical aberration should be corrected for two different colours, such as the visually and the photographically brightest light.

The correction of coma calls for the 'breaking' of one of the contact surfaces, just as in aplanatic two-lens objectives. The simplest and most effective way of securing spherical correction for two colours then consists in adding also a moderate air-space between the lenses at the broken contact, as in the previous section.

The independent bending of the two parts of the separated objective plus the variation of the air-space provide the three liberties necessary for the simultaneous fulfilment of three conditions. The air-space required is quite small—about 1/250 of the focal length of the complete objective will usually be its approximate size. The difference between the curvatures at the air-space is also usually quite small.

In preparing for the solution, it must be borne in mind that we are not bound to adhere to the sequence of the three lenses used in the TL,Chr. solution, for the achromatism of a thin system, as well as its power, is independent of sequence. As it is desirable that the central lens should be approximately equiconvex or equiconcave and as the first surface of any aplanatic objective tends to be more strongly convex than the last, it is likely to prove beneficial if, in a system with a *concave* central lens, we place the *stronger* of the two convex lenses in front, whilst in a system like example (3) of section [41], with a *convex* central lens, it will be advisable to place the *weaker* concave lens in front.

There will also be an initial doubt as to which one of the two contacts should be selected for the air-space. As a rule, it will be best to select the contact that causes the lens or cemented pair in front of it to assume *convex* properties; therefore in the case of the usual concave middle lens, we have the arrangement shown in Fig. 115 (a), but when the middle lens is convex, we have the arrangement of Fig. 115 (b). The reason for this choice is that it leads to converging rays in the air-space and that in consequence the aperture of the second part will be smaller than that of the first. With the reverse arrangements, in case the air-space is comparatively large, the second part would demand a larger clear aperture than the [88] first, and in a large telescope objective, that would mean a serious increase in cost! These two suggestions as to the sequence of lenses and the choice of the broken contact must not be regarded as hard-and-fast laws, however, but merely as advice which will prove correct in the majority of cases.

We will take example (2) of section [41] for the numerical work.   The TL,Chr. data for this example are, correct as to sequence,

$$c_a = 1.54 \quad c_b = -2.26 \quad c_c = 0.935 \quad \text{for} \quad f' = 8,$$

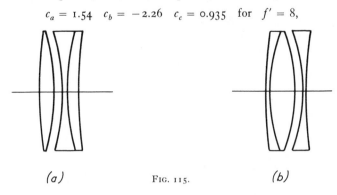

(a)   FIG. 115.   (b)

and in accordance with the advice just given, the broken contact will be between lenses (a) and (b) whilst (b) and (c) will be cemented.†

The properties of the three kinds of glass are:

	Nd	Nf−Nc	V	Pg'f	Pfd
(a) Chance 6493	1.5160	0.00809	63.8	0.561	0.701
(b) Chance 3389	1.5376	0.01069	50.3	0.576	0.707
(c) Chance 466	1.5833	0.01251	46.6	0.592	0.711

In judging the progress in correcting the aberrations it must be borne in mind that the focal length '8' will in actual instruments be at least in *feet* and may be in *metres*, hence the fourth decimal in the final intersection lengths should be reasonably dependable and six-figure logs will be needed in actual practice.

The beginner at any rate will be wise to make a TL solution first for correction of spherical aberration and of coma; it will prove to be a decidedly rough approximation on account of the deep curves and considerable thicknesses, but it will be near enough to facilitate greatly the work of trigonometrical correction.   When experience has sufficiently developed the instinct for a form which such an objective is likely to take, then it *may* save time to begin at once with trigonometrical work on a guessed prescription.

† It may here be emphasized that, for large lenses, such as occur in telescope objectives for serious astronomical work, the cement must be *very soft* as otherwise the lenses would be distorted by unequal thermal expansion with changes of temperature.   Soft Canada balsam, practically unbaked, will be safe for a considerable period but will eventually harden to a prejudicial extent.   Castor oil or glycerine is safer.

[88]   In order that the TL solution may fit in with the subsequent trigonometrical work, it is desirable to estimate the admissible aperture.   It may be taken as practically certain that in large telescope objectives a sufficiently close reduction of the secondary and tertiary aberrations will prove impossible if the clear aperture is equal to the shortest radius of curvature, for that would make the $(U+I)$ of the surface of deepest curvature $30°$ and either I or I' still larger.   This criterion gives us a first means of estimating the greatest possible aperture.   Our middle lens has $c_b = -2.26$; if we were lucky enough to arrive at an exactly equiconcave form of this lens, the two surfaces would have curvatures of ∓ 1.13 or radii of $1/1.13 = 0.885$.   Therefore 0.885 is an ultra-optimistic estimate for the maximum clear aperture.   But our first lens also has the high total curvature of 1.54, and nearly all of this is likely to be on its second surface, for photovisual objectives tend to be very flat on their first surfaces.   We must therefore be prepared for, say, $c_2 = -1.50$ or $r_2 = -0.667$, and that brings the permissible aperture down to 0.667; but whilst this *would* give $U_2 + I_2 = 30°$, the marginal ray would come out with a strong convergence and I'_2 would therefore greatly exceed the risky value of $30°$. Simple arguments of this type lead to the conclusion that it will be wise to stop at an aperture of 0.500, and this we will adopt, giving $y = 0.250$ for use in the TL(10)** formulae by the G-sums.   This choice is in good accordance with predominating practice, giving a ratio of focal length to aperture (or f-number) of 16; existing objectives in fact more often exceed the ratio 16 than they fall short of it.

We have now settled the form of our objective (Fig. 115a) and the clear aperture to be aimed at, and are now ready to work out the spherical aberration and coma by the G-sums.   This work will be only slightly different from that exemplified for the usual two-component aplanatic objective.   For the single front lens, it will be best to work by the 'original G-sums' in terms of v_1 and c_1.   For telescope objectives (and also for nearly all photographic objectives) the objects are assumed at infinity, so $v_1 = 0$ and that reduces the spherical G-sum to three terms and the coma G-sum to two terms.   Hence we have only to put $y = 0.25$, $l'_k = 8$ (namely, the intended focal length of the complete system), and $c_a = 1.54$ into the usual computing scheme, with the G-values taken out for $Nd = 1.5160$.   The scheme then gives the spherical contribution of the front lens in the form of a quadratic equation in c_1 and the coma contribution as a linear equation in c_1.   In the latter we use $H'_k = 1$ for the sake of simplification.

For the cemented second component, it will be best to express the aberrations in terms of the curvature and intersection-length at the contact; therefore we must use the 'alternative G-sums' for the lens (b) and the 'original G-sums' for the last lens (c).   That necessitates calculating the convergence (or in this case the divergence) $v'_4 = v_5$ with which the rays would pass through the assumed infinitesimal air-space between lenses (b) and (c), for this is the v-value which must be used for both lenses.   By the universal thin-lens formula $v' = (N-1)c + v$, the first lens will—with $v_1 = 0$—give $v'_2 = v_3 = (N_a - 1)c_a = 0.5160 \times 1.54 = 0.793$ by slide rule.   The middle lens then gives $v'_4 = v_5 = (N_b - 1)c_b + v_3 = 0.5376 \times (-2.26) + 0.793 = -1.215 + 0.793 = -0.422$, again by slide rule, and this is the v' for lens (b) and the v for lens (c) to be used in the G-sums.   As the objective is considered to be *thin*, again $y = 0.25$, $l'_k = 8$, and $H'_k = 1$.   The contributions are [88] of course calculated separately for (b) and (c), but as both will be in terms of the contact curvature, they can be algebraically added when found and give the contribution of the cemented component to the total spherical aberration as a quadratic equation in c_4 and the coma contribution as a linear equation in c_4. Adding the corresponding equations found for the front lens, we obtain the total spherical aberration at the final focus as a quadratic in c_1 and c_4, and these two simultaneous equations are solved precisely like those of an ordinary aplanatic objective.   There will be no possible doubt as to which is the useful solution and which is the freak.   From the values of c_1 and c_4 thus found, the curvature of the remaining surfaces will be obtained by the usual application of the total curvatures c_a, c_b, and c_c.

From our numerical example, the TL solution for correction of coma and spherical aberration gives the following contributions:

Lens	(a)	$SC' =$	$3.686c_1{}^2 - 9.870c_1 + 8.662;$	$CC' = 0.04121c_1 - 0.05797$
	(b)		$-5.590c_4{}^2 + 15.608c_4 - 18.196;$	$-0.06265c_4 + 0.08949$
	(c)		$2.468c_4{}^2 - 1.245c_4 + 0.744;$	$0.02780c_4 - 0.00631$
or (b+c) cemented			$-3.122c_4{}^2 + 14.363c_4 - 17.452;$	$-0.03485c_4 + 0.08318$

Adding the aberrations of the first lens we find

Total $SC' = 3.686c_1{}^2 - 9.870c_1 - 3.122c_4{}^2 + 14.363c_4 - 8.790$

Total $CC' = 0.04121c_1 - 0.03485c_4 + 0.02521$

These two equations have to be solved for the coordinated values of c_1 and c_4 which will produce the desired state of correction, in our case zero for Total SC' and Total CC'.   By the linear equation, we find $c_4 = [0.0728]c_1 + [9.8594]$, and using this to eliminate c_4 from the quadratic equation, we find the equation for c_1 to be

$$0.679c_1{}^2 - 1.771c_1 = -0.034$$

and this gives the two solutions for $c_1 =$ either 2.59 or 0.0194.   The first solution is manifestly out of the question as it would give $r_1 < 0.4$!   Hence we adopt the second solution.   Introduced into the equation of condition $c_4 = [0.0728]c_1 + [9.8594]$, it gives the corresponding $c_4 = 0.7463$ and the total curvatures then give the complete prescription:

$$c_1 = 0.0194 \quad \text{whence} \quad r_1 = 51.55$$
$$c_2 = -1.5206 \qquad\qquad\quad r_2 = -0.6576$$
$$c_3 = -1.5137 \qquad\qquad\quad r_3 = -0.6606$$
$$c_4 = 0.7463 \qquad\qquad\quad r_4 = 1.340$$
$$c_5 = -0.1887 \qquad\qquad\quad r_5 = -5.30$$

and this is the prescription which has to be corrected trigonometrically.

The first step should always be the fixing of proper thicknesses by a sufficiently

[88] large and accurate scale drawing, using the radii obtained from the TL solution and a full aperture exceeding the intended clear aperture by from 5 to 10 per cent. according to size. Whilst a sharp edge on a convex component should always be avoided, it will be necessary to be content with a comparatively thin edge (perhaps 1/50 or even 1/100 of the aperture) to eke out the thickness of the plates as usually supplied by the glass makers. For the same reason, the concave element will probably have to be restricted to 1/20 or even less of the clear aperture as its central thickness, although this increases the difficulty of producing first-class surfaces. Whenever possible, concave lenses should have a thickness of 1/10 or at least 1/15 of their aperture. But with good mounting of lenses on the grinding and polishing machines much less can be tolerated; as an example, the flint lens of the 36-inch Lick objective (one of the best in the world!) has a central thickness of a little less than *one inch*!

Since r_3 is longer than r_2, no air-space need be provided between the front lens and the back combination. A scale drawing gives as quite sufficient $d'_1 = 0.07$, $d'_3 = 0.02$, $d'_4 = 0.05$, the first and last of which would bear reducing by another 0.01 if necessary. The edge thickness of the biconcave lens is in excess of 0.10, which means difficulty in obtaining a disk as the usual maximum is 1/7 of the diameter. The value of d'_3 could not safely be diminished below 0.015, so it may be necessary to lengthen the focus in order to eke out a given disk. The thicknesses stated above will be adopted in our example. It will be understood that, in making the scale drawing, the first and last radii should be used to *calculate* the depth of curvature because a long beam-compass is not usually available and is inconvenient even when at hand. The calculated sag X_1 shows that the curvature of surfaces of long radius may safely be treated as negligible in comparison with the arbitrary selection of the minimum edge thickness which may be regarded as still sufficient. For the same reason all the X may be *calculated* by the first-approximation formula $Y^2/2r$ in telescope objectives of every kind; but this is not safe in deeply curved photographic or microscope objectives.

An accurate ray-tracing in C and F light through the analytical form up to the last surface gives $l'_{4c} = 101.0413$ and $l'_{4f} = 119.9743$. Following the recommended procedure of section [87] F, the last radius is next calculated for C-F achromatism by the *paraxial* form of Chr. (1), giving $r_5 = -5.2894$, whence we find for $Y = 0.25$:

$$LA'_c = -0.00871 \qquad LA'_f = -0.02419$$
$$OSC'_c = +0.00012 \qquad OSC'_f = +0.00026$$
$$f'_c = 7.9907$$

As it is desirable to correct the spherical aberration in the two colours C and F, a small air-space is now introduced between the first and second lenses, the width of which is calculated by equation (8) on page 575. This formula gives $d' = 0.0325$, and by applying equation (9), the necessary bending δc_b of the second component to remove the spherical aberration caused by the introduction of the air-space is found to be $\delta c_b = -0.05366$. The necessary further bending δc of the whole system to remove coma is then found, by repeated applications of equation

(5), to be $\delta c = +0.040$.   Applying both these bendings, the improved prescription   [88]
becomes:

$$r_1 = \quad 2.3844$$
$$0.07$$
$$r_2 = -0.8923$$
$$(0.0325)$$
$$r_3 = -0.8566$$
$$0.02$$
$$r_4 = \quad 0.9152$$
$$0.05$$
$$r_5 = \quad 4.5848 \text{ by Chr. (1)}$$

An accurate tracing of the marginal and paraxial rays in C light through this system reveals an under-corrected residual of $LA'_c = 0.0580$.   This is far too large to be tolerated in view of the fact that the 'unit' to which this lens is constructed may well be in the foot, in which case the Rayleigh limit for spherical aberration will be only 0.0086.

To remove this residual of spherical aberration, the rear component alone must be given a further small bending.   A few trials show that $\delta c_b = +0.0057$ will be satisfactory, giving the improved prescription for the rear component:

$$r_3 = -0.8524$$
$$r_4 = \quad 0.9200$$
$$r_5 = \quad 4.7104 \text{ by Chr. (1)}$$

With this additional bending, the aberrations of the whole objective become:

$$LA'_c = -0.0012 \qquad LA'_f = -0.0018$$
$$OSC'_c = +0.00044 \qquad OSC'_f = +0.00044$$
$$f'_c = 9.0730$$

Thus the air-space d' and the bending δc of the whole lens have satisfactorily accomplished their respective duties of equalizing the spherical aberration in C and F light and correcting the offence against the sine condition.

It remains now to investigate the residual zonal aberration, and to ascertain how well the original aim of uniting the three colours C, F, and G' at a common focus has been achieved.   Tracing of 0.7 zonal rays in C and F light reveals the presence of the following zonal residuals of the unusual *over-corrected* form:

$$LZA'_c = -0.0074 \qquad LZA'_f = -0.0087.$$

The Rayleigh tolerance of $6\lambda/\sin^2 U'_m$ being 0.0132 if the unit is the foot, we see that the lens will be fully acceptable so far as the spherical and zonal aberrations in C and F light are concerned.

The chromatic aberration residuals are found by tracing paraxial rays in the four colours C, D, F, and G' through the objective, with the results:

$$l'_c = 8.6643$$
$$l'_d = 8.6608$$
$$l'_f = 8.6643$$
$$l'_{g'} = 8.6656$$

[88] The C, F, and G' lines are therefore very closely united, as is the aim in a photo-visual objective, but it should be noticed that the D focus falls short of the united C-F focus by 0.0035, which is about 1/2600 of the focal length.   For an ordinary two-lens achromat, it was shown on page 157 of Part I that the D line falls short of the C-F focus by 1/2500 of the focal length, so we conclude that the *tertiary* spectrum of our photovisual objective for the D line has about the same magnitude as the *secondary* spectrum of an ordinary achromat for that line.   The great advantage of the photovisual lens appears, of course, in the blue portion of the spectrum, where the normal chromatic over-correction has been virtually eliminated.

CHAPTER XII

OPTICAL PATH DIFFERENCES

BEFORE undertaking the study of aberrations and their tolerances on the basis [89] of the wave nature of light, which will form the substance of the next four chapters, the student is urged to re-read Chapter III of Part I, in which the undulatory theory of light and the physical aspect of optical images are discussed. The following important conclusions to be drawn from that discussion will form the basis of the present study:

(1) That light travels in straight lines with a velocity which is an absolute constant for any transparent medium and for any colour of light.

(2) That in free space this velocity is about 186,000 miles per second, but in a denser medium the velocity is reduced in proportion to the refractive index N. Thus light takes as long to travel through a geometrical distance L in a medium of index N as to travel an *optically equivalent* distance $N \cdot L$ in empty space. We shall employ this law to reduce all path-lengths to their 'equivalent optical paths' in air by multiplying the geometrically measured path in any medium by the refractive index of the medium.

(3) That light is of an undulatory character and is propagated by waves. The number of vibrations per second is an absolute constant for any one colour in the spectrum. Division of the distance travelled in one second by the number of vibrations per second gives the *wave-length*, or distance from crest to crest. As the velocity of light is less in dense media than in air, the wave-length is also lower in such media. Wave-lengths stated without further specification are always understood to be those in air of an agreed standard density. The most usual unit of wave-length is the Ångstrom Unit, equal to 10^{-7} mm. In this unit, wave-lengths come out in four-figure numbers, ranging for visible light from about 7600 Å in the deep red to about 3900 Å in the near ultraviolet. However, in optical calculations it is more convenient to use the micron (0.001 mm.) as a unit; the visual range of wave-lengths then runs from 0.76 to 0.39μ, leading to convenient numerical coefficients.

(4) That in accordance with the principle of interference, or the superposition of wave motions, two portions of light from the same source when meeting will produce a maximum of intensity if they meet in the same phase of vibration, say crest and crest, and a minimum of intensity if they meet in opposite phases, crest and trough. In optical instruments we nearly always seek to produce the brightest and smallest possible image of each point in the object. In accordance with the principle of interference just stated, this end will be attained if all the light passing from the object-point through different parts of the clear aperture to the image-point does so along paths of precisely the same equivalent optical length, for all the light will then reach the image-point after the same time interval and therefore in the same phase. We thus arrive at the chief fundamental principle of the physical treatment of optical problems: *The optical paths from object-point to image-point should all be equal within a small fraction of a wave-length.*

[89]   Although the object-point may be very distant, nearly the whole of the total paths are necessarily and precisely equal on strictly Euclidean principles, for as the object-point sends out spherical waves (in an isotropic medium), the paths are equal up to a spherical surface having the object-point B as centre and touching the first surface of the lens either on the axis or at the margin according to the sign and length of the first lens radius (Fig. 116).   On the image side, the same applies with reference to the path-lengths from a spherical surface touching the last lens-surface and having B' as centre in the image-point.   Consequently the only paths which require watching as to equality are those between these two auxiliary spheres, and even these paths become further reduced by similar auxiliary concentric spherical surfaces which can be fitted within each lens or space.   The part of the total optical paths which depend on the skill of the designer and the maker of a lens system is thus reduced—with rare exceptions—to a small fraction of an inch, and the precision called for sinks to something of the order of one part in ten thousand.

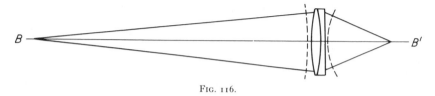

FIG. 116.

(5) As residuals of aberration are inevitable in optical instruments, the most important question from the optical designer's point of view is: To what extent may the optical paths leading to an image-point be allowed to vary without sensible, or at any rate without serious, detriment to the brightness and sharpness of the image?   The answer depends on rather laborious calculation of the light-distribution at a focus in the presence of residual aberrations.   The late Lord Rayleigh was the first to tackle this extremely important problem and arrived at the conclusion (Part I, section [38]) that an extreme variation of the optical paths to the extent of one-quarter of a wave-length was a suitable limit.   This is the value which we shall here adopt, but we shall add conclusions as to departures from Lord Rayleigh's value in both directions when the occasions arise.

The Method of Optical Path Differences

The procedure indicated by the above principles obviously is that we must first determine the geometrical path-length of all light to be considered in a given problem, separately for each medium traversed; that in the second place we multiply each geometrical path by the index of the medium in which it lies in order to turn it into the equivalent optical path in vacuo—or, in practical optics, in air; that thirdly we form the algebraic sum of all the separate pieces of each total track from object-point to image-point; and finally that we compare the various optical pathlengths to determine the conditions under which their maximum difference will not exceed the Rayleigh limit.

As a first example, we take the refraction of a pencil of rays (in geometrical [89] optical language) at a single refracting surface. In order that as many quantities as possible shall have a positive sign, we take for our diagram (Fig. 117) the case of a pencil of rays which would form a perfect focus at a given point F but is intercepted by the refracting surface AP. This surface separates a medium of index N at its left from a medium of index N' at its right and would, geometrically, produce a more or less perfect focus F' of the refracted rays in the second medium. By the geometrical method, we should have to trace each ray by the law of refraction and then seek the point of closest union of the refracted rays and claim that as the best possible image. By the physical method which we are now learning to apply, we begin by noting that, since F is given as a perfect focus, all the light (along all the rays of the pencil) would arrive at F in the same phase of vibration or without any

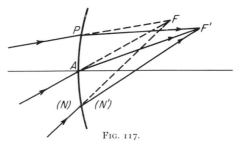

FIG. 117.

difference of optical paths. On interposing the medium of index N' with the bounding surface AP and enquiring as to what will happen at any arbitrarily chosen point F' in the new medium, we note that the light from P would have to travel along PF' instead of along PF. As the index of the medium in which the focus F would be formed is N, the geometrical path PF corresponds to an equivalent optical path $N \cdot PF$. Similarly the actual path PF' of the light in the medium of index N' at the right of the surface corresponds to an equivalent optical path $N' \cdot PF'$. It follows that the optical path from P to F' is longer than the optical path from P to F by the difference of the two or by ·

$$N' \cdot PF' - N \cdot PF$$

It might of course happen that $N' \cdot PF'$ came out shorter than $N \cdot PF$: that possibility is covered by sign-conventions, for the formula we have written down will in such a case give a negative value of the difference and will thus point out that there is really a shortening instead of a lengthening of the path. Similarly the formula covers the case when F is a real object-point at the left of AP, for in accordance with our universal sign-conventions FP would then have the negative sign and this would automatically turn the algebraic difference represented by our formula into an arithmetical sum, as the case would obviously call for.

Therefore, provided our sign-conventions are correctly observed and interpreted, the formula for the lengthening of the path is universally applicable.

In accordance with principles 4 and 5, the only thing of interest is the *differences*

[89] of optical path with which light from different points (like P) of the clear aperture arrives at P'. The most convenient way to determine these differences is to adopt the path through one particular point of the clear aperture as a standard with which the others are compared. Naturally our choice for this standard path will nearly always fall upon the central or principal ray of the pencil. Let A therefore be the point where the principal ray of the pencil pierces the refracting surface. Merely putting A in place of P in our formula, we conclude that the lengthening of the optical paths of light passing from A to F' will be

$$N' \cdot AF' - N \cdot AF.$$

Now it is evident that because light from both A and P would arrive at F simultaneously and as the optical paths have been lengthened by the amounts expressed in the two formulae, F' would be reached simultaneously by the light coming from A and from P if $(N' \cdot PF' - N \cdot PF)$ was found to be equal to $(N' \cdot AF' - N \cdot AF)$. It is equally evident that if the two lengthenings of optical paths are found to be unequal, their difference will measure the degree of imperfection of the union of the corresponding light at F'. This difference therefore is the *difference of equivalent optical paths* which we seek to determine. It only remains to decide in which sense the difference is to be taken. We shall find before long that ordinary spherical under-correction causes the marginal light to arrive at the paraxial focus along a path which is optically shorter than the axial path. As it appears desirable to retain the positive sign for spherical under-correction, we therefore decide to take the difference in the sense of axial lengthening minus marginal lengthening of path. Introducing further the symbol OPD for the equivalent optical path difference, with the usual dash for the medium to the right of a surface, we arrive at the fundamental formula

$$OPD' = (N' \cdot AF' - N \cdot AF) - (N' \cdot PF' - N \cdot PF).$$

This formula is more convenient to use if we make a slight transposition, by collecting the terms which have N' and N as respective factors, and write

$$OPD' = N'(AF' - PF') - N(AF - PF).$$

We are now ready to attack our first problem in exact mathematical form.

A NARROW BEAM OF LIGHT INCIDENT OBLIQUELY AT A SINGLE REFRACTING SURFACE

[90] WE choose the method of coordinate solid geometry, and retaining the horizontal X-axis and vertical Y-axis in the plane of the diagram as used in all our two-dimensional investigations, we add a Z-axis at right angles to the other two and count the Z-coordinate positive behind the plane of the diagram and negative in front of that plane. In the perspective drawings the Z-coordinate, if positive, will be sloped upward and to the right, as in this simple diagram of a rectangular prism occupying the corner of the all-positive space (Fig. 118). The corner P of the prism opposite to the origin O then has the positive coordinates x, y, z as marked, and P really lies behind the plane of the diagram. These conventions as to the coordinates and the method of projection should be carefully noted and remembered

in order to minimize the well-known difficulty of clearly visualizing the three- [90] dimensional objects represented in such a flat diagram. In the optical-path studies, we shall have to deal almost entirely with the distances (like PF in our general discussion) of points, say of P_1, from P. The formula for such a distance

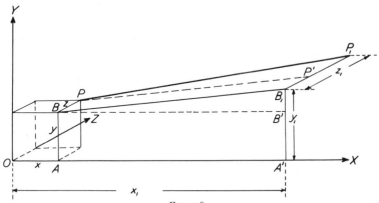

FIG. 118.

can be read off the diagram. Drawing BB' parallel to AA', we have $BB' = AA' = x_1 - x$, and $B'B_1 = y_1 - y$. Therefore by the theorem of Pythagoras

$$BB_1{}^2 = (x_1 - x)^2 + (y_1 - y)^2,$$

and drawing PP' parallel to BB_1 in the plane figure PBB_1P_1, we have $PP' = BB_1$ and therefore

$$(PP')^2 = (x_1 - x)^2 + (y_1 - y)^2.$$

We also have $P'P_1 = z_1 - z$ and so obtain for the square of the required distance PP_1, again by the Pythagorean theorem,

$$(PP_1)^2 = (x_1 - x)^2 + (y_1 - y)^2 + (z_1 - z)^2.$$

This formula is universally true, no matter what the relative position of P and P_1 may be, provided that the sign of the several coordinates is carefully observed. The sign to be given to the distance PP_1 itself is not decided by the formula on account of the ambiguity attaching to any square root in that respect. The sign to be given to the distance found by the formula must therefore be deduced in each problem by appropriate considerations. If any coordinate entering into the formula is zero, it naturally contributes nothing to the result and may be omitted; in particular, all three coordinates are zero at the origin O. The distance of any point x, y, z from the origin is therefore given by

$$OP^2 = x^2 + y^2 + z^2.$$

As the differences $(x_1 - x)$ &c. are only used *squared*, we may set either $(x_1 - x)$ or $(x - x_1)$ at pleasure in working with the general equation, but whichever form may be chosen must be worked out with the correct individual signs of x and x_1.

[90]   For our optical problem we now refer to Fig. 119. The origin of coordinates, A, shall be the point where the principal ray of the entering pencil pierces the refracting surface, and AX shall be the normal of that surface. The latter therefore has the YZ-plane as its tangent at the point A. The point B_h in the XY-plane shall be the virtual object-point, its rectangular coordinates being l in the X-direction and H in the Y-direction. The XY-plane thus represents the plane of incidence for the ray AB_h. The image-point for which the differences of optical paths are to be determined is not to be restricted in any way by preconceived notions as to where it may eventually be located, and for the present we therefore do *not* assume it to be found in the XY-plane but give it the general coordinates l', H', and z'. Similarly, we give to the point P at which the second ray is to pierce the refracting

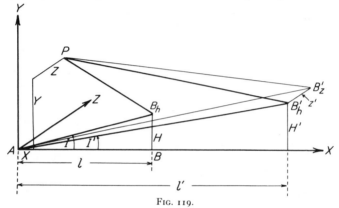

FIG. 119.

surface the perfectly general coordinates X, Y, Z. The use of the distance-formula then gives us

$$PB_h{}^2 = (l-X)^2+(H-Y)^2+Z^2$$
$$AB_h{}^2 = l^2+H^2$$
$$(PB'_z)^2 = (l'-X)^2+(H'-Y)^2+(Z-z')^2$$
$$(AB'_z)^2 = l'^2+H'^2+z'^2$$

We shall work out the formulae in this form subsequently (page 603); for our present purpose of searching for the conditions under which the isolated point B_h is best depicted in the image space, it is desirable to define B_h in terms of its distance AB_h from the point of incidence and of the angle I between AB_h and the normal AX. Evidently, from the plane triangle ABB_h we have

$$l = AB_h \cdot \cos I \quad \text{and} \quad H = AB_h \cdot \sin I.$$

Defining B'_h in the same way by the distance AB'_h and the angle I' with the normal, gives

$$l' = AB'_h \cdot \cos I' \quad \text{and} \quad H' = AB'_h \cdot \sin I'.$$

The four distances which enter into the OPD-formula then become, for the present [90] investigation

$$PB_h{}^2 = (AB_h \cdot \cos I - X)^2 + (AB_h \cdot \sin I - Y)^2 + Z^2$$
$$AB_h{}^2 = AB_h{}^2$$
$$(PB'_z)^2 = (AB'_h \cdot \cos I' - X)^2 + (AB'_h \cdot \sin I' - Y)^2 + (Z - z')^2$$
$$(AB'_z)^2 = (AB'_h)^2 + z'^2.$$

As the expressions for the first two distances differ from the corresponding last two only by not having the dash and by having $z = 0$, it is only necessary to work out the last two; the first two can then be written down by analogy. As $\cos^2 + \sin^2$ of any angle equals unity, the working out of the equation for $(PB'_z)^2$ gives

$$(PB'_z)^2 = (AB'_h)^2 - 2X \cdot AB'_h \cdot \cos I' - 2Y \cdot AB'_h \cdot \sin I' + X^2 + Y^2 + Z^2$$
$$- 2zz'Z + z'^2.$$

Here $X^2 + Y^2 + Z^2$ is, as was shown, the square of the distance of P from the origin, or PA^2. Introducing this, extracting $(AB'_h)^2$ as a common factor, and then taking the square root† gives

(a) $$PB'_z = AB'_h \cdot \sqrt{1 - 2\frac{X \cdot \cos I'}{AB'_h} - 2\frac{Y \cdot \sin I'}{AB'_h} + \frac{PA^2}{(AB'_h)^2} - 2\frac{z'Z}{(AB'_h)^2} + \left(\frac{z'}{AB'_h}\right)^2}$$

as a perfectly exact expression. For the purpose of further development and discussion, we must however eliminate the root. This root is of the form $\sqrt{(1 + a)}$ if we put a as a simple symbol for the total of the five algebraic terms. Such a root can be converted into a series of rational terms, either by extracting the root by the usual rule of algebra or more conveniently by Taylor's theorem, with the result

$$\sqrt{(1 + a)} = 1 + \frac{a}{2} - \frac{a^2}{8} + \frac{a^3}{16} - \frac{5}{128}a^4 \quad \&c.$$

However, if this series is to be convergent and fit for discussion, 'a' must be numerically smaller than unity, and in fact it must be a decidedly small fraction if the convergence is to be rapid. If we introduce 'a' into equation (a) instead of the five algebraic terms which it represents, then square and transpose, we find $a = 1 - (PB'_z/AB'_h)^2$, and we see that 'a' will reach the upper limit $a = +1$ if $PB'_z/AB'_h = 0$, and that it will sink to the lower limit $a = -1$ when PB'_z reaches the value $1.4142\ AB'_h$. From the study of oblique pencils by the geometrical method of Part I, it will be clear that the only cases of practical interest are those for which B'_z falls very close to B'_h as otherwise an enormously expanded image-patch would be implied. A few diagrams of oblique pencils, neglecting a possible slight value of z' by drawing the outer ray from P directly towards B'_h, will quickly convince the student that either of the limiting values of PB'_h which would cause the series to cease to be convergent is almost impossible from the practical optical

† The root obviously must be taken with the positive sign as we must obtain $PB'_z = AB'_h$ when P is taken very close to A and B'_z very close to B'_h.

[90] point of view as it would call for excessive obliquity of the pencil or curvature of the surface or both.   Hence we can safely draw a first and highly important conclusion that our development of the square root into a series is always legitimate and that therefore no exceptions can arise with regard to the validity of the resulting equations.   But even when we demand that ' a ' shall be within the much narrower limits of $+1$ and -1 in order to assure swift convergence of the series, our equation $a = 1 - (PB'_z/AB'_h)^2$ shows that this condition will be satisfied when PB'_z is greater or less than AB'_h by not more than 5 per cent.; a few more diagrams will demonstrate that even this much narrower limit covers most of the cases which are likely to arise in practical designs.   We can therefore confidently assert that our series will not only be convergent in all conceivable cases but that the convergence will usually be decidedly rapid.

The order of importance of the terms in the series will depend upon the values which we desire to admit for the several variables contained in equation (a), and we now stipulate that the angles I and I' shall not be restricted in any way but shall be allowed to assume any magnitude up to the obvious limit of $\pm 90°$.   With a finite value of AB'_h this stipulation causes X, Y, Z, PA, and z' to be the only variables which can make the terms small. Y and Z can become very nearly equal to PA, the semi-aperture of the pencil, and although we have already noticed that only very small values of z' can be of practical interest, we will for the present treat z' as being also of the same order of magnitude as PA. X, however, represents the depth of curvature of our refracting surface, and as we laid the X-axis into the normal of the surface at A, X will to a first approximation (strictly for a spherical surface!) be proportional to PA^2 for any continuous surface and will be small of the second order when Y, Z and PA are small of the first order.   We shall limit the present first discussion to first- and second-order terms; hence we must write down all terms of the $\sqrt{(1+a)}$ series arising from $\frac{1}{2}a$ and must add the term in Y^2 which arises from $-a^2/8$.   The expression for PB'_z then becomes, with a slightly altered sequence of the terms:

$$PB'_z = AB'_h\left\{ 1 - Y\frac{\sin I'}{AB'_h} - \frac{Z \cdot z'}{(AB'_h)^2} - X\frac{\cos I'}{AB'_h} + \frac{1}{2}\left(\frac{PA}{AB'_h}\right)^2 \right.$$
$$\left. - \frac{1}{2}Y^2\frac{\sin^2 I'}{(AB'_h)^2} + \frac{1}{2}\left(\frac{z'}{AB'_h}\right)^2 + \text{terms of third and higher orders}\right\}.$$

The expression for AB'_z is very easily treated in the same way and gives:

$$AB'_z = AB'_h\left\{ 1 + \frac{1}{2}\left(\frac{z'}{AB'_h}\right)^2 + \text{terms of fourth and higher orders}\right\}.$$

We can now work out our fundamental formula for OPD' by noting that the image-point which was called F', is now called B'_z whilst the object-point, formerly called F, is now called B_h.   For the first part of the OPD'-formula, we therefore have to determine $N'(AB'_z - PB'_z)$ by putting in the above values of the two paths, and having done so we can write down the second part: $N(AB_h - PB_h)$ by analogy by omitting the dashes and also those terms containing z'.

Carrying out the multiplication by AB'_h in the separate expressions for AB'_z and [90] PB'_z, then forming the difference and multiplying this by N', we obtain

$$N'(AB'_z - PB'_z) = Y \cdot N' \sin I' + \frac{N'z'Z}{AB'_h} + X \cdot N' \cos I' - \tfrac{1}{2} \frac{PA^2}{AB'_h} N' + \tfrac{1}{2} Y^2 N' \frac{\sin^2 I'}{AB'_h}$$

and by analogy

$$N(AB_h - PB_h) = Y \cdot N \cdot \sin I \qquad + X \cdot N \cdot \cos I - \tfrac{1}{2} \frac{PA^2}{AB_h} N + \tfrac{1}{2} Y^2 \cdot N \cdot \frac{\sin^2 I}{AB_h}.$$

Subtracting the second equation from the first gives the difference of optical paths with which light from A meets that from P at the point B'_z, in the form†

$$\text{OP(1)} \quad OPD' = \left\{ \begin{array}{l} Y(N' \sin I' - N \sin I) + N' \cdot \dfrac{Z \cdot z'}{AB'_h} + X(N' \cos I' - N \cos I) \\[2mm] - \tfrac{1}{2}PA^2\left(\dfrac{N'}{AB'_h} - \dfrac{N}{AB_h}\right) + \tfrac{1}{2}Y^2\left(\dfrac{N' \sin^2 I'}{AB'_h} - \dfrac{N \sin^2 I}{AB_h}\right) \\[2mm] + \text{third and higher order terms.} \end{array} \right\}$$

and this equation we must now discuss. In this discussion we must bear in mind firstly that OPD' measures the differences of optical paths with which light passing through any outer point P of the aperture meets that through the central point A at any assumed point B'_z, and secondly that a good image can result only if we succeed in making these differences very small, if possible smaller than the Rayleigh limit of $\tfrac{1}{4}$ wave-length. Hence the point of chief interest is to discover under what conditions the successive terms may be reduced to zero, and as we have shown that our series may practically always be relied upon to converge quickly, we must discuss the first terms first because they will be capable of reaching the highest numerical value.

The first term $Y(N' \sin I' - N \sin I)$ has Y as a factor, and as Y can reach all values from zero to plus or minus the full semi-aperture of the refracting surface, this term can become unconditionally zero only if the bracketed factor is reduced to zero. That gives the condition $N' \sin I' - N \sin I = 0$ or, transposed, $N' \sin I' = N \sin I$, and we recognize the law of refraction as the necessary and sufficient condition for the disappearance of the first term *at any finite value of* Y; obviously we need not worry about infinite values of Y as the aperture of optical instruments is inevitably restricted to at most a few yards! The form of the second term, $Z \cdot N' \cdot z'/AB'_h$, in the same way leads to the conclusion that this term can be made to disappear unconditionally at any value of Z, which again means any finite aperture, only if the factor of Z is made zero. Since N' is necessarily finite and since an infinite value of the image-distance AB'_h is only a very remote and improbable special case, the removal of the second term of our series for OPD' calls for $z' = 0$, or for an image-point in the XY-plane, which is the *plane of incidence* of the arriving principal ray of the pencil. We thus reach the first highly important conclusion that the first two terms of our series for OPD' will be

† The letters OP will be prefixed to the numbers of all equations referring to optical paths.

[90] unconditionally zero if, and only if, we seek the image-point on the principal ray traced in strict accordance with the complete law of refraction.

Whilst this conclusion is rigidly correct for any possible aperture, we can make further highly important deductions by assuming that we restrict the aperture to so small a value that the higher terms of our series may be regarded as very small. In that case, the whole of the light from the small aperture surrounding the principal ray will arrive at any point of the latter without any sensible differences of optical path on the sole condition that the principal ray has been traced in strict accordance with the law of refraction. Hence the light from the small aperture will produce its highest possible intensity on the principal ray, with reduced intensity in the vicinity, because the first two terms of our series will have a sensible value at points not on the principal ray owing to non-fulfilment of the law of refraction there. Therefore all the points of maximum intensity of light lie on the trigonometrically traced ray, and as the latter represents a straight line, we have now proved the 'rectilinear propagation' of the light transmitted by any small aperture. Nevertheless, we must note the important reservation that, while all the points of *maximum intensity* lie on a strictly straight line—the ray of geometrical optics—this line of maximum intensity is surrounded by an extended region within which there is only a gradual reduction of intensity. If the aperture is very small, this region becomes quite extensive, and in fact it covers the entire space beyond the aperture if the latter measures only a small fraction of a wave-length across. This is the proof of the fictitious nature of the isolated rays of light with which we usually deal in geometrical optics.

This property of thin isolated pencils, which obviously holds for any number of successive refractions, supplies the justification of the Hartmann method of determining the course of rays through a lens system. This method consists in covering the clear aperture with an opaque screen pierced with a number of suitably distributed small holes and then photographing the resulting patches of light in two planes at known locations on either side of the focal point. We can also deduce at once that the method will give reliable results only if the holes are sufficiently small. Moreover, the photographs must be taken at such a distance from the focus that there is no suspicion of overlapping of any of the light-patches; for if this were to happen, the patches would almost certainly be drawn away from their proper positions by interference effects and would also assume unsymmetrical forms.

Another property of the geometrical ray, and for our purposes the most important of all, becomes manifest when we consider the waves of light on their passage through a refracting surface (Fig. 120). The light from point B travels in vibrations. As long as there is no change of medium, there will at any given instant be a succession of concentric spherical surfaces—wave-fronts—at perfectly equal intervals of one wave-length, on each of which the light is in the same phase of vibration, say at the crest stage. On passing the refracting surface and entering the medium beyond (supposed in Fig. 120 to be denser), the wave-length becomes shortened in the proportion of N/N'. The obvious result is that the axial part of each wave is retarded as compared with its outer part, which is still in the lighter medium, and a change of curvature results. For our present purpose, we have only to note that there will be a continuation of the wave surfaces in the new [90] medium, that the light over the whole of these refracted waves will again be in the same state of vibration, and therefore that all optical paths from *B* to any of those *refracted* waves will be absolutely equal. We may now regard the thin pencils of rays drawn from *B*, Fig. 120, through the upper part of the refracting surface as an equivalent in all essential respects of the pencils of small aperture considered in connection with Fig. 119. For the latter case, we found that the path assigned to the geometrical ray PB'_h by the law of refraction was the one that made the optical paths leading to B'_h from all closely neighbouring points around *P* equal.

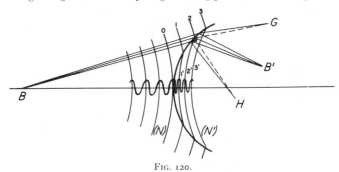

FIG. 120.

We can now ask in what direction the refracted central ray of the thin pencil in Fig. 120 must lie with reference to the wave-fronts in order to give equal optical paths for the whole light to points at some finite distance to the right of the refracting surface.

We have just shown that all optical paths from *B* to the refracted wave 2′ are equal. Therefore we have only to determine the direction in which a point must lie in order to be equidistant from all parts of the small segment of, say, wave 2′ which is enclosed by our thin pencil.

If we look at a point like *H*, we see at once that the lower part of the wave segment is nearer to *H* than the upper part, because the wave segment is tilted with reference to a line joining *H* to its centre. For point *G* we have the opposite state of affairs, but again no equality of the various paths. The conclusion is obvious and irresistible that the true point *B′* through which the refracted ray from the centre of the small pencil must go must lie on the *normal* erected at the centre of the small wave segment, for there is no one-sided tilt of the latter with reference to its normal. We thus reach the conclusion that, in an isotropic medium:

'The rays of geometrical optics are always and unconditionally normals of the wave surface.'

The importance of this theorem arises from the fact that it establishes a simple connection between rays and light waves which will enable us to draw surprisingly far-reaching conclusions from the combined evidence of a few trigonometrically traced rays and one or two determinations of differences of optical paths.

[90]   We have now established the significance of the first two terms in equation OP(1). They are both unconditionally zero for a pencil of rays of any aperture surrounding a principal ray traced in accordance with the strict law of refraction. For such a pencil, our equation therefore reduces itself to

$$OPD' = X(N' \cos I' - N \cos I) - \tfrac{1}{2}PA^2\left(\frac{N'}{AB'_h} - \frac{N}{AB_h}\right) + \tfrac{1}{2}Y^2\left(\frac{N' \sin^2 I'}{AB'_h} - \frac{N \sin^2 I}{AB_h}\right)$$

+ third and higher order terms.

This equation now determines the difference of optical paths with which the light from any point P of the clear aperture meets the light from A at any point on the trigonometrically-traced principal ray through A. It represents the most general proof of one of the most useful theorems of applied physical optics, namely, Fermat's theorem of the 'minimum optical path' as it is still often called, although we shall see that the 'minimum' characteristic is frequently absent. One correct way in which we can state the theorem is the negative form, which very usually is unsatisfactory but which in the case of Fermat's theorem will often prove to be its most valuable form. We arrived at the last equation by proving that, for points on a trigonometrically-traced principal ray, the two first-order terms of the original complete equation, in Y and in Z respectively, disappear absolutely and unconditionally. Thus we have not yet found it necessary to restrict the two lengths AB_h and AB'_h in any way, our last equation holding for light passing from any point on the arriving principal ray to any point on the refracted principal ray of an oblique pencil passing through a continuous surface separating a medium of index N from another of index N'. The absence of first-order terms in the resulting differences of optical paths therefore justifies the statement:

'*Fermat's Theorem:* The differences of optical paths with which light through the outer parts of a given aperture meets that travelling along the trigonometrically-traced ray contain no linear terms, or, more explicitly, contain no terms which are directly proportional to the aperture.'

The value of this negative statement arises from the fact that it implies that OPD' can only contain terms in the second or still higher powers of the aperture and that, as we first noted in the case of longitudinal spherical aberration, such higher-order terms have extremely small values for a fairly large central part of the aperture. The statement also has the advantage that it is unconditionally true in every conceivable case (in an isotropic medium) and therefore, unlike the positive forms of the theorem, requires no restrictive qualifications.

We obtain the most useful form of the theorem by noting that in the last equation the X in the first term is, to a first and usually sufficient approximation, proportional to PA^2 and therefore is small and of the second order, as the other two terms obviously are. Hence if, for a given point P the three stated terms give a sum differing from zero, we can say that the OPD' will grow as the square of the aperture. But there is the possibility of a zero sum of the three stated terms for particular points P or—as we shall see—even for all possible points P; in that case, we should

have to evaluate the third and higher order terms of the series to find the law of   [90]
OPD'. Hence the positive statement must take the form:

> '*Fermat's Theorem:* All optical paths between any two points on a trigono-
> metrically-traced ray differ from the optical path along the latter by quantities
> of the second or higher order, or by differences which are proportional to the
> second or higher powers of the aperture.'

The theorem means that if we computed all neighbouring paths of a given
trigonometrically-determined one, and plotted the *OPD* with which the latter is
met by the former against the distances *AP* at which the neighbouring paths
penetrate the refracting surface, we should obtain graphs which in the vicinity of *A*
would be ordinary vertical parabolas with their poles at *A*. Fig. 121 (a) and (b)
show the two possible cases of this kind, in which the trigonometrically-determined
path through *A* represents respectively a minimum and a maximum value with
reference to its neighbouring paths. If in a particular case the second-order terms

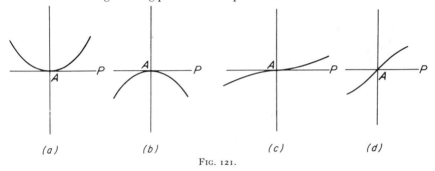

(a)   *(b)*   *(c)*   *(d)*

FIG. 121.

in our last equation added up to zero, a still greater approach to equality of the
paths near AB'_h would result; we should have to evaluate the hitherto omitted
higher terms of the series. If we then found the third-order term to be of sensible
magnitude, a graph like Fig. 121 (c) or its reflection in a mirror would result,
namely, a cubic parabola, and the optical path of the central ray would be neither
a maximum nor a minimum; it would be described as having a stationary value.
On the other hand, a similar graph for the optical path differences for a point not
situated on the central path AB'_h found by the law of refraction would give a graph
like Fig. 121 (d), that is, an *inclined* straight line or curve. This would indicate a
very rapid variation of the optical paths from the neighbourhood of *A* to an assumed
image-point not fulfilling the law of refraction. This last graph will show how
important it is to limit applications of Fermat's theorem to the neighbourhood of a
central path determined in strict accordance with the law of refraction. Fig. 121
(a), (b), and (c) demonstrate that our objection to calling Fermat's theorem the law
of 'minimum optical path' is fully justified, for the trigonometrically-determined
path may correspond to a minimum or a maximum or a stationary value according
to the position of the two points B_h and B'_h, the curvature of the refracting surface,

[90] the angles of incidence and refraction, and the orientation of the normal plane in which the comparison is made.

We have proved the theorem directly for one surface, but it will be evident that, as it holds for any continuous surface, it must also hold for the combined effect of any number of continuous surfaces.

By proving that the light passing in the immediate vicinity of a 'ray' traced in accordance with the law of refraction harmonizes as closely as possible in phase and thus produces a maximum intensity along the ray, Fermat's theorem supplies strong evidence in favour of another theorem due to Helmholtz, namely:

'In the interior of a wave train the energy of light is propagated in the direction of the rays of geometrical optics.'

By the 'interior of a wave train' is meant all that light which is neither very close to the extreme outside mantle of a cone of rays nor close to a focus. For more rigorous details, reference must again be made to books on theoretical optics, and Drude's text-book may be specially mentioned in this connection. An American translation of it is available. This book also deals with another difficult side-issue of physical optics, namely, the so-called boundary conditions, which are here merely mentioned because they are occasionally trotted out as objections to physical methods of dealing with lens problems. The apertures of even the smallest lens systems are such large multiples of the wave-length that the percentage of the total transmitted light which could possibly come under the ban of these boundary conditions is negligibly small. It is different in the case of fine diffraction gratings or the minute structure of delicate microscopic objects.

Helmholtz' theorem means for us that we must trace the paths of light through any system in strict accordance with the laws of reflection and refraction, no matter whether the result at the final focus is to be discussed by the geometrical or by the physical method. It will therefore now be clear to the student that he will not have to unlearn anything taught in Part I; we shall merely apply a different and more accurate interpretation to the *results* of our ray-tracing.

We now reach the final stage of the discussion of our *OPD* equation. We ask under what condition the second-order terms, which we have so far regarded as expressive of Fermat's theorem, can become zero and so indicate a still nearer approach to the perfect union of the light reaching B'_h through the clear aperture of our refracting surface.

We shall first discuss our equation for the case (almost the only one of practical interest) when the refracting surface is truly spherical and therefore has the same radius of curvature, which we call simply r, in every normal plane. Putting $X = PA^2/2r$ into the first term of the last equation and collecting the terms depending on PA^2, we find

$$OPD' = \tfrac{1}{2}PA^2 \left(\frac{N}{AB_h} + \frac{N' \cos I' - N \cos I}{r} - \frac{N'}{AB'_h} \right) + \tfrac{1}{2}Y^2 \left(\frac{N' \sin^2 I'}{AB'_h} - \frac{N \sin^2 I}{AB_h} \right).$$

Evidently the simplest case is that corresponding to normal or radial incidence of our pencil, that is, when H and H' in Fig. 119 are zero and B_h and B'_h become B and

B', for in that case we have $I = I' = 0$ and our pencil becomes a paraxial one. [90] The second term of the equation then disappears because $\sin I$ and $\sin I'$ are both zero and in the first term the cosines become unity and disappear as factors. Consequently we have for a paraxial pencil

$$OPD' = \tfrac{1}{2}PA^2\left(\frac{N}{AB}+\frac{N'-N}{r}-\frac{N'}{AB'}\right)+\text{higher terms}$$

and the first term of this will become zero for *any* value of PA if the bracketed term is zero. Using our old symbols l and l' for the intersection-lengths AB and AB', the condition under which all the light reaches B' without any differences of optical paths arising from the second-order term of OPD' is therefore

$$\frac{N}{l}+\frac{N'-N}{r}-\frac{N'}{l'} = 0.$$

This is merely a transposed form of our old paraxial equation (6p), which thus shows that, in looking for the conditions under which the second-order terms of OPD' are zero, we are really searching for the true focus of our thin pencil.

We now drop the assumption that I and I' are zero and we have therefore to discuss the general equation for the second-order terms of OPD' at a spherical surface, and not the last simplified form. The general equation has two terms depending upon PA^2 and upon Y^2 respectively, and as for any zone of the aperture its fixed value of PA becomes associated with all possible values of Y between zero in the XZ-plane and very nearly PA in the XY-plane, it is clear that OPD' can become zero for every point in the aperture only on condition that the factors of PA^2 and of Y^2 in our equation are separately zero. That gives the two conditions

$$\frac{N}{AB_h}+\frac{N'\cos I'-N\cos I}{r}-\frac{N'}{AB'_h} = 0$$

and

$$\frac{N'\sin^2 I'}{AB'_h}-\frac{N\sin^2 I}{AB_h} = 0.$$

Taking the latter condition first and extending the first term by N'/N', the second by N/N, we can take out $N^2\sin^2 I$ as a common factor because $N'\sin I' = N\sin I$ by the law of refraction and the condition becomes

$$\frac{1}{N'\cdot AB'_h}-\frac{1}{N\cdot AB_h} = 0 \quad \text{or} \quad AB'_h = \frac{N}{N'}\cdot AB_h.$$

This then is one indispensable condition for the formation of a perfect focus by a thin oblique pencil at any *one* spherical refracting surface. We can find out what radius r would be required to produce this result for a given intersection-length AB_h by putting $AB_h\cdot N/N'$ in place of AB'_h in our first condition, which then becomes

$$\frac{N}{AB_h}+\frac{N'\cos I'-N\cos I}{r}-\frac{N'^2}{N\cdot AB_h} = 0.$$

[90] Solving for r we then obtain

$$r = AB_h \frac{N(N' \cos I' - N \cos I)}{N'^2 - N^2}.$$

On comparing these results with those obtained in Part I for the case of aplanatic refraction at a spherical surface, namely, $L' = L \cdot N/N'$ and $r = L \cdot N/(N' + N)$, we see a very close resemblance; our first condition for an oblique pencil is really identical to the corresponding earlier result and the second would become so if I and I' were so small that their cosines were sensibly equal to unity. We have therefore learnt that the case of aplanatic refraction on the optical axis is also a very favourable one for the corresponding oblique pencils.

Having shown that an oblique pencil refracted at a spherical surface can have a perfect focus only when the above conditions are fulfilled, we realize that, in the general case of some other ratio between r and AB_h, there can be no definite

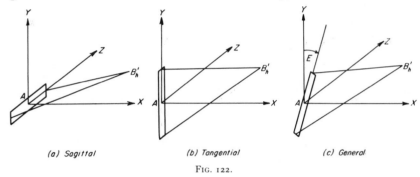

(a) Sagittal   (b) Tangential   (c) General

Fig. 122.

focus for a complete oblique pencil of circular aperture and that the latter will require further restriction in order to secure a close approach to the desired zero value of OPD' over the whole of the restricted aperture. Two particular restricted apertures of this type are easily singled out and are almost the only ones of practical interest. As the second term of the OPD' equation depends upon Y^2 it obviously vanishes for $Y = 0$, that is, for a slit aperture lying in the XZ-plane which has already been referred to in Part I as a sagittal fan of rays. But Y^2 will be very small as long as Y itself is reasonably small, and as its factor contains $\sin^2 I$ and $\sin^2 I'$, which are also small for all but exceptionally large angles of obliquity, we see that it is not necessary to restrict our sagittal fan to *infinite* thinness in the Y-direction; it will retain a close approach to a definite focus as long as we restrict ourselves to a rectangular aperture the length of which in the Z-direction is a considerable multiple of its width in the Y-direction (Fig. 122a). Under these conditions, the Y^2-term in the OPD' equation may be regarded as negligible and the equation becomes for a sagittal fan:

$$OPD' = \tfrac{1}{2}PA^2\left(\frac{N}{AB_h} + \frac{N' \cos I' - N \cos I}{r} - \frac{N'}{AB'_h}\right).$$

By the reasoning already employed, we find the condition under which all light [90] reaches B'_h in the same phase by equating the bracketed factor to zero. If we introduce the symbol s for AB_h, the intersection-length of the incident fan, and s' for AB'_h, the condition yields the equation

$$\frac{N'}{s'} = \frac{N' \cos I' - N \cos I}{r} + \frac{N}{s},$$

which is a very simple and widely used one for the tracing of sagittal fans of rays and was proved in Part I (page 410) by the geometrical method.

The second important special case is that of a tangential fan of rays extending in the Y-direction (Fig. 122b). For such a fan, Y^2 is very little different from $PA^2 = X^2 + Y^2 + Z^2$ because for the relatively small values of PA to which we are still restricted (as otherwise the omitted third- and higher-order terms in the OPD' equation would become important) X is small compared to PA, and we can make Z small by keeping down the width of the rectangular aperture. For a tangential fan, the equation can therefore be simplified by putting $Y^2 = PA^2$ and yields

$$OPD' = \tfrac{1}{2}PA^2\left(\frac{N}{AB_h} - \frac{N \sin^2 I}{AB_h} + \frac{N' \cos I' - N \cos I}{r} - \frac{N'}{AB'_h} + \frac{N' \sin^2 I'}{AB'_h}\right)$$

or the condition to be fulfilled by the tangential intersection-lengths $AB_h = t$ and $AB'_h = t'$:

$$\frac{N}{t}(1 - \sin^2 I) + \frac{N' \cos I' - N \cos I}{r} - \frac{N'}{t'}(1 - \sin^2 I') = 0.$$

By setting $(1 - \sin^2) = \cos^2$ and transposing, we obtain the tangential companion to the above equation for sagittal fans:

$$\frac{N' \cos^2 I'}{t'} = \frac{N' \cos I' - N \cos I}{r} + \frac{N \cos^2 I}{t}.$$

It will be noted that in both equations the same 'astigmatic constant' $(N' \cos I' - N \cos I)/r$ appears; this renders simultaneous calculation of the s and t pencils very convenient and rapid, as has already been demonstrated in Part I.

It is easy to deduce a more general equation for a fan of rays at any angle E with the XY-plane as shown in Fig. 122c. For such a fan we have very nearly $Y = PA \cdot \cos E$, and by putting this value into the general equation and reducing, the equation for the intersection-lengths l_E and l'_E of the general fan is found to be

$$\frac{N'(1 - \sin^2 I' \cdot \cos^2 E)}{l'_E} = \frac{N' \cos I' - N \cos I}{r} + \frac{N(1 - \sin^2 I \cdot \cos^2 E)}{l_E}.$$

This gives the s-formula by putting $E = 90°$ and the t-formula by putting $E = 0$, and also the old paraxial formula (6p)** by putting $I = I' = 0$. But the general equation is valid for a fan at any angle E and may occasionally be found useful for tracing a pencil through a decentred system of refracting surfaces.

We obtain a further interesting result by noting that the general equation for a slit aperture at any position-angle E contains this angle only in the form $\cos^2 E$.

[90] This square necessarily has the same value for $-E$ as for $+E$, or for the slit shown in Fig. 122c for a positive E and for a corresponding one forming the same angle with the Y-axis but in the opposite clock-sense. Hence OPD' will vanish, as far as second-order terms are concerned, not merely for the single slit shown in Fig. 122c but for a cross-shaped slit $\times$ at the distance l'_E calculated by the last equation. In the light of this conclusion, we must regard the horizontal slit which yields the sagittal focus as the superposition of the two slits of our cross when E has attained $\pm 90°$ and the vertical slit as a corresponding superposition of the two slits for $E = 0$.

We need at present refer only briefly to the still more general case of a non-spherical refracting surface. As has already been pointed out, such a surface will in general have a different radius of curvature in the normal planes laid through A at different angles E. It is easily seen that all the formulae given above will remain valid if we replace the fixed r of the spherical surface by the proper value of the variable r_E of the non-spherical surface. But as in practical optics we only employ,

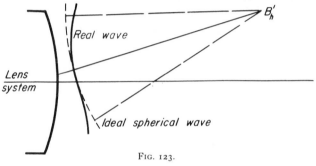

FIG. 123.

at most, surfaces with very slight departures from a true spherical form produced by local polishing or 'figuring' and as these can be allowed for in a simpler way, it is not worth while to follow up the subject further along the lines taken in the present chapter.

We must, however, add the most generally useful interpretation of the OPD' determined by our present method and by its modification which will follow later. We deduced at the beginning the conclusion that the ideal state of correction of any optical system calls for the exact equality of all the paths along which the light from a given object-point travels towards the conjugate image-point; that is evidently equivalent to a demand that the differences of optical paths, our OPD', should be zero for the entire clear aperture. In that ideal case, we should have truly spherical waves emerging from the system and converging accurately upon the image-point, which would be their centre. The ideal wave form is therefore that of a segment of a sphere with the image-point as centre, shown by Fig. 123. Our equations for OPD' are put into such a form as to give zero value for the *immediate* neighbourhood of the adopted principal ray; hence the *real* wave will cross the ideal one, or touch it, on the principal ray. But if OPD' yields finite values for the

outer regions of the clear aperture, a positive value means, according to our selection [90]
of signs, that the path along the principal ray is longer than that along the particular
outer ray; hence the light along the outer ray will at any given moment be closer to
the end of its journey at B'_h than the central light.   Therefore a positive value of
OPD' signifies that the outer part of the real wave to which it refers is ahead of
the ideal spherical wave, whilst a negative OPD' means that in that particular
region the real wave is lagging behind the ideal wave.   The heavy line drawn in
Fig. 123 as a possible section of a real wave thus indicates positive OPD' above the
principal ray and negative OPD' below.   The geometrical separation of the two
waves at any point, however, does not represent the calculated OPD' itself but
OPD'/N', where N' is the index of the medium in which the waves are proceeding.
Since our equivalent optical paths are defined as the geometrically measured length
of the path *multiplied by* the refractive index, it is evident that we must reverse this
operation when interpreting the geometrical value of a given OPD'.

Our equations give OPD' in general algebraical terms for any point of the
clear aperture; they may therefore be regarded as defining the separation between
the real and the ideal spherical waves in a perfectly continuous way.   In other
words, these equations are the equation of the true wave form, although referred
not to the usual rectilinear coordinates but to the ideal spherical wave, with OPD'/N'
to be measured along the radii drawn toward the adopted image-point.

Fig. 123 gives a first idea of another highly important aspect: our equations in
their original form are valid for any point B'_h, and the image received at any such
point will be the sharper and brighter the smaller the variation of OPD'.   Looking
at the tilted real wave in Fig. 123 (which corresponds roughly to a case of *coma*),
it is at once obvious that a point some distance below the B'_h as drawn would be
more nearly equidistant from all parts of the true wave than the adopted point and
would therefore yield a sharper and brighter image.   Hence an important part of
our work on the physical basis will consist in searching for the point that is most
nearly equidistant from all parts of the true emerging waves.

OPTICAL PATH DIFFERENCES AT SMALL APERTURE AND FIELD

THE discussions in the previous section were based on the consideration of an [91]
oblique pencil without direct reference to a corresponding axial pencil.   For
general investigations of the properties of lens systems, it is desirable to study the
system as a whole, and we shall therefore now develop the first expression which
we found in the preceding section (page 590) for the optical paths to be compared.
But as we have proved that the image of an object-point can only be found in or
close to the incidence plane of the principal ray of the oblique pencil, we shall
omit z' and work with the simplified equations:

$$PB_h{}^2 = (l-X)^2 + (H-Y)^2 + Z^2$$
$$AB_h{}^2 = l^2 + H^2$$
$$PB'_h{}^2 = (l'-X)^2 + (H'-Y)^2 + Z^2$$
$$AB'_h{}^2 = l'^2 + H'^2.$$

[91] The equations for the plain and dashed quantities are quite analogous. We therefore work out only the first two and then write out the result for the last two by analogy. Developing the squares of $(l-X)$ and of $(H-Y)$, putting $X^2 + Y^2 + Z^2 = PA^2$, and then extracting l^2 as a common factor, we find the exact expressions:

$$PB_h = l\sqrt{1 + \left[-2\frac{X}{l} + \left(\frac{H}{l}\right)^2 - 2\frac{YH}{l^2} + \left(\frac{PA}{l}\right)^2 \right]}$$

$$AB_h = l\sqrt{1 + \left(\frac{H}{l}\right)^2}.$$

By introducing these values and the corresponding 'dashed' ones into the general equation for OPD', the latter becomes

$$OPD' = N'(AB'_h - PB'_h) - N(AB_h - PB_h)$$

We could *calculate* the resulting differences of phase with any desired degree of accuracy for given values of the data; but—if l and l' were large—we might have to use seven- or even ten-figure logs to secure the desirable accuracy to small fractions of a wave-length, and in any case we should only obtain isolated results for the particular rays submitted to calculation. We shall subsequently develop more convenient formulae for the exact determinations of individual phase differences. For the present, we must deduce general analytical results which—whilst we shall not often use them for numerical calculations—will put us into a position to deduce general conclusions as to the degree of perfection of a given system from the tracing of a small number of rays, coupled when necessary with a few direct determinations of differences of optical paths. We shall obtain this general analytical result by developing the roots in our expressions for the optical paths by the $\sqrt{(1+a)}$ series used in the previous section. We shall first transform X. For the strictly spherical surfaces which are almost exclusively employed in precision optical work, we shall have rigorously $X = PA^2/2r$. But as non-spherical lens surfaces are of considerable theoretical interest and as they may in the near future become worthy of consideration in actual commercial practice, we will include such surfaces in the discussion.

The equations which we are going to develop are intended for application to *centred* lens systems which have the axis of Fig. 119 as their optical axis. As symmetry with reference to the optical axis is the chief feature of centred lens systems, it is at once clear that their separate surfaces should also have this symmetry. Therefore we can, without loss of general validity, restrict our investigation to refracting surfaces which are *surfaces of rotation* with reference to the optical axis. For such surfaces X can only depend on powers of $(Y^2 + Z^2)$ and therefore also of PA^2. We can reach the same conclusion optically by noting that the root to be developed in our present work contains only terms in PA^2 and that the use of non-spherical surfaces can only be justified if it leads to cancellation or a favourable modification of some of the terms resulting from the development. We therefore generalize the expression for the X of a strictly spherical surface by adding to it terms in PA^4, PA^6, &c., and adopt the form

$$X = PA^2/2r + k_4 PA^4 + k_6 PA^6 + \&c.,$$

but as we shall find that even the sixth-order terms become very complicated, [91] we shall break off with that order. The k_4 and k_6 may be called 'figuring constants', for in the present state of the art they will have to be realized by zonal polishing of the surfaces.†

Putting the generalized value of X into the root expression for PB_h, the latter becomes

$$PB_h = l\sqrt{1 + \left[-\frac{PA^2}{l}\left(\frac{1}{r}-\frac{1}{l}\right) + \left(\frac{H}{l}\right)^2 - 2\frac{YH}{l^2} - 2k_4\frac{PA^4}{l} - 2k_6\frac{PA^6}{l} \right]},$$

and this root can now be developed by the $\sqrt{(1+a)}$ series by calling the long square-bracket expression 'a'. By breaking off the development of a^2 and a^3 with terms of the sixth order (that is, those in which the power-indices of H, Y and PA add up to six) the result will be

$$PB_h = l - \tfrac{1}{2}PA^2\left(\frac{1}{r}-\frac{1}{l}\right) + \tfrac{1}{2}\frac{H^2}{l} - \frac{YH}{l} - k_4 PA^4 - k_6 PA^6$$

(represents $-\tfrac{1}{8}a^2$)

$$\left[\begin{array}{l} -\tfrac{1}{8}\dfrac{PA^4}{l}\left(\dfrac{1}{r}-\dfrac{1}{l}\right)^2 - \tfrac{1}{8}\dfrac{H^4}{l^3} - \tfrac{1}{2}\dfrac{Y^2H^2}{l^3} + \tfrac{1}{4}\dfrac{PA^2H^2}{l^2}\left(\dfrac{1}{r}-\dfrac{1}{l}\right) \\[2mm] -\tfrac{1}{2}\dfrac{PA^2YH}{l^2}\left(\dfrac{1}{r}-\dfrac{1}{l}\right) - \tfrac{1}{2}k_4\dfrac{PA^6}{l}\left(\dfrac{1}{r}-\dfrac{1}{l}\right) + \tfrac{1}{2}\dfrac{YH^3}{l^3} + \tfrac{1}{2}k_4\dfrac{PA^4H^2}{l^2} \\[2mm] -k_4\dfrac{YHPA^4}{l^2} \end{array} \right.$$

(represents $+\tfrac{1}{16}a^3$)

$$\left[\begin{array}{l} -\tfrac{1}{16}\dfrac{PA^6}{l^2}\left(\dfrac{1}{r}-\dfrac{1}{l}\right)^3 + \tfrac{1}{16}\dfrac{H^6}{l^5} - \tfrac{1}{2}\dfrac{Y^3H^3}{l^5} + \tfrac{3}{16}\dfrac{PA^4H^2}{l^3}\left(\dfrac{1}{r}-\dfrac{1}{l}\right)^2 \\[2mm] -\tfrac{3}{8}\dfrac{PA^4YH}{l^3}\left(\dfrac{1}{r}-\dfrac{1}{l}\right)^2 - \tfrac{3}{8}\dfrac{H^5Y}{l^5} + \tfrac{3}{4}\dfrac{H^4Y^2}{l^5} - \tfrac{3}{16}\dfrac{PA^2H^4}{l^4}\left(\dfrac{1}{r}-\dfrac{1}{l}\right) \\[2mm] -\tfrac{3}{4}\dfrac{PA^2Y^2H^2}{l^4}\left(\dfrac{1}{r}-\dfrac{1}{l}\right) + \tfrac{3}{4}\dfrac{PA^2H^3Y}{l^4}\left(\dfrac{1}{r}-\dfrac{1}{l}\right). \end{array} \right.$$

The development of the root for AB_h gives

$$AB_h = l + \tfrac{1}{2}\frac{H^2}{l} - \tfrac{1}{8}\frac{H^4}{l^3} + \tfrac{1}{16}\frac{H^6}{l^5},$$

and if we now form the difference $(AB_h - PB_h)$ which the expression for OPD' calls for, we shall find

$$AB_h - PB_h = \tfrac{1}{2}PA^2\left(\frac{1}{r}-\frac{1}{l}\right) + \frac{YH}{l}$$

† As an example, we see that for a paraboloidal surface

$$X = \frac{Y^2+Z^2}{2r} = \frac{PA^2-X^2}{2r} = \frac{PA^2}{2r} - \frac{PA^4}{8r^3} + \frac{PA^6}{16r^5} - \cdots.$$

Hence for this particular non-spherical surface, the figuring constants become $k_4 = -1/8r^3$ and $k_6 = 1/16r^5$.

4th-order terms
$$\left\{ \begin{array}{l} + \tfrac{1}{8}PA^4\left[8k_4 + \dfrac{1}{l}\left(\dfrac{1}{r} - \dfrac{1}{l}\right)^2\right] \\[2mm] + \tfrac{1}{2}PA^2YH\,\dfrac{1}{l^2}\left(\dfrac{1}{r} - \dfrac{1}{l}\right) \\[2mm] - \tfrac{1}{4}PA^2H^2\,\dfrac{1}{l^2}\left(\dfrac{1}{r} - \dfrac{1}{l}\right) \\[2mm] + \tfrac{1}{2}Y^2H^2\,\dfrac{1}{l^3} - \tfrac{1}{2}YH^3\,\dfrac{1}{l^3} \end{array} \right.$$

6th-order terms
$$\left\{ \begin{array}{l} + \tfrac{1}{16}PA^6\left[16k_6 + 8k_4\,\dfrac{1}{l}\left(\dfrac{1}{r} - \dfrac{1}{l}\right) + \dfrac{1}{l^2}\left(\dfrac{1}{r} - \dfrac{1}{l}\right)^3\right] \\[2mm] + \tfrac{3}{8}PA^4YH\left[\tfrac{8}{3}k_4\cdot\dfrac{1}{l^2} + \dfrac{1}{l^3}\left(\dfrac{1}{r} - \dfrac{1}{l}\right)^2\right] \\[2mm] - \tfrac{3}{16}PA^4H^2\left[\tfrac{8}{3}k_4\,\dfrac{1}{l^2} + \dfrac{1}{l^3}\left(\dfrac{1}{r} - \dfrac{1}{l}\right)^2\right] \\[2mm] + \tfrac{3}{4}PA^2Y^2H^2\,\dfrac{1}{l^4}\left(\dfrac{1}{r} - \dfrac{1}{l}\right) \\[2mm] - \tfrac{3}{4}PA^2YH^3\,\dfrac{1}{l^4}\left(\dfrac{1}{r} - \dfrac{1}{l}\right) + \tfrac{3}{16}PA^2H^4\,\dfrac{1}{l^4}\left(\dfrac{1}{r} - \dfrac{1}{l}\right) \\[2mm] + \tfrac{1}{2}Y^3H^3\,\dfrac{1}{l^5} - \tfrac{3}{4}H^4Y^2\,\dfrac{1}{l^5} + \tfrac{3}{8}H^5Y\,\dfrac{1}{l^5}. \end{array} \right.$$

By analogy $(AB'_h - PB'_h)$ will give a precisely similar set of terms with H' and l' in place of H and l. The expression for OPD' can then be written down by multiplying $(AB'_h - PB'_h)$ by N' and subtracting from it $(AB_h - PB_h)N$ with the result—omitting the terms of the sixth order after the first as of no present interest—

OP(2) $OPD' = \tfrac{1}{2}PA^2\left[N'\left(\dfrac{1}{r} - \dfrac{1}{l'}\right) - N\left(\dfrac{1}{r} - \dfrac{1}{l}\right)\right] + Y\left[\dfrac{H'N'}{l'} - \dfrac{HN}{l}\right]$

Seidel aberrations
$$\left\{ \begin{array}{l} + \tfrac{1}{8}PA^4\left[8k_4(N'-N) + \dfrac{N'}{l'}\left(\dfrac{1}{r} - \dfrac{1}{l'}\right)^2 - \dfrac{N}{l}\left(\dfrac{1}{r} - \dfrac{1}{l}\right)^2\right] \\[2mm] + \tfrac{1}{2}PA^2Y\left[\dfrac{H'N'}{l'^2}\left(\dfrac{1}{r} - \dfrac{1}{l'}\right) - \dfrac{HN}{l^2}\left(\dfrac{1}{r} - \dfrac{1}{l}\right)\right] \\[2mm] - \tfrac{1}{4}PA^2\left[\dfrac{H'^2N'}{l'^2}\left(\dfrac{1}{r} - \dfrac{1}{l'}\right) - \dfrac{H^2N}{l^2}\left(\dfrac{1}{r} - \dfrac{1}{l}\right)\right] \\[2mm] + \tfrac{1}{2}Y^2\left[\dfrac{H'^2N'}{l'^3} - \dfrac{H^2N}{l^3}\right] \\[2mm] - \tfrac{1}{2}Y\left[\dfrac{N'H'^3}{l'^3} - \dfrac{NH^3}{l^3}\right] \end{array} \right.$$

$$\left.\begin{array}{l}\text{Secondary}\\\text{spherical}\\\text{aberration}\end{array}\right\{\begin{array}{l}+\tfrac{1}{16}PA^6\left\{16k_6(N'-N)+8k_4\left[\dfrac{N'}{l'}\left(\dfrac{1}{r}-\dfrac{1}{l'}\right)-\dfrac{N}{l}\left(\dfrac{1}{r}-\dfrac{1}{l}\right)\right]\right.\\\\\qquad\qquad\qquad\qquad\left.+\dfrac{N'}{l'^2}\left(\dfrac{1}{r}-\dfrac{1}{l'}\right)^3-\dfrac{N}{l^2}\left(\dfrac{1}{r}-\dfrac{1}{l}\right)^3\right\}\end{array}$$

[91]

As was indicated in the opening paragraph of this section, the equation OP(2) differs from OP(1) by referring the *OPD'*, not to the distances of the object- and image-points measured along the principal ray of an isolated oblique pencil, but by referring them to the object- and image-heights H and H' and to the axial distances l and l'; the new equation therefore establishes a direct correlation between associated axial and oblique pencils, besides embodying in a directly useful way the possible inclusion of non-spherical or figured surfaces. Equation OP(2) thus corresponds exactly to the geometrical treatment of the Seidel aberrations in Part I.

As the equation was again obtained by developing roots of the $\sqrt{(1+a)}$ type into series, we must begin by discussing the limits within which this development is valid. The two roots developed in the present work were

$$PB_h = l\sqrt{(1+a)} \quad\text{and}\quad AB_h = l\sqrt{(1+a)},$$

and in both cases 'a' must again be restricted within the limits of plus and minus unity. In the same way as in Section [90], we find that both AB_h/l and PB_h/l must lie between zero and 1.4142; moreover, we inferred similar roots for the 'dashed' data, and for these the same limiting values must also be respected if the series is to be valid. If these limits are examined—most easily by drawing some sketches—it will be seen that our present equation is not valid for object- and image-planes which coincide with the pole of the refracting surface. The reason is simply that l and l' would then be zero while AB_h and AB'_h would be equal to the respective H and H' and therefore finite, so that the ratio AB_h/l would be infinite and utterly beyond the limiting value 1.4142. It will also be found that for other, less extreme, cases, the range of apertures and fields within which OP(2) is valid is more restricted than it is for OP(1), although OP(2) is valid for rather large apertures and fields except when l and l' approach zero. It will be remembered that in the method used in Chapter VI of Part I the breakdown occurred when $l = l' = r$, which is perfectly admissible for OP(2); on the other hand, the method of Part I is admissible for $l = l' = 0$. It will thus be seen that the old and the new methods cover each other's weak points, and hence that together they cover every conceivable case except when the aperture or the field or both are excessive.

We now begin the discussion of OP(2) and as we desire to have a converging series and a zero value, we take the terms in succession and ask whether they can be brought to the vanishing point by introducing suitable special values for such data as we can control. The first term,

$$\tfrac{1}{2}PA^2\left[N'\left(\dfrac{1}{r}-\dfrac{1}{l'}\right)-N\left(\dfrac{1}{r}-\dfrac{1}{l}\right)\right],$$

[91] will evidently become zero for any conceivable value of the semi-aperture PA if the bracketed factor is zero, or when the condition is fulfilled that

$$N'\left(\frac{1}{r}-\frac{1}{l'}\right) = N\left(\frac{1}{r}-\frac{1}{l}\right).$$

It is at once seen that this condition represents the paraxial equation

(6p)** $$\frac{N'}{l'} = \frac{N'-N}{r}+\frac{N}{l}.$$

with the sequence of its terms altered.   Hence we have a first conclusion that the first term vanishes unconditionally if the distances l and l', at which the object- and image-planes are erected, are determined by the ordinary paraxial ray-tracing equations.   Therefore we henceforth adopt this location of these planes and are then permanently rid of the first term in OP(2).

The second term,

$$Y\left[\frac{H'N'}{l'} - \frac{HN}{l}\right],$$

having the necessarily finite aperture-measure Y as an outside factor, also vanishes unconditionally if $H'N'/l' = HN/l$ or $H'/H = Nl'/N'l = m'$ of Part I, Chapter I.   This value of the linear magnification m' was deduced in Part I by the theorem of Lagrange.   Consequently the two first terms of OP(2) are unconditionally zero and may be omitted if, in addition to locating the object- and image-*planes* at the successive paraxial focal points, we also determine all the object- and image-*heights* by the theorem of Lagrange, or equivalently by the paraxial formulae for the linear magnification.   We should note also that this proof of the Lagrange theorem is valid for $l = l' = r$, where the proof in Part I breaks down.   We again see the remarkable amount of optical information contained in these easily deduced optical-path equations.   We must also note that the two conditions which we have adopted in order to remove the first two terms of OP(2) are identical with those applied in Part I, Chapter VI to the treatment of the geometrical Seidel aberrations; hence the analogy between the two entirely different methods is becoming closer.

CHAPTER XIII

OPTICAL PATH DIFFERENCES AT AN AXIAL IMAGE-POINT

EQUATION OP(2) with the first two terms omitted represents the OPD' with [92] which the light from any point P of the refracting surface meets that from its pole A at any point B'_h in the paraxial focal plane. We shall for the present limit the discussion to the *axial* image-point B'. For this point and the conjugate object-point B, H and H' are of course absolutely zero; hence all terms of OP(2) containing H and H' as factors vanish unconditionally for the axial image-point, and we have for this paraxial focal point

$$OPD' = \tfrac{1}{8}PA^4\left[8k_4(N'-N)+\frac{N'}{l'}\left(\frac{1}{r}-\frac{1}{l'}\right)^2-\frac{N}{l}\left(\frac{1}{r}-\frac{1}{l}\right)^2\right]$$
$$+\tfrac{1}{16}PA^6 \text{ [Factor stated explicitly in OP(2)]}.$$

It is obvious that these two terms must represent the primary and secondary spherical aberration of the refracting surface, with inclusion of the effects of possible 'figuring' according to the values of the 'figuring constants' k_4 and k_6. With regard to the latter, it is evident that whatever value the second and third terms in the factor of PA^4 may have, we can find a value of the figuring constant k_4 which will cancel them. The primary spherical aberration can therefore be removed by using the non-spherical surface departing from the sphere by a PA^4-law. Similarly, when k_4 has thus been determined, the factor of the secondary spherical aberration shows that it will be always possible to choose a value of k_6 which will simultaneously remove that aberration. This is one of the attractive properties of non-spherical surfaces: but the practical difficulties of removing glass according to the laws and to the extent demanded by the figuring constants are almost insuperable and put a less alluring aspect upon the idea unless the amount of glass to be removed is very small.

For a *spherical* surface, the factor of PA^4 becomes reduced to the second and third terms as k_4 is zero for such a surface. Because at the paraxial focus $N'(1/r-1/l') = N(1/r-1/l)$, the factor can be written

$$N^2\left(\frac{1}{r}-\frac{1}{l}\right)^2\left(\frac{1}{N'l'}-\frac{1}{Nl}\right),$$

and this shows at once that a spherical surface will be free from spherical aberration if $N'l' = Nl$, which is the condition we reached in Part I by the ray-tracing method. It is easily seen that the same condition will also cause the PA^6 term to disappear, so that our physical method of attack confirms the previous geometrical result as to the existence of a pair of aplanatic points for a spherical surface.

It is in fact easily shown that the first term of the equation for OPD', when worked out for small aperture and therefore for its value as primary spherical aberration, gives a result bearing a close resemblance to those obtained in Part I

[92] for the geometrical primary spherical aberration in its various measures. Omitting the figuring term in k_4 and writing y instead of PA for the now small aperture, we have

$$\text{Primary } OPD' = \tfrac{1}{8}y^4\left[\frac{N'}{l'}\left(\frac{1}{r}-\frac{1}{l'}\right)^2 - \frac{N}{l}\left(\frac{1}{r}-\frac{1}{l}\right)^2\right]$$

or, on taking $y \cdot y^2$ into the square bracket,

$$\text{Primary } OPD' = \tfrac{1}{8}y\left[N'\frac{y}{l'}\left(\frac{y}{r}-\frac{y}{l'}\right)^2 - N\frac{y}{l}\left(\frac{y}{r}-\frac{y}{l}\right)^2\right].$$

Now we have by the usual paraxial formula $y/l = u$; $y/l' = u'$; $y/r = u+i = u'+i'$, and on introducing these we find

$$\text{Primary } OPD' = \tfrac{1}{8}y[N'u'i'^2 - Nui^2].$$

As by the law of refraction $Ni = N'i'$, we can take out a factor and get

$$\text{Primary } OPD' = \tfrac{1}{8}yN'i'[u'i'-ui],$$

and if we now set $u' = u+i-i'$, we find

$$\text{Primary } OPD' = \tfrac{1}{8}yN'i'[ui'+ii'-i'^2-ui]$$
$$= \tfrac{1}{8}yN'i'(i'-u)(i-i')$$

for the four terms in the square brackets. It will be noticed that these two factors are exactly those appearing in all our geometrical expressions for primary spherical aberration. The closest relation is that with the angular aberration, for equation (10)**** of Part I (page 117) gives for the *new* angular aberration arising at a surface

$$AA'_p = \tfrac{1}{2}i'(i'-u)(i-i'),$$

which gives at once the simple relation

$$\text{Primary } OPD' = \tfrac{1}{4}yN'\cdot AA'_p.$$

As by definition $AA'_p = TA'_p/l' = LA'_p \cdot u'/l'$, we have the complete set, on putting $y/l' = u'$:

$$\text{Primary } OPD' = \tfrac{1}{4}yN'\cdot AA'_p = \tfrac{1}{4}N'u'\cdot TA'_p = \tfrac{1}{4}N'u'^2\cdot LA'_p.$$

It is extremely significant that TA'_p appears with the factor $N'u'$ and LA'_p with the factor $N'u'^2$, for these are exactly the transfer factors with which the transverse and longitudinal aberrations persistently entered into the summations of Part I, Chapter VI. Evidently this amounted to adding them up as their equivalent in optical path difference, and then re-converting them into longitudinal or transverse aberrations by the final division by $N'_k u'_k{}^2$ or $N'_k u'_k$!

SUMMATION FOR A COMPLETE SYSTEM

We must now inquire how spherical aberration expressed as OPD' is added up for a whole system of centred surfaces. To a first approximation—for primary aberration—the question is most easily answered. For a first surface we can [92] calculate OPD' by the formulae already given; the result will tell us, assuming OPD' to come out positive, that the marginal light would arrive at the paraxial focus of the first surface a corresponding number of wave-lengths ahead of the axial light. If aperture and aberrations are small, the contraction of the cone of rays at the second surface can be determined in the manner regularly employed in Part I and we can then calculate in the first instance the new OPD' which would be given by the second surface if the waves *received* by it were truly spherical 'ideal' waves. If OPD' again came out positive, we should have another gain of the marginal light, and simple common sense at once demands that the previous gain at the first surface must be purely additive, for if in walking from A to C via B a second time, we improve on the previous performance by, say, two minutes between A and B and by a further three minutes between B and C, we inevitably shall do the whole distance in five minutes less time than before. It is equally obvious that if there are sometimes gains and sometimes losses, then the final result must be still the simple algebraic sum. That is indeed the law by which the primary OPD' combine. But if we now discuss in a stricter manner what happens when we apply the more exact complete equation for primary and secondary OPD' to a complete system of considerable aperture and try to maintain the identity of the primary and secondary parts, that is, those growing as PA^4 and those growing as PA^6, so as to be able to discuss them at the final focus for the aperture as a whole, then a whole host of awkward corrections come in and render this otherwise attractive method extremely tedious.

Our formula for the OPD' at the paraxial focus may be put into the abbreviated form for a first surface receiving light from an original object-point:

$$OPD'_1 = \tfrac{1}{8}v_1 PA_1{}^4 + \tfrac{1}{16}w_1 PA_1{}^6 + \text{higher terms},$$

in which v_1 is a short symbol for the factor of the first term and w_1 for the factor of the second term. It is, however, preferable and more in accordance with the usual practice to measure the semi-aperture by the ordinate Y instead of the chord PA. For the present case of an axial pencil, we have $PA^2 = X^2 + Y^2$, therefore $PA^4 = Y^4 + 2Y^2X^2 + X^4$ and $PA^6 = Y^6 + 3Y^4X^2 + 3Y^2X^4 + X^6$. On the other hand, to a first approximation which is sufficient for sixth-order accuracy, $X = Y^2/2r$ or $X^2 = Y^4/4r^2$, so for this degree of accuracy, we may put

$$PA^4 = Y^4 + Y^6/2r^2 \quad \text{and} \quad PA^6 = Y^6.$$

The omitted terms are in the eighth or higher powers and would only be appreciable if tertiary aberrations were to be included. If these sufficiently close values of PA^4 and PA^6 are put into the equation for OPD', it becomes

$$OPD'_1 = \tfrac{1}{8}v_1 Y_1{}^4 + \tfrac{1}{16}Y_1{}^6(w_1 + v_1/r_1{}^2).$$

We must note that the coefficient of the fourth-order term remains as in the PA form of the equation, but the sixth-order term becomes more complicated even for a first surface.

For the second surface the above equation will again apply formally, that is, the new OPD' produced by this surface will be

$$OPD'_2 = \tfrac{1}{8}v_2 Y_2{}^4 + \tfrac{1}{16}Y_2{}^6(w_2 + v_2/r_2{}^2).$$

[92] However, to express the combined effect in a single expression clearly separated into fourth- and sixth-order parts, we must work out Y_2 in terms of some convenient Y of the system, and Y_1 will for our present purpose be the most convenient. If we now refer to Fig. 124, we shall have for the paraxial region the usual contraction of the cone of rays proportional to l_2/l'_1; hence to a first approximation $Y_2 = Y_1 \cdot l_2/l'_1$. But for the more remote rays this will not be sufficiently exact, firstly because P_1 and P_2 will be closer together or in other cases more widely separated than A_1 and A_2, and secondly because the marginal light will, according to the theorem of Helmholtz, proceed towards the marginal focus B'_{m_1} and not towards the paraxial focus B'_{p_1}. It is not difficult to show directly that the correction of Y_2 is to a first and sufficient approximation proportional to $Y_1{}^3$, but we arrive at that conclusion more easily by the symmetry argument first employed in Chapter II of Part I: The correction must have the same magnitude but the

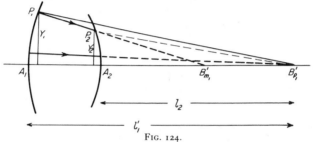

Fig. 124.

opposite sign for $+Y$ and for $-Y$ and that can only be if it is expressed in odd powers of Y_1. Hence we must use for the second surface

$$Y_2 = Y_1 \cdot l_2/l'_1 + p \cdot Y_1{}^3,$$

p being a constant, and if we form the fourth and sixth powers of this, breaking off with sixth-order terms, we find

$$Y_2{}^4 = Y_1{}^4(l_2/l'_1)^4 + 4Y_1{}^6 \cdot p \cdot (l_2/l'_1)^3; \qquad Y_2{}^6 = Y_1{}^6(l_2/l'_1)^6$$

and introduction of these into the formal equation for OPD'_2 gives

$$OPD'_2 = \tfrac{1}{8}Y_1{}^4 v_2(l_2/l'_1)^4 + \tfrac{1}{16}Y_1{}^6(w_2 + v_2/r_2{}^2)(l_2/l'_1)^6 + \tfrac{1}{2}Y_1{}^6 v_2 p(l_2/l'_1)^3.$$

The first term still retains the simple value of the primary aberration, but the second term is further complicated. In adding OPD'_1 and OPD'_2 there finally arises yet another complication, for the OPD' calculated for the first surface assumes that the light from P_1 travels directly toward B'_{p_1} whilst the theorem of Helmholtz directs it toward B'_{m_1}; at the second surface, we assume that the light arriving at P_2 has a direction toward $B_{p_2} = B'_{p_1}$ and the OPD' obtained from the above equation is only correct on that assumption. Evidently this brings in a third correction: we must evaluate the difference between the broken track $P_1P_2B'_{p_1}$

and the direct track $P_1 B'_{p_1}$, and this correction amounts to a lengthening of the [92] marginal path since the sum of two sides of a triangle is always larger than the third side. The result is a subtractive correction of OPD' that is proportional to the square of the angular aberration at P_1 or again to the sixth power of Y; the correction therefore further complicates the total secondary aberration.

We have thus amply proved that a direct determination of the primary *and* secondary OPD' of a centred lens system would be a task of extreme difficulty by means of Equation OP(2). But we have also learnt that all these complications affect only the secondary term. Hence we shall use OP(2) exclusively for the determination of the primary differences of optical path and shall find it very convenient and useful for that purpose. Another valuable conclusion from the discussion is that, although even the second term becomes extremely complicated in the summing up, it remains proportional to $Y_1{}^6$; hence the OPD' of any centred system for the axial pencil follows the law

$$OPD' = v_4 Y^4 + v_6 Y^6 + \&c.$$

if v_4 and v_6 are looked upon as constant factors to be determined indirectly by ray-tracing methods.

THE RELATION BETWEEN THE GEOMETRICAL AND PHYSICAL REPRESENTATION OF SPHERICAL ABERRATION

THE relations between our new measure of spherical aberration as a difference of [93] optical path and the more usual geometrical measures will play a very important part in subsequent work, and we shall therefore work them out more closely without restricting ourselves to purely primary aberrations. Re-drawing Fig. 123 of Chapter XII for the present case of an axial pencil and of the paraxial focus of that pencil as our adopted reference point, we obtain Fig. 125. We shall draw the ideal spherical wave with B'_p as centre and with l' as radius so that it touches the refracting surface at its pole. The real wave will depart from the ideal one everywhere by OPD'/N', OPD' being calculated by the proper form of OP(2) for the points for which the gap between the two curves is to be determined.

We proved in Chapter XII that the rays of geometrical optics are always exact normals of the corresponding light waves in isotropic media, hence the true marginal ray corresponding to the ideal ray PB'_p must be at right angles with the real wave. This relation enables us to locate the marginal ray PB'_m and with it the longitudinal spherical aberration $B'_m B'_p$ and the angular aberration $B'_p PB'_m = AA'$.

Before putting these relations into mathematical form, we must note that they find their chief, and almost only important, application at the final focus of complete systems, where only small residual aberrations can be tolerated. Hence the OPD' that come into question will usually amount to at most a few wave-lengths and the geometrical aberrations will be correspondingly small. While it is highly important to know these final and usually incurable residuals within a moderate percentage, say within 5 or 10 per cent. to be able to decide whether they will have a serious effect on the image, it is not of the slightest interest to determine them

[93] within fractions of one per cent., not only because such minute variations could not possibly be appreciated by even the most delicate tests, but mainly because the technical imperfections of the actual lens system may be expected to be of the same order of magnitude as the usual tolerances, but entirely erratic in sign and tendency. Hence there is no need to split hairs over these relations between OPD' and the geometrical aberrations.   Another important deduction from the smallness of the residuals at which we are aiming is that our diagrams necessarily will have to be drawn with enormous exaggeration of the actual defects; this exaggeration inevitably leads to apparent absurdities, for when the really minute 'gaps' between the ideal and the real waves and the corresponding faint differences in curvature are magnified about 10,000 times, the real wave may assume the appearance of reversed curvature (corresponding to diverging instead of converging normals or geometrical rays), which in reality would be quite unthinkable.   This exaggeration of the diagrams must therefore perpetually be borne in mind.

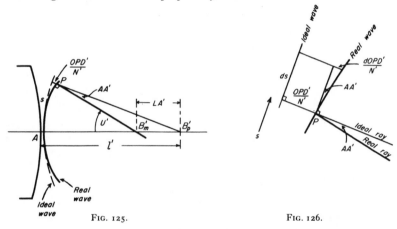

FIG. 125.   FIG. 126.

We obtain our fundamental relation between the geometrical and the physical aberration by assuming in the first instance that we know OPD' as a function of s, the length of the *arc* of the ideal wave between the central ray and any outer ray. If we then consider two closely neighbouring points of the two waves in the immediate neighbourhood of P in Fig. 125, and draw a hugely magnified diagram of this region which will enable us to represent the OPD' at something like its true relative size, we shall have at P a separation between the two waves equal to OPD'/N' calculated for the distance s from the central ray.   For an increase ds of this distance there will be an increase (positive or negative) of OPD'/N' by $d \cdot OPD'/N'$ which could be calculated by our postulated function giving OPD' in terms of s.   Bearing in mind the minuteness of OPD' compared to any likely distance of B'_p, it is clear that the radii drawn from the ends of ds toward the distant B'_p cannot sensibly approach each other between the two waves.   Therefore the parallel to ds in Fig. 126 may be regarded as sensibly equal to ds and the small

triangle gives the angle of tilt AA' of the real wave with reference to the ideal wave [93]
as

$$\tan AA' = \frac{d(OPD')}{N' \cdot ds}.$$

This equation *is* the general solution of our problem, because the rays of geometrical optics are the normals of the corresponding waves. The ray PB'_p in Fig. 125, which is the ideal marginal ray that we should have in the absence of aberration, is the normal of the ideal wave, and the real marginal ray, affected by aberration and drawn as PB'_m in Fig. 125, must be the true normal of the distorted real wave. The angle between two lines in a plane is equal to that between the normals of these lines; therefore the angle AA' represents the angle between the direction (toward B'_p) which the marginal ray ought to take and that which it really takes (toward B'_m) and is the ' angular aberration' of the marginal ray with reference to the desired focus at B'_p. AA' being determined, we now obtain $B'_mB'_p = LA'$ from the triangle $PB'_pB'_m$ in which we know $PB'_p = l'$, the angle at $P = AA'$, and the angle at $B'_m = U'$, the latter being given by the trigonometrical ray-tracing. The trigonometrical sine relation then gives

$$B'_mB'_p = LA' = l' \sin AA' / \sin U'.$$

Our equation for the determination of AA' really gives the tangent; but as the angle obviously cannot exceed a small number of minutes in any case that could be of interest, we can neglect the difference between tangent and sine and substitute the value of $\tan AA'$ by our first solution for $\sin AA'$ in the second, thus finding the practically exact equation

$$LA' = \frac{l'}{N' \sin U'} \cdot \frac{d(OPD')}{ds}.$$

The equations for LA' and AA' are practically exact when OPD' is expressed as a function of the arc s of the ideal wave between A and P. Since this arc has B'_p as its centre and l' as radius, we have $s = l' \times$ (angle at B'_p), and since l' is a constant, the equations remain exact if OPD' is expressed as a function of this angle, which can either be directly calculated from the data of the usual ray-trace or can be set equal to $U' - AA'$, in which case approximations will have to enter into the solution. Nevertheless, the latter procedure is the better if the aberrations are large and if at the same time a considerable degree of precision is desired.

On account of the very limited precision that is required in nearly all cases, as was shown above, and because the aberrations to be dealt with as well as the angles U' are usually quite small, we shall nearly always use an approximation which consists in noting that under these conditions the arc s will be very little different from either the PA or the Y at the refracting surface itself. This means that we may express the OPD' as a function of PA or Y in the usual way without introducing a serious inaccuracy. We now collect the formulae:

OP(3)   (1) If OPD' is expressed as a function of the angle at $B'_p = U' - AA'$, then with high precision

$$\text{Tan } AA' = \frac{d(OPD')}{N' \cdot d(U' - AA')}; \quad LA' = \frac{l'}{N' \cdot \sin U'} \cdot \frac{d(OPD')}{d(U' - AA')};$$

[93] (2) In nearly all cases occurring in practice, OPD' may be expressed in terms of the usual Y, when

$$AA' = \frac{d(OPD')}{N' \cdot dY}; \quad LA' = \frac{l'}{N' \sin U'} \cdot \frac{d(OPD')}{dY}.$$

Examples of the practical application of these equations will be found in the following chapter.

TRIGONOMETRICAL CALCULATION OF OPD'_m

[94] FOR direct determinations of the OPD with which the light travelling along any given ray reaches the final focus, we shall rely entirely on the following exact trigonometrical process, illustrated in Fig. 127.

We assume that a ray meeting a spherical refracting surface at P and aiming,

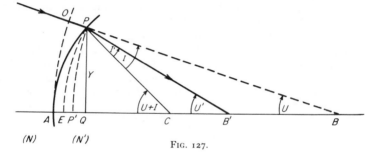

FIG. 127.

before refraction, at B, has been trigonometrically traced through the surface to its intersection at B', and that all its trigonometrical data are known.

Assuming for the present that B is a perfect focus of all the arriving light, strictly spherical waves will converge upon B as their centre, and the circle AO with centre at B in the diagram will represent the axial section of such a wave. While the marginal part of this wave traverses the distance OP in the medium of index N and therefore covers the optical path $N \cdot OP$, its axial part has entered the new medium but necessarily covers the same *optical* path. The ray-tracing has determined B' as the new intersection-point of the refracted ray. If no spherical aberration were to arise at the surface, we should have a new spherical wave EP with B' as centre, and the axial light would reach E with the optical path $N' \cdot AE$ at the same instant as the marginal light reaches P from O with the optical path $N \cdot OP$. Consequently, if $N \cdot OP$ and $N' \cdot AE$ are separately computed and compared, their equality proves the absence of spherical aberration at the surface. Conversely, any difference between them must be due to spherical aberration and must in fact be the OPD which we set out to determine. Obviously the axial light will be too far from B' at the instant when the marginal light reaches P if $N' \cdot AE > N \cdot OP$, and as we have decided to take the OPD' as positive when the axial path is the longer, we conclude that OPD'_m (suffix m to indicate that it refers to the [94] marginal focus B') will be determined by $OPD'_m = N' \cdot AE - N \cdot OP$.

To determine OP, we transfer it to the optical axis by a circle PP' with B as centre, so that $AP' = OP$. Dropping also the perpendicular $PQ = Y$ from P upon the optical axis, we have $AP' = AQ - P'Q$. Now AQ is the X of our refracting surface, with angle $(U+I)$ at the centre C of curvature; $P'Q$ is the corresponding depth of curvature of the arc PP' with centre at B and the angle U at that centre. Using formula (8)d for these X-values, we find

$$OP = AP' = Y[\tan \tfrac{1}{2}(U+I) - \tan \tfrac{1}{2}U].$$

AE can be evaluated in the same manner, as $AQ - EQ$, and gives

$$AE = Y[\tan \tfrac{1}{2}(U'+I') - \tan \tfrac{1}{2}U'],$$

$U'+I'$ being used instead of $U+I$ (to which it is unconditionally equal) for the sake of symmetry. We therefore have

$$OPD'_m = N'Y[\tan \tfrac{1}{2}(U'+I') - \tan \tfrac{1}{2}U'] - NY[\tan \tfrac{1}{2}(U+I) - \tan \tfrac{1}{2}U].$$

This equation can be converted into a convenient computing formula by first putting $\tan = \sin/\cos$. This gives for the factor of the second term, by bringing to a common denominator,

$$\tan \tfrac{1}{2}(U+I) - \tan \tfrac{1}{2}U = \frac{\sin \tfrac{1}{2}(U+I) \cdot \cos \tfrac{1}{2}U - \sin \tfrac{1}{2}U \cdot \cos \tfrac{1}{2}(U+I)}{\cos \tfrac{1}{2}(U+I) \cdot \cos \tfrac{1}{2}U}$$

$$= \frac{\sin [\tfrac{1}{2}(U+I) - \tfrac{1}{2}U]}{\cos \tfrac{1}{2}(U+I) \cdot \cos \tfrac{1}{2}U} = \frac{\sin \tfrac{1}{2}I}{\cos \tfrac{1}{2}(U+I) \cdot \cos \tfrac{1}{2}U}.$$

The factor of the first term in OPD'_m obviously will give the same expression with 'dashes' throughout; therefore

$$OPD'_m = N'Y \frac{\sin \tfrac{1}{2}I'}{\cos \tfrac{1}{2}(U'+I') \cdot \cos \tfrac{1}{2}U'} - NY \frac{\sin \tfrac{1}{2}I}{\cos \tfrac{1}{2}(U+I) \cdot \cos \tfrac{1}{2}U}.$$

If we now extend the first fraction by $2 \cos \tfrac{1}{2}I'/2 \cos \tfrac{1}{2}I'$ and the second by $2 \cos \tfrac{1}{2}I/2 \cos \tfrac{1}{2}I$, the numerators become $\sin I'$ and $\sin I$ respectively, and as $N' \sin I' = N \sin I$, we can extract a common factor (note also that $\cos \tfrac{1}{2}(U'+I') = \cos \tfrac{1}{2}(U+I)$!) and obtain

(a) $$OPD'_m = \frac{N'Y \sin I'}{2 \cos \tfrac{1}{2}(U+I)} \left[\frac{1}{\cos \tfrac{1}{2}I' \cdot \cos \tfrac{1}{2}U'} - \frac{1}{\cos \tfrac{1}{2}I \cdot \cos \tfrac{1}{2}U} \right].$$

Bringing the square bracket to a common denominator, this becomes

$$OPD'_m = \frac{N'Y \sin I'}{2 \cos \tfrac{1}{2}(U+I)} \cdot \frac{\cos \tfrac{1}{2}I \cdot \cos \tfrac{1}{2}U - \cos \tfrac{1}{2}I' \cdot \cos \tfrac{1}{2}U'}{\cos \tfrac{1}{2}U \cdot \cos \tfrac{1}{2}I \cdot \cos \tfrac{1}{2}U' \cdot \cos \tfrac{1}{2}I'},$$

and if we now introduce in the first term of the final numerator the obvious identity $\cos \tfrac{1}{2}U = \cos [\tfrac{1}{2}(U - U') + \tfrac{1}{2}U']$ and resolve this by the $\cos (a+b)$ formula, and in the second term of this numerator $\cos \tfrac{1}{2}I' = \cos [\tfrac{1}{2}(U - U') + \tfrac{1}{2}I]$ (which follows from $U+I = U'+I'$) also resolved by the $\cos (a+b)$ formula, the numerator is obtained in the form of four terms of which two cancel each other while the

[94] remaining two have $\sin \frac{1}{2}(U - U')$ as a common factor. The final formula thus obtained is

$$\text{OP(4)} \qquad OPD'_m = \frac{N'Y \sin I' \cdot \sin \frac{1}{2}(U - U') \cdot \sin \frac{1}{2}(I - U')}{2 \cos \frac{1}{2}U \cdot \cos \frac{1}{2}I \cdot \cos \frac{1}{2}U' \cdot \cos \frac{1}{2}I' \cdot \cos \frac{1}{2}(U + I)}.$$

On account of the advisability of using the PA-check in all trigonometrical ray-tracing, it will usually be preferable to replace the Y in this formula, which would have to be computed by (8)* on page 29 of Part I, by the PA which is obtained as a by-product in the check calculation. As $Y = PA \cos \frac{1}{2}(U + I)$ the formula then becomes

$$\text{OP(4)} \qquad OPD'_m = \frac{N' \cdot PA \cdot \sin I' \cdot \sin \frac{1}{2}(U - U') \cdot \sin \frac{1}{2}(I - U')}{2 \cos \frac{1}{2}U \cdot \cos \frac{1}{2}I \cdot \cos \frac{1}{2}U' \cdot \cos \frac{1}{2}I'}$$

and has one term less than the formula in Y because $\cos \frac{1}{2}(U + I)$ cancels out.

The original formula, in Y, is easily transformed to meet the one special case of a plane surface. By putting, according to Pl.(1) $I = -U$ and $I' = -U'$, we obtain

$$\text{OP(4) Plano:} \qquad OPD'_m = \frac{N'Y \cdot \sin U' \cdot \sin \frac{1}{2}(U + U') \cdot \sin \frac{1}{2}(U - U')}{2 \cos^2 \frac{1}{2}U \cdot \cos^2 \frac{1}{2}U'},$$

in which $Y = L \tan U = L' \tan U'$ by Pl.(2*).

Both computing formulae imply less work than their length suggests because they give the small OPD directly and not as a difference of two large numbers; therefore they are almost invariably calculated with ample accuracy by four-figure logs taken from a five- or six-figure table and involve only a small amount of interpolation, if any. At glass-air surfaces, one line of figures can be saved by noting that of the alternative $N' \sin I' = N \sin I$, one will be simply the sine of the angle in air. That one should therefore be selected.

A check on OP(4) may nevertheless be considered desirable, for the OPD's calculated by it form the final crucial test of all microscope-objective designs and an error in them would therefore be fatal. Such a check is easily derived from the intermediate equation (a). If we put $Y = PA \cdot \cos \frac{1}{2}(U + I)$ and split the equation into two parts by remembering that $N' \sin I' = N \sin I$, it assumes the form

$$\text{OP(4) check:} \qquad OPD'_m = \frac{N' \cdot PA \cdot \sin I'}{2 \cos \frac{1}{2}I' \cdot \cos \frac{1}{2}U'} - \frac{N \cdot PA \cdot \sin I}{2 \cos \frac{1}{2}I \cdot \cos \frac{1}{2}U}$$

and should be computed exactly as written by extracting PA, $\sin I$, and $\sin I'$ from the main calculation, and N and N' from as near the fountain-head as possible, and taking out the cosines with the utmost accuracy, preferably to six decimal places. Thus computed, the check should agree closely with the standard OP(4) result, but the latter will be the more accurate one and should be retained. For a plane, the check takes the form

$$\text{OP(4) Plano check:} \qquad OPD'_m = \frac{N \cdot Y \cdot \sin U}{2 \cos^2 \frac{1}{2}U} - \frac{N' \cdot Y \cdot \sin U'}{2 \cos^2 \frac{1}{2}U'}$$

and should be computed in the same manner advised for the general case. The reversed sequence of the plain and 'dashed' terms in the plano check arise from the relation $I = -U$ and $I' = -U'$, which reverses the signs of the sines.

Both these checks also amount to a final and very searching test of the correct [94] application of the refractive indices in the trigonometrical ray-tracing, and an obstinate disagreement between the results of OP(4) and its check may find its explanation there.

As the *OPD* are very small quantities and have the wave-length as the proper unit of length by which their importance is to be estimated, it is desirable (and it saves many zeros after the decimal point) to calculate them in approximate wave-lengths. One inch contains 50,000 and one millimetre 2000 wave-lengths of some kind of green light. As all the OPD'_m formulae have a numerical factor $\frac{1}{2}$, they will therefore give their results in approximate wave-lengths by replacing the factor $\frac{1}{2}$ by the factor 25,000 if inch measure is used for radii and thicknesses, or by the convenient factor 1000 if the millimetre is the adopted unit.

The trigonometrically determined OPD'_m has the great advantage that the contributions of the successive surfaces to the final total combine strictly by direct algebraic addition, simply because the marginal light is tracked along the true marginal ray. This means that the correction pointed out above as arising in the case of OP(2) owing to the paraxial focus being its reference point is entirely avoided. On the other hand, OPD'_m shares with all strictly trigonometrical processes the limitation that it gives only the one piece of information, namely, the exact difference of optical path with which the light along the particular marginal ray meets the light travelling along the optical axis at the marginal focus. OPD'_m, therefore, requires supplementary calculations before we can draw conclusions as to the value of OPD' for other zones of the complete aperture. In many cases the primary OPD' determined above is valuable in this respect, and we shall therefore add its formula, together with its remarkable relations with the three measures of the geometrical aberration:

OP(4p)
$$OPD'_p = \tfrac{1}{8}yN'i'(i'-u)(i-i')$$
$$= \tfrac{1}{4}yN' \cdot AA'_p = \tfrac{1}{4}N'u' \cdot TA'_p = \tfrac{1}{4}N'u'^2 \cdot LA'_p$$

It follows from the earlier discussion of the addition of the OPD' by OP(2) that the primary OPD' is simply additive without any corrections. The relations between the four different measures of primary spherical aberration hold not only for any one surface, but for the total aberrations after passage through several or all the surfaces of a system.

As a numerical example we shall calculate OPD'_m for the marginal light of the telescope objective forming Example 2 in Chapter I of Part I. Taking the data of the objective as being in terms of inches, the formulae will be

OP(4)
$$OPD'_m = 25,000 \frac{N' \cdot PA \cdot \sin I' \cdot \sin \tfrac{1}{2}(U-U') \cdot \sin \tfrac{1}{2}(I-U')}{\cos \tfrac{1}{2}U \cos \tfrac{1}{2}I \cos \tfrac{1}{2}U' \cos \tfrac{1}{2}I'}$$

to be calculated by four-figure logs, and

OP(4) check:
$$OPD'_m = 25,000 \frac{N' \cdot PA \cdot \sin I'}{\cos \tfrac{1}{2}I' \cos \tfrac{1}{2}U'} - 25,000 \frac{N \cdot PA \cdot \sin I}{\cos \tfrac{1}{2}I \cos \tfrac{1}{2}U}$$

to be calculated with the exact values of N, N', $\sin I$, and $\sin I'$ found or used in the trigonometrical calculation. In the denominator of the second formula, the

[94] cosines are to be taken out to the exact half-angles of the angle register, with six decimals of the log if possible, and the anti logs of the two separate terms must also be found with the utmost accuracy.

If the dimensions of the objective were in millimetres, the numerical factor would be 1000, or if the dimensions were in centimetres, it would be 10,000.

The numerical work begins with forming a register of the halved angles (excepting $U + I$), but as this is also required for the PA-check, only the formation of $\frac{1}{2}(U - U')$ and of $\frac{1}{2}(I - U')$ represents additional work.   If the OP(4) check is to be included, then the half-angles should preferably be put down exactly, not rounded off to the full second.

	First surface	Second surface	Third surface
$\frac{1}{2}U$	0- 0- 0	2-49-42.5	2- 3-24
$\frac{1}{2}I$	8-10-50	−10-59-54	−2-31-19.5
$\frac{1}{2}I'$	5-21- 7.5	−10-13-35.5	−4- 6-31
$\frac{1}{2}U'$	2-49-42.5	2- 3-24	3-38-35.5
$\frac{1}{2}I - \frac{1}{2}U$	(Required for PA-check)		
$\frac{1}{2}I' - \frac{1}{2}U'$			
$\frac{1}{2}U - \frac{1}{2}U'$	−2-49-42.5	0-46-18.5	−1-35-11.5
$\frac{1}{2}I - \frac{1}{2}U'$	5-21- 7.5	−13- 3-18	−6- 9-55

We then extract from the main calculation:

$\log N$	0.00000	0.18087	0.21101
$\log N'$	0.18087	0.21101	0.00000
$\log \sin I$	9.44977	9.57351n	8.94410n
$\log \sin I'$	9.26890	9.54337n	9.15511n
$\log PA$	0.00444	0.00388	9.98884

When the whole calculation is on one sheet, the above need not of course be written out; they are brought down directly from the main calculation.

The OPD'_m formula is then calculated thus:

$\log 25{,}000$	4.3979	4.3979	4.3979
$+ \log N'$ or $\log N$		0.2110	
$+ \log \sin I'$ or $\log \sin I$	9.4498	9.5434n	9.1551n
$+ \log PA$	0.0044	0.0039	9.9888
$+ \log \sin \frac{1}{2}(U - U')$	8.6933n	8.1294	8.4423n
$+ \log \sin \frac{1}{2}(I - U')$	8.9698	9.3539n	9.0310n
$+ \operatorname{colog} \cos \frac{1}{2}U$	0.0000	0.0005	0.0003
$+ \operatorname{colog} \cos \frac{1}{2}I$	0.0044	0.0081	0.0004
$+ \operatorname{colog} \cos \frac{1}{2}I'$	0.0019	0.0070	0.0011
$+ \operatorname{colog} \cos \frac{1}{2}U'$	0.0005	0.0003	0.0009
$\log OPD'_m =$	1.5220n	1.6554	1.0178n
$OPD'_m =$	−33.27	+45.23	−10.42

OPD'_m at final focus $= 45.23 - 33.27 - 10.42 = +1.54$ wave-lengths (app.).

For OP(4) check, we then calculate:

log 25,000	4.39794	4.39794	4.39794
$+\log N'$	0.18087	0.21101	0.00000
$+\log \sin I'$	9.26890	$9.54337n$	$9.15511n$
$+\log PA$	0.00444	0.00388	9.98884
$+\operatorname{colog} \cos \frac{1}{2}I'$	0.001898	0.006955	0.001117
$+\operatorname{colog} \cos \frac{1}{2}U'$	0.000529	0.000280	0.000879
log (first part) =	3.854577	$4.163435n$	$3.543886n$
log 25,000	4.39794	4.39794	4.39794
$+\log N$	0.00000	0.18087	0.21101
$+\log \sin I$	9.44977	$9.57351n$	$8.94410n$
$+\log PA$	0.00444	0.00388	9.98884
$+\operatorname{colog} \cos \frac{1}{2}U$	0.00000	0.000529	0.000280
$+\operatorname{colog} \cos \frac{1}{2}I$	0.004442	0.008051	0.000421
log (second part) =	3.856592	$4.164780n$	$3.542591n$
First part	7154.47	− 14569.17	− 3498.53
− Second Part	− 7187.73	+ 14614.37	+ 3488.12
Check OPD'_m =	− 33.26	45.20	− 10.41

The check values agree within a few hundredths of a wave-length with those by OP(4) and the calculation may therefore be accepted as correct.

The result of our calculation is that, although in this objective the marginal rays were brought very accurately to the paraxial focus, the marginal light does not meet the paraxial light in the same phase of vibration. The marginal light has a path shorter by approximately 1.54 wave-lengths (or strictly by 1.54 units of 0.000 02 inch) than the axial light, and as the Rayleigh limit is one quarter of a wave-length, it looks as if our objective were a very bad one. The proof, to be given in Chapter XIV, that this is incorrect is one of the most important deductions from a study of the differences of optical path.

EFFECTS OF A SHIFT OF THE IDEAL IMAGE-POINT

THIS proof, and many other valuable deductions, depend on the possibility [95] briefly referred to at the end of the last chapter that the point to which the optical path-differences were referred in the first instance on account of convenience in the numerical calculation may not be the location of the best image. On the contrary, there may be another point, naturally not very far away, at which the light from all parts of the clear aperture arrives with a smaller variation of the optical paths. To find the change in the differences of optical path resulting from any small shift of the point at which the light is received, we refer to Fig. 128. We

[95] assume that it is known with what differences of optical path the light from an optical system would arrive at a point B' for which it was convenient to determine those differences. We ask: by what amount will these differences be changed if we receive the light at a point B'_δ which is shifted in the X-direction by $\delta l'$, in the Z-direction by $\delta z'$, in the Y-direction by $\delta H'$? The X-axis will as a rule be represented by the originally adopted principal ray for an oblique pencil, or by the principal optical axis for axial pencils. We now strike an ideal spherical wave with B' as centre and with a radius l'. It will nearly always be convenient and advisable to make l' equal to the distance from the refracting surface to B', but we shall see that moderate departures from that value alter the final result by such small amounts that they are usually negligible. We consider the light proceeding

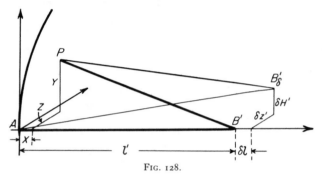

FIG. 128.

from A and from any point P of our auxiliary sphere towards the final region. All the light will necessarily have exactly equal paths (i.e., radii of the sphere) to traverse in getting to B'. We now determine the lengths of the new paths AB'_δ and PB'_δ, for *their* difference obviously will be the additional difference of optical path by which the total OPD' will be changed at B'_δ. When point P is defined by the rectangular coordinates X, Y, Z, the distance formula gives

$$(AB'_\delta)^2 = (l'+\delta l')^2 + \delta H'^2 + \delta z'^2,$$
$$(PB'_\delta)^2 = (l'+\delta l' - X)^2 + (Y-\delta H')^2 + (Z-\delta z')^2.$$

The latter gives, when $(l'+\delta l')$ is treated as a simple quantity,

$$(PB'_\delta)^2 = (l'+\delta l')^2 - 2X(l'+\delta l') - 2Y\delta H' - 2Z\delta z' + \delta H'^2 + \delta z'^2 + PA^2.$$

As P lies on the sphere of radius l', we have $X = PA^2/2l'$, and when this is put into the second term of the last equation, $-2Xl'$ is seen to cancel the final $+PA^2$, hence the equation simplifies to

$$(PB'_\delta)^2 = (l'+\delta l')^2 - PA^2\delta l'/l' - 2Y\delta H' - 2Z\delta z' + \delta H'^2 + \delta z'^2.$$

Finally, if $(l'+\delta l')^2$ is taken out as a common factor and the root is extracted, we find

$$PB'_\delta = (l'+\delta l')\sqrt{1 - \frac{\delta l'}{l'}\left(\frac{PA}{l'+\delta l'}\right)^2 - 2\frac{Y\delta H'}{(l'+\delta l')^2} - 2\frac{Z\delta z'}{(l'+\delta l')^2} + \frac{\delta H'^2 + \delta z'^2}{(l'+\delta l')^2}}.$$

We can at once write down the corresponding equation [95]

$$AB'_\delta = (l'+\delta l')\sqrt{1+\frac{\delta H'^2+\delta z'^2}{(l'+\delta l')^2}}.$$

In developing these roots into series, we can save useless elaboration by noting that obviously the $\delta l'$ will in practice be comparable with the ordinary longitudinal aberration and therefore small of the second order. Moreover, the $\delta H'$ and $\delta z'$ will be of the transverse-aberration type or small of the third order, and Y, Z, and PA are to be treated as small of the first order. It follows that all the algebraic terms under the first root are small of the fourth order with exception of the last term, which is of the sixth order, and the algebraic term under the second root is small of the sixth order. Hence the a^2 term in $\sqrt{(1+a)}$ would produce nothing below the eighth order, and as that corresponds to tertiary aberration in the OPD' measure, we can disregard everything beyond the first term of the root development. We thus find

$$PB'_\delta = (l'+\delta l')-\tfrac{1}{2}\frac{\delta l'PA^2}{l'(l'+\delta l')}-\frac{Y\delta H'}{l'+\delta l'}-\frac{Z\delta z'}{l'+\delta l'}+\tfrac{1}{2}\frac{\delta H'^2+\delta z'^2}{l'+\delta l'},$$

$$AB'_\delta = (l'+\delta l') \qquad\qquad\qquad\qquad\qquad +\tfrac{1}{2}\frac{\delta H'^2+\delta z'^2}{l'+\delta l'}.$$

The difference between these two will be the additional $\delta OPD'$ resulting from the shift of focus if the latter is formed in the air, but for the sake of universal validity we will add the usual factor N' to cover the possibility of an image formed in a medium other than air. Since we have stipulated that the OPD' are to be taken as axial lengthening of the path, minus the marginal, we have to form the difference $N'(AB'_\delta-PB'_\delta)$, which leads to the formula

OP(4)* $$\delta OPD' = \tfrac{1}{2}N'\frac{\delta l'PA^2}{l'(l'+\delta l')}+N'\frac{Y\delta H'}{l'+\delta l'}+N'\frac{Z\delta z'}{l'+\delta l'}.$$

This means that, at the shifted focus B'_δ, the total OPD' will be equal to OPD' at $B'+\delta OPD'$ by OP(4)*. It is a highly exact formula for it omits nothing below the eighth order of magnitude, and it is universal for it covers a shift of focus in any conceivable direction.

As a rule we shall not require the full precision, and then we can treat $\delta l'$ as small compared with l' and simplify the formula by replacing $(l'+\delta l')$ in the denominators by l' itself. When the convergence of the pencil is small, say when marginal U' is not more than $10°$, we can in many cases use another simplification by noting that then PA/l' is very nearly equal to $\sin U'$. This means that the first term becomes $\tfrac{1}{2}N'\delta l' \sin^2 U'$, which fits in with the trigonometrical calculations.

We shall make considerable use of this equation in the following chapter and subsequently.

A PATH-DIFFERENCE PROOF OF THE OPTICAL SINE THEOREM†

[96]   IN Fig. 129 let BX be the optical axis of a system of lenses or mirrors, or both combined, no other restriction being imposed than that all the refracting or reflecting surfaces shall be continuous surfaces of rotation with BX as axis, and the system therefore a 'centred' one.   The rays from B shall be allowed to enter the system through a narrow zone marked by the circle P with A (on BX) as centre; after passing through the system, they are assumed to emerge by the zone defined by the circle P' with A' as centre, and to come to a focus at B'.   That all rays of such a zone must come to a common focus follows from the symmetry of the entire system round the axis BX.   For the same reason any one ray, such as $BPP'B'$, must always remain in the plane containing the axis BX in which it started from B; the angles U and U' formed by the incident and emergent rays with the optical

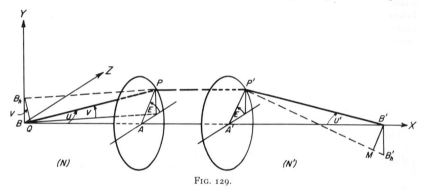

FIG. 129.

axis must be the same for all rays of the zone under consideration, and the paths of all such rays must be equal, both as to their entire length and as to their corresponding constituent parts.

Owing to these properties of a centred optical system, it is possible to discuss, without computing the paths of the separate rays through the system, the formation of the image of any point B_h near the optical axis.   By 'near' we mean so close to B that, to any ray passing from B through the system, we can find a corresponding ray from B_h which forms only small angles of the first order with it, and which does not become separated from it by more than small quantities of the first order anywhere within the system.

This generalization depends on Fermat's theorem that any path which agrees with a known optically determined path within small quantities of the *first* order is of the same length within small quantities not exceeding the *second* order, and is therefore sensibly equal to it.   We are, therefore, justified in assuming that the

---

† On page 375 of Part I, a proof of the sine theorem not depending on the auxiliary optical axis was promised for Part II.   The proof referred to was probably that given by Conrady in his paper 'The Optical Sine-condition', Mon. Not. R.A.S., **65**, 501–509, 1905, the essential part of which is reproduced here by kind permission of the Royal Astronomical Society.

paths traversed *within* the optical system by light from B_h are sensibly equal to [96] those followed by light from B, and we can base our study of the image of B_h on a comparison of the paths *outside* the system.

In order that B'_h may be the image of B_h, all optical paths between B_h and B'_h must be of equal length, for that is the condition under which all light from B_h will arrive at B'_h in the same phase of vibration so as to form a bright image. We know that all paths between B and B'—via the zone under consideration—*are* equal; our task, therefore, reduces itself to comparing any ray from B_h with its corresponding ray from B, and by Fermat's theorem we need not consider the portion within the system.

Now BP and B_hP are two such corresponding rays, and by dropping a perpendicular B_hQ from B_h upon BP we see that B_hP is shorter than BP by BQ. On reference to the rectangular system of coordinates shown in the figure, B_h lying on the axis BY, it is seen that, V being the angle between BP and the XZ-plane, we have

$$BQ = BB_h \cdot \sin V.$$

Calling the angle between radius AP and the XZ-plane E, we further have

$$\sin V = \sin E \cdot \sin U$$

and therefore

(1) $$BQ = BB_h \cdot \sin U \sin E.$$

The angle E being the angle between the XZ-plane and the plane in which the ray BP proceeds through the system, it is constant for this ray; hence, by analogy, we can at once write down the quantity B'_hM by which $P'B'_h$ is longer than $P'B'$:

(2) $$B'_hM = B'B'_h \cdot \sin U' \sin E.$$

Equations (1) and (2) give us the *geometrical* difference between the paths $BPP'B'$ and $B_hPP'B'_h$. But we must compare optical paths, and each of the geometrical paths must, therefore, be multiplied by the index of the medium in which it lies in order to be commensurable. Taking the indices surrounding object and image respectively as N and N', we therefore get the difference of the optical paths BB' and $B_hB'_h$ as

$$OPD = N \cdot BQ - N' \cdot B'_hM = \sin E\{N \cdot BB_h \cdot \sin U - N' \cdot B'B'_h \cdot \sin U'\}.$$

This difference varies, therefore, in general according to the value of E. But it vanishes, and all optical paths between B_h and B'_h become equal, when the bracketed term becomes zero; hence a narrow zone of *any* centred optical system will yield sharp images of points near but not in the optical axis, the position of the image being determined by the condition:

(3) $$N \cdot H \cdot \sin U = N' \cdot H' \cdot \sin U',$$

where H and H' refer to BB_h and $B'B'_h$ respectively.

One further remark must be added to complete this proof. The point B'_h has been considered to represent the equiphase-focus of a *single zone* of the lens aperture. In the presence of coma, as is shown on page 743, the equiphase point of a single zone falls at the sagittal focus of the zone; hence the image height $B'B'_h$ represents the quantity called h'_s in equation (a) on page 368 of Part I.

CHAPTER XIV

OPTICAL TOLERANCES

[97] WE can now prove those formulae for optical tolerances which refer either to cases when there is no aberration at the image-point originally adopted or when there is only spherical aberration or pure astigmatism. If these restrictions are satisfied, the tolerances will apply to oblique as well as to axial pencils.

THE RAYLEIGH LIMIT

IN 1878 the late Lord Rayleigh published a series of papers (Collected Researches, Vol. I, pp. 415–453) on the resolving power of spectroscopes and other optical instruments which would have been immediately recognized as epoch-making and as the obvious starting-point of a new line of optical research if applied optics had been in a healthy state. Unfortunately, optics had for generations remained in a most deplorable traditional groove as a mere stalking-horse for complicated algebraic exercises, and so it came about that Lord Rayleigh's brilliant lead was practically ignored.

Lord Rayleigh stated that an optical instrument would fall very little short of its utmost possible theoretical perfection if all the light arrived at the focus with differences of optical paths not exceeding one quarter of a wave-length.

Lord Rayleigh drew this conclusion because a number of test cases, including spherical aberration at the paraxial focus, coma, and astigmatism, showed that, with 1/4 wave-length of extreme difference of optical path (i.e., between the longest and the shortest path), the maximum brightness of the image of a point of light was only 20 per cent. or thereabouts less than it would be if the image were absolutely perfect. There could therefore only be a very small increase in the size of the 'spurious disk' and a correspondingly small loss of resolving power.

Recently the amount of material upon which the discussion of the Rayleigh limit can be based has been very largely extended. The brightness of the image throughout its extent to the third diffraction ring and even beyond has been calculated for a number of cases at and near the Rayleigh limit, not only for the paraxial focus but also for neighbouring points. The results bear out Lord Rayleigh's conclusion as to the maximum brightness of the image, but show variations in the percentage of loss at the limit from about 27 per cent. to as little as 11 per cent. A most important and significant fact becomes apparent when the complete light distribution is studied. All the light received by the lens system from a luminous object-point must of course (on the principle of the conservation of energy) reappear in the image, apart from losses by reflection and absorption. Lord Rayleigh's paper leaves it an open question whether the diminished brightness of the image in the presence of differences of optical path is compensated by an enlarged area of the central diffused spurious disk or whether the light unaccounted for in the central disk is scattered over a wide surrounding area and thus rendered practically invisible. The complete integrations bear out the latter alternative.

Up to the Rayleigh limit, and even up to the doubled limit, the central disk—which in the majority of cases is the only part of the image bright enough to be seen and appreciated—has a *practically constant diameter*. Consequently there is *no* loss of resolving power for close isolated points (say double stars) up to the *doubled* Rayleigh limit. But the light lost in the central disk (at the doubled limit, about 60 per cent. of the maximum light possible with a perfect instrument) appears as an extended halo of low luminosity on which are superimposed more or less marked diffraction rings surrounding the central disk. This scattered light necessarily diminishes the *contrast* between the lighter and darker parts of an extended object, such as an ordinary microscopic object or the surface of the moon or a planet, and if the details are of low contrast, they may be completely masked by this scattered light; for the human eye appreciates differences of brightness of adjoining small and ill-defined areas only if they exceed about 5 per cent., that is, if the relative brightness of the two is in the ratio of 19:20. The integrations thus bear out the experience of all keen observers that the resolution of strongly marked detail even right up to the theoretical limit is *not* a proof of excellence; it merely proves that the instrument is *not bad*. But if, after bearing this test of mere resolving power, an instrument is also found to be good at showing delicate contrasts in decidedly coarser objects, then, and not till then, may it be passed as good.

Up to the original Rayleigh quarter-wave limit, the loss of contrast is not very serious; in fact, only a very small proportion of existing microscope and telescope object-glasses are corrected within that limit. We may therefore adopt it as a safe guide, but we should never draw upon it to save a little extra trouble in perfecting a permanent design. It should be used only as a criterion for fixing the maximum allowable aperture or field of any given system, when the best obtainable state of correction has been determined by a careful and complete study of the possibilities.

From the designer's point of view, the chief requirement is to know the relation between the Rayleigh limit and the residuals of aberration found in the usual methods of calculation so that any result may be interpreted with the least amount of trouble. These relations will now be given.

THE FOCAL RANGE OF PERFECT OPTICAL SYSTEMS

A perfect optical system has no differences of optical path at the exact focus at which the image of the conjugate object-point is formed. At that point it therefore gives a spurious disk of the full intensity following from Airy's calculations for such a perfect system. There will then be a certain range of focal adjustment near the exact focus within which the differences of optical path do not exceed the Rayleigh limit, and this is the focal range which we have already found very useful in Part I.

As in this case there are no differences of optical path at the exact focus, it is evident that the whole of the imperfection at any changed focus is represented by the $\delta OPD'$ of OP(4)* on page 623 and that we can find the focal range by determining the limits at which the $\delta OPD'$ reaches the Rayleigh limit of $1/4$ wave-length. Moreover, our assumed perfectly-corrected pencil will be symmetrical around its central ray and we need therefore only to inquire about longitudinal shifts of

[97] the focus and can omit the terms in $\delta H'$ and $\delta z'$ which refer to transverse shifts. Hence we have to solve

$$\delta OPD' = \tfrac{1}{2}N' \frac{\delta l' \cdot PA^2}{l'(l'+\delta l')}$$

for $\delta l'$ and find

$$\delta l' = \frac{2 \cdot \delta OPD'}{N'\left(\dfrac{PA}{l'}\right)^2 - \dfrac{2 \cdot \delta OPD'}{l'}}.$$

As the value of chief interest is that for $\delta OPD' = 1/4$ wave-length, it is obvious that the second term in the denominator will practically always be very small compared with the first term and may be neglected. Putting also $PA/l' = \sin U'$, which will nearly always be near enough, and $\delta OPD' = 1/4$ wave-length, we find

$$\delta l' = \pm \frac{\tfrac{1}{2}\text{ wave-length}}{N'\sin^2 U'}$$

as the distance from the exact focus at which the differences of optical path reach the Rayleigh limit. As this distance applies in either direction from the exact focus, the whole focal range will be $2\delta l'$ or

OT(1) Focal Range $= \dfrac{1\text{ wave-length}}{N'\sin^2 U'}.$

It will be convenient to discuss this equation together with several following ones, and we shall therefore proceed to the determination of the tolerances for spherical aberration.

TOLERANCE FOR PRIMARY SPHERICAL ABERRATION

[98] WE learnt in Part I that in many cases we have to admit a considerable amount of spherical aberration, chiefly of the primary type, either because simple systems do not allow it to be corrected or more often, as in practically all eyepieces and in many forms of photographic objectives, because a moderate amount of spherical aberration makes it possible to correct the oblique pencils much better. We shall now determine the tolerance corresponding to the Rayleigh limit on the assumption that pure *primary* aberration is present. We shall then have at the paraxial focus of such a system the relation determined in OP(4p), namely

(a) $$OPD'_p = \tfrac{1}{4}N'u'^2 \cdot LA',$$

which is convenient because it gives OPD' in terms of the longitudinal aberration, which we usually calculate in these cases. Again there will be symmetry of the pencil of rays with reference to the central or principal ray and only longitudinal shifts of the focus can come into question. Hence OP(4)* is again reduced to its first term:

$$\delta OPD' = \tfrac{1}{2}N'\delta l' \frac{PA^2}{l'(l'+\delta l')}$$

and here we may replace PA by the usual Y in spherical-aberration formulae and omit the $\delta l'$ in the denominator. At a point at distance $\delta l'$ from the paraxial focus, we shall therefore have [98]

$$OPD' = OPD'_p + \delta OPD' = \tfrac{1}{4}N'\left[u'^2 LA' + 2\delta l'\left(\frac{Y}{l'}\right)^2\right].$$

In the first term within the square bracket, we may safely put u' as sensibly equal to Y/l' because the systems to which the present investigation refers always have cones of rays of very moderate angular extent. The equation then becomes

(b) $$OPD' = \tfrac{1}{4}N'(Y/l')^2(LA' + 2\delta l').$$

It must be carefully remembered that LA' is the longitudinal aberration of the rays from the zone of semi-aperture Y, also that LA' is considered positive when the paraxial focus lies at the right of the marginal focus, while $\delta l'$ follows the sign-convention of analytical geometry and is therefore positive to the right of the paraxial focus. Equation (b) immediately yields two highly important conclusions:

(1) OPD' becomes zero if $\delta l' = -\tfrac{1}{2}LA'$, that is, midway between the paraxial and the marginal foci; at that point marginal and paraxial light therefore meet without any difference of phase.

(2) In order to reach the marginal focus, we evidently must make $\delta l' = -LA'$. This value put into (b) gives

$$OPD'_m = \tfrac{1}{4}N'(Y/l')^2(-LA') = -OPD'_p$$

by equation (a). This means that at the marginal focus the marginal light meets the paraxial light with precisely the same difference of optical path as at the paraxial focus, but of the opposite sign. In the case of primary spherical aberration, the paraxial and the marginal foci are therefore equally good from the physical point of view, that is, judged by the residual differences of optical path; but the first conclusion renders it evident that the midway point gives a very much better union of all the light than either the paraxial or the marginal focus.

Figure 130 will render the significance of these results perfectly clear. If B'_p represents the paraxial focus of a lens system, then in the absence of aberration we should have strictly spherical waves like p converging upon B'_p. If primary aberration is present, the real wave will depart from this ideal one according to a Y^4-law or will be, roughly, of the shape denoted by the heavy line in the figure. The real marginal ray will be normal to the real wave and will reach, say, B'_m. The ideal wave form for this focus would be that of spherical waves—like m—converging upon the marginal focus. Our second conclusion simply means that for any given aperture of a lens system afflicted with primary spherical aberration, the margin of the real wave always lies midway between the margins of the ideal waves for the paraxial and marginal foci respectively. And conclusion (1) means that the axial and marginal zones of the real wave are equidistant from the point midway between the two chief foci. Evidently the ideal wave for the mid-focus is a far closer fit to the real wave than either of the others and must lead to far smaller residual OPD-values. We can in fact easily conclude, without elaborate mathematics, that the mid-focus gives the absolute minimum of residual OPD. For if

[98] we struck a circle from a point a little to the left (in our diagram) of the mid-focus, that circle would obviously pass farther to the right of the hump in the intermediate zone of the real wave and would thus lead to increased *OPD* for that zone. If, on the other hand, we struck the circle from a point a little to the right of the

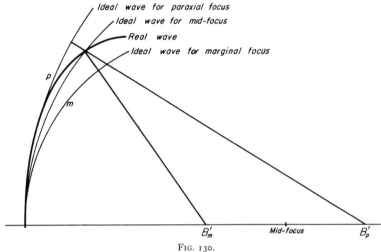

FIG. 130.

mid-focus, the circle (i.e., the ideal wave for that point) would certainly approach the 'hump' more closely, but it would simultaneously fall to the left of the margin of the real wave. As the gap between two circles increases in proportion to the square of the distance from the axis on which they are tangent to each other, the approach to the 'hump' will be less than the recession behind the margin and the maximum *difference* of optical path will again be greater than at the mid-focus. We therefore recognize the latter as the best possible compromise for a system suffering from primary spherical aberration, and it only remains to determine the maximum *OPD'* (the height of the hump of the real wave) for that mid-focus. For that purpose we return to the *general* equation (b).

If we call the longitudinal spherical aberration of our system for its extreme marginal zone LA'_m, we shall have to make $\delta l' = -\frac{1}{2}LA'_m$ to obtain the *OPD'* at the best possible focus midway between the paraxial and marginal ones. Hence the general equation will be, for the mid-focus.

$$OPD' = \tfrac{1}{4}N'(Y/l')^2(LA' - LA'_m),$$

and we have to determine the maximum or minimum (for negative *OPD'*) value of this function. The aberration LA' for any intermediate value Y of the semi-aperture grows as Y^2; hence we have, if the full marginal semi-aperture is denoted by Y_m,

$$LA' = LA'_m \cdot Y^2/Y_m{}^2$$

or

$$OPD' = \tfrac{1}{4}N'(Y/l')^2 \cdot LA'_m(Y^2/Y_m{}^2 - 1),$$

which by a slight rearrangement becomes   [98]

(c) $OPD' = \frac{1}{4}N'(LA'_m/l'^2)(Y^4/Y_m^2 - Y^2).$

In this equation Y is the only variable, hence

$$d(OPD')/dY = \frac{1}{4}N'(LA'_m/l'^2)(4Y^3/Y_m^2 - 2Y) = 0$$

for maximum or minimum, and we have the condition

$$4Y^2/Y_m^2 - 2 = 0 \quad \text{or} \quad Y^2 = \frac{1}{2}Y_m^2$$

as the solution.

Putting this value of Y^2 into equation (c) we find the maximum difference of optical path at the mid-focus to be

Maximum $OPD' = \frac{1}{4}N'(LA'_m/l'^2)(\frac{1}{4}Y_m^2 - \frac{1}{2}Y_m^2) = -\frac{1}{16}N'(Y_m/l')^2 \cdot LA'_m,$

which is exactly one quarter of the difference of optical path at either the paraxial or the marginal focus. It is very remarkable that this great reduction in the value of OPD' at the mid-focus completely escaped previous investigators, including Lord Rayleigh, and that as a consequence they put the limit of admissible spherical aberration at one quarter of its true value.

Putting the maximum OPD' equal to $1/4$ wave-length and introducing again Y/l' as being nearly enough equal to $\sin U'_m$, we find the true tolerance from

$$1/4 \text{ wave-length} = \frac{1}{16}N' \sin^2 U'_m \cdot LA'_m,$$

whence

OT(2)   Permissible primary $LA'_m = 4$ wave-lengths$/N' \sin^2 U'_m.$

There is no need to retain the minus sign as only the absolute value of the residual OPD' is of interest; U'_m has been used to emphasize the fact that the extreme marginal convergence angle must be used in working out the tolerance.

TOLERANCE FOR SECONDARY AND ZONAL SPHERICAL ABERRATION

IN most types of instruments it is possible to correct the longitudinal spherical [99] aberration for the marginal rays. But owing to the presence of higher aberration the intermediate rays then display the zonal aberration, usually in the sense of under-correction, which we first detected in the $f/4$ telescope objective of section [2] in Part I. From our new point of view, this means that the higher terms of the OPD-series are of sensible amount. Except in the case of the higher powers of microscope objectives, it is usually safe to assume that only the term in Y^6 is of sufficient magnitude to be taken into consideration, and therefore we shall now discuss the case when the distortion of the waves emerging from an optical system is sufficiently nearly represented by the equation

(a) $OPD'_p = v_4Y^4 + v_6Y^6.$

[99] We can find the corresponding longitudinal aberration by OP(3) (page 615), but as for these small residuals of aberration it would be perfectly futile to aim at extreme precision, we shall use the second form. With $\sin U'$ set as being nearly enough equal to Y/l', the equation then becomes

$$LA' = \frac{l'^2}{N'Y} \cdot \frac{d(OPD')}{dY},$$

or with the above value of OPD',

$$LA' = \frac{l'^2}{N'} (4v_4 \cdot Y^2 + 6v_6 \cdot Y^4).$$

Systems in which the zonal aberration calls for attention are nearly always corrected so as to bring the extreme marginal rays exactly to the paraxial focus, and we can limit the subsequent discussion to that case without appreciably sacrificing the general utility of the results. This assumption means that LA' must be zero for the extreme marginal rays with semi-aperture Y_m and gives the condition

$$4v_4 \cdot Y_m^2 + 6v_6 \cdot Y_m^4 = 0,$$

which can be transposed into $v_4 = -1\tfrac{1}{2}v_6 Y_m^2$. Hence the equations for LA' and OPD'_p become, *for trigonometrically corrected systems only*,

(b) $$OPD'_p = -1\tfrac{1}{2}v_6 Y_m^2 Y^4 + v_6 Y^6,$$

(c) $$LA' = \frac{6l'^2}{N'} (-v_6 Y_m^2 Y^2 + v_6 Y^4).$$

We arrive at a full understanding of the present case most quickly and most instructively if we determine the wave form resulting from equation (b), after transforming the latter by the identity $Y \equiv Y_m \cdot \dfrac{Y}{Y_m}$ into

$$OPD'_p = -v_6 Y_m^6 [1\tfrac{1}{2}(Y/Y_m)^4 - (Y/Y_m)^6].$$

Here, as comparison with equation (a) shows, we have outside the square bracket the full amount of the secondary wave distortion $v_6 Y^6$ at the margin, but with reversed sign; consequently the square bracket alone indicates the *fraction* of the full marginal secondary wave distortion which is effective at every point of the clear aperture. We easily find from Barlow's tables by looking up the square and cube of $(Y/Y_m)^2$:

(Y/Y_m)	= 0.2	0.4	0.5	0.6	0.7	0.8	0.9	1.0	1.1
$1\tfrac{1}{2}(Y/Y_m)^4$	= 0.002	0.038	0.094	0.194	0.360	0.614	0.984	1.500	2.196
$-(Y/Y_m)^6$	=		−0.004	−0.016	−0.047	−0.118	−0.262	−0.531	−1.000	−1.772

| $1\tfrac{1}{2}(Y/Y_m)^4 -$ $(Y/Y_m)^6$ | = 0.002 | 0.034 | 0.078 | 0.147 | 0.242 | 0.352 | 0.453 | 0.500 | 0.424 |

These are the amounts to which the OPD' at the respective values of Y/Y_m is proportional. Plotting these radially from an ideal spherical wave struck with the combined paraxial and marginal foci at $B'_p = B'_m$ as centre, we find the true wave form shown, with a maximum departure equal to $-\frac{1}{2}$ of the full marginal secondary [99] aberration from the ideal wave at the extreme margin. Fig. 131. The general forward tilt of this real wave at once suggests that there must be a point closer to the real wave which is more nearly equidistant from all zones of it. It can be

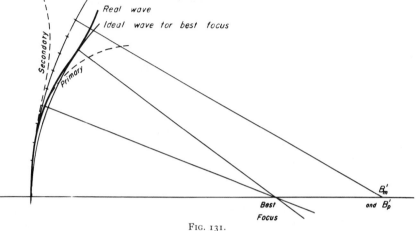

FIG. 131.

shown that the best focus corresponds to a term $-\frac{9}{16}(Y/Y_m)^2$ added to the above square bracket, which gives for the respective values of Y/Y_m

$$-\frac{9}{16}(Y/Y_m)^2$$
$$= -0.022 \quad -0.090 \quad -0.141 \quad -0.202 \quad -0.276 \quad -0.360 \quad -0.456 \quad -0.562 \quad -0.681$$

Hence $-\frac{9}{16}(Y/Y_m)^2 + 1\frac{1}{2}(Y/Y_m)^4 - (Y/Y_m)^6$
$$= -0.020 \quad -0.056 \quad -0.063 \quad -0.055 \quad -0.034 \quad -0.008 \quad -0.003 \quad -0.062 \quad -0.257$$

This shows that at the best focus thus selected, only 1/16 of the full marginal secondary wave-distortion is effective as the residual difference of optical path.

We can prove this directly by discussing the complete formula

$$\text{Best-focus } OPD' = -v_6 Y_m{}^6 \left[-\frac{9}{16}(Y/Y_m)^2 + 1\frac{1}{2}(Y/Y_m)^4 - (Y/Y_m)^6 \right]$$

in the usual way by determining maximum and minimum *and* the value at the margin for $Y/Y_m = 1$. We easily find that it gives $OPD' = \frac{1}{16}v_6 Y_m{}^6$ for $Y = \frac{1}{2}Y_m$ (highest numerical value at half aperture) and the same value at full aperture, and that it gives the lowest numerical value $OPD' = 0$ for $Y = 0$ and again at $Y = \sqrt{\frac{3}{4}} \cdot Y_m$, where the best ideal wave touches the true distorted wave. It follows that the best focus lies at the point where the true rays from half aperture and from $\sqrt{\frac{3}{4}} = 0.866$ of the full aperture meet the optical axis. This once more shows a contradiction between physical and geometrical optics, for it was shown in Chapter II of Part I that the geometrical disk of least confusion in this case lies at the common focus of the rays from 0.346 and 0.938 of the full aperture. In the

[99] same way as in the case of the primary spherical aberration, it is easily shown that any departure in either direction from the best focus marked in Fig. 131 gives an increased difference between the longest and the shortest optical path and that we are therefore justified in claiming it to be the best possible one.

The 'tolerance' corresponding to the Rayleigh quarter-wave limit takes different forms in the case of zonal aberration according to the method of determining the aberration:

(1) We may determine the OPD'_m by the strict trigonometrical equation, OP(4). As we are restricting our present discussion to the case when the marginal ray is brought accurately to the paraxial focus, i.e., when $LA'_m = 0$, we have now $OPD'_m = OPD'_p$ and equation (b) gives for the marginal rays, or $Y = Y_m$,

$$OPD'_m = OPD'_p = -\tfrac{1}{2}v_6 Y_m{}^6.$$

This means that, the calculated marginal OPD' represents one half of the total secondary wave-distortion. On the other hand, we found that at the best possible focus 1/16 of the total secondary wave-distortion appears as the residual difference of optical path which under the Rayleigh limit may reach 1/4 wave-length. As the OPD' calculated for the united paraxial and marginal foci is eight times as large, it may reach $8 \times 1/4 = 2$ whole wave-lengths; hence

Permissible $OPD'_m = 2$ wave-lengths.

In section [94] we calculated OPD'_m for the 2-inch telescope objective as 1.54 wave-lengths approximately, hence by this method the objective is working at $1.54/2 = 0.77$ of the Rayleigh limit.

(2) We may calculate the *primary* OPD'_p by the much more convenient equation OP(4p) with our usual nominal paraxial angles resulting from beginning with $u_1 = \sin U_1$ or $y_1 = Y_1$. By equation (b) this will represent the first or primary term for $Y = Y_m$ or

Primary $OPD'_p = -1\tfrac{1}{2}v_6 Y_m{}^6.$

We see by comparison with the trigonometrical OPD'_m evaluated above that the primary OPD'_p is three times as large as the OPD'_m, hence the tolerance will also be three times as large or

Permissible primary $OPD'_p = 6$ wave-lengths.

Direct calculation of primary OPD'_p by equation OP(4p) with the data of section [21] of Part I will be found to give approximately 4.07 wave-lengths, so by this criterion the 2-inch object-glass would be working at $4.06/6 = 0.68$ of the Rayleigh limit. The imperfect agreement with the value of 0.77 found by computing OPD'_m is directly explained by the fact that the 3-to-1 ratio of primary OPD'_p to OPD'_m holds strictly when primary and secondary aberration *alone* are present in a system. In reality, systems which display considerable zonal aberration always have also very sensible amounts of tertiary aberration, as was proved for the 2-inch object-glass in section [31] of Part I, and the disagreement referred to may always be taken as indicative of sensible tertiary aberration. In such cases the OPD'_m method is the safer one.

(3) We can determine the zonal aberration geometrically by tracing a ray through [99] 0.7071 of the full aperture and finding the maximum LZA'. To find the tolerance for this measure, we have to discuss equation (c). In the first place, determination of the maximum value gives, in agreement with the result obtained in Part I, 0.7071 Y_m as the zone for which LA' becomes largest. If we now put $Y = 0.7071 Y_m$, or $Y^2 = \frac{1}{2} Y_m{}^2$ into equation (c), we find

$$LZA' = 6 \frac{l'^2}{N'}(-\tfrac{1}{2}v_6 Y_m{}^4 + \tfrac{1}{4}v_6 Y_m{}^4) = -1\tfrac{1}{2}\frac{l'^2}{N'} v_6 Y_m{}^4 = -1\tfrac{1}{2}\frac{l'^2}{N'Y_m{}^2} v_6 Y_m{}^6,$$

and if we now reintroduce $Y_m/l' = \sin U'_m$ we obtain

$$LZA' = -1\tfrac{1}{2}v_6 Y_m{}^6/N' \sin^2 U'_m.$$

The numerator agrees with the value found above for primary OPD'_p and may accordingly reach 6 wave-lengths, whence

OT(3)   Permissible $LZA' = 6$ wave-lengths$/N' \sin^2 U'_m$.

For the 2-inch telescope objective, $\sin^2 U'_m$ was found to be 0.127, whence

       Permissible $LZA' = 0.00012/0.0162 = 0.0074$.

By the five-figure calculation in section [21] of Part I we found $LZA' = 0.0050$ or 0.675 of the Rayleigh limit. The more precise six-figure result used in section [31] gave $LZA' = 0.0056$ or 0.76 of the Rayleigh limit, in very close agreement with the result by the totally independent OPD'_m method.

(4) Finally, we can detect the zonal aberration by calculating the primary longitudinal spherical aberration from (10)** or one of its modifications. We can derive the corresponding tolerance directly from the relations of the different measures of the primary aberration collected in OP(4p), which include

$$\text{Primary } OPD'_p = \tfrac{1}{4}N'u'^2 \cdot LA'_p$$

or, transposed:

$$LA'_p = 4OPD'_p/N'u'^2.$$

We proved in (2) that primary OPD'_p may reach 6 wave-lengths, and as lens systems in which the zonal aberration is worth discussing are sure to satisfy the sine condition quite closely, we may also replace u'^2 by $\sin^2 U'_m$ and obtain the tolerance equation

OT(3)*   Permissible primary $LA'_p = 24$ wave-lengths$/N' \cdot \sin^2 U'_m$.

For the 2-inch telescope objective this gives, with $\sin U'_m = 0.127$

       Permissible $LA'_p = 0.0048/0.0162 = 0.0297$.

The actual calculation of the primary aberration in section [28] of Part I gave 0.0205, well inside the limit, for $0.0205/0.0297 = 0.69$ of the Rayleigh limit. This agrees almost exactly with the 0.68 found by primary OPD'_p, as indeed must be the case on account of the fixed relations OP(4p) between all measures of the primary aberrations.

[99]   The first method, by direct trigonometrical calculation of OPD'_m, is the most dependable one, and as it is less laborious than the calculation of LZA', it may well supersede this method, which in other respects comes next in reliability.   In all ordinary cases of telescope and low-power microscope objectives, sufficiently close results will be obtained by the primary OPD'_p method because the division by $N'\sin^2 U'_m$ is saved.

TOLERANCE FOR TERTIARY SPHERICAL ABERRATION

[100]   IN all ordinary lens systems, it is unnecessary to take aberrations higher than the secondary into account if the method consistently adhered to in the present work is adopted, that is, if the trigonometrical determination of the exact longitudinal spherical aberration of the extreme marginal ray is made the principal basis of the discussion.   The only exception, but a highly important and emphatic one, is represented by high-power microscope objectives exceeding about o.5 in NA. In these objectives, angles of incidence and of convergence exceeding 60° are reached at certain surfaces, and tertiary or even higher aberrations are likely to reach a decidedly serious magnitude and must therefore be searched for and corrected within the Rayleigh limit.

If a system of this exceptional type is submitted to our usual tests for zonal aberration according to the preceding subsection, the need of higher correction will reveal itself by an inadmissible magnitude of the zonal aberration; and we thus realize that the real problem is to correct or at least to greatly diminish the zonal aberration.   The usual purely trigonometrical way of meeting this requirement consists in tracing a paraxial, a marginal, *and a zonal* ray through the system and in modifying the design until these *three* rays come to a common focus.   This method is not only most emphatically a laborious one, but it leaves a grave uncertainty; for while we may confidently claim that a system thus corrected must be superior to one in which the geometrical zonal aberration is left at some considerable magnitude, we cannot be sure whether the residual differences of optical path for the *whole* of the clear aperture are below the Rayleigh limit.   For these reasons, we will adopt another criterion which has stood the test of a quarter of a century of practical application to every type of microscope objective.   We shall stipulate that in addition to correction of the longitudinal spherical aberration for the marginal zone there shall also be no difference of optical path at the common focus of the paraxial and the marginal rays.   Our conditions therefore are:

$$\text{Trigonometrical } LA'_m = \text{o}$$

$$\text{Trigonometrical } OPD'_m = \text{o};$$

together they assure that the light from the central part of the clear aperture and that from the marginal zone cooperate perfectly.   But before accepting even this correction as sufficient, we must search for the maximum residual differences of optical path in the intermediate zones.

A reasonably simple discussion is possible only if we assume that only Y^8 aberration is effectively present in addition to the primary and secondary aberration. As this is certainly not correct, especially in systems of the highest NA, our results will require caution when they are applied in practice, for it is easily shown by a modification of the arguments employed in the following paragraphs that, if the uncontrollable higher aberration is made up of a mixture of Y^8, Y^{10}, &c. terms, then the residual phase differences at the focus will tend to be larger than those which we shall find under our simplifying assumption.

We accordingly assume that the first three terms of the aberration series are of sensible magnitude and that the OPD's are represented by the equation

$$OPD'_p = v_4 Y^4 + v_6 Y^6 + v_8 Y^8.$$

The condition that the longitudinal aberration of the marginal ray shall be zero then again leads to the demand that $d(OPD'_p)/dY$ shall be zero for $Y = Y_m$ or, from the preceding equation,

$$4v_4 Y_m^3 + 6v_6 Y_m^5 + 8v_8 Y_m^7 = 0.$$

The condition that the difference of optical path with which the marginal light meets the paraxial light shall be zero gives the second equation,

$$v_4 Y_m^4 + v_6 Y_m^6 + v_8 Y_m^8 = 0,$$

and dividing the first equation of condition by $4Y_m^3$ and the second by Y_m^4 we obtain

$$v_4 + 1\tfrac{1}{2}v_6 Y_m^2 + 2v_8 Y_m^4 = 0$$
$$v_4 + v_6 Y_m^2 + v_8 Y_m^4 = 0.$$

These two equations can be solved for v_4 and v_6 in terms of v_8 and Y_m to give

$$v_4 = v_8 Y_m^4, \qquad v_6 = -2v_8 Y_m^2,$$

and putting these values of v_4 and v_6, which embody our two conditions to be fulfilled, into the general equation for OPD'_p, we obtain

$$OPD'_p = v_8 Y_m^4 \cdot Y^4 - 2v_8 Y_m^2 \cdot Y^6 + v_8 \cdot Y^8$$

as the general equation of the residual OPD with which light from all parts of the clear aperture reaches the combined marginal and paraxial focus when our *two* conditions are fulfilled.

The equation is easily discussed with regard to the maximum and minimum values of OPD'_p by the usual method of forming $d(OPD'_p)/dY$ and equating this to zero. This gives

$$4v_8 Y_m^4 \cdot Y^3 - 12v_8 Y_m^2 \cdot Y^5 + 8v_8 \cdot Y^7 = 0$$

or on division throughout by $8v_8 Y_m^4 \cdot Y^3$:

$$\tfrac{1}{2} - 1\tfrac{1}{2}\left(\frac{Y}{Y_m}\right)^2 + \left(\frac{Y}{Y_m}\right)^4 = 0.$$

The solution of this quadratic equation in $(Y/Y_m)^2$ gives the values for which OPD'_p is a maximum or a minimum as $(Y/Y_m)^2 =$ either 1 or $\tfrac{1}{2}$, and the forming of

[100] the second differential coefficient of OPD'_p shows that, with a positive value of v_8, there will be a maximum at $Y = 0.7071\ Y_m$ and a minimum at $Y = Y_m$, and the reverse for a negative v_8. In either case, the value of OPD'_p for $Y = 0.7071$ Y_m is $\frac{1}{16}v_8 Y_m{}^8$ or $1/16$ of the whole eighth-order aberration present in the marginal zone, and of course zero for the marginal zone itself. The biggest departure from the ideal spherical wave form is therefore at $Y = 0.7071\ Y_m$; as in accordance with the Rayleigh limit this biggest departure may reach $1/4$ wave-length and as it was found to represent $1/16$ of the total Y^8 aberration in the margin, it follows that the latter may be allowed to reach 4 whole wave-lengths.

A diagram again gives an immediate and complete idea of the state of affairs (Fig. 132) and shows that, while there is no possibility in this case of finding a shifted focus at which the residual OPD would be smaller, there is the possibility of increasing the working aperture to the marked point, at which the distorted wave, on

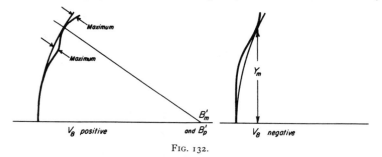

FIG. 132.

bending away from the ideal one in accordance with the existence of a maximum or minimum at Y_m, regains the same distance from the ideal wave, and in the same sense, as at $Y = 0.7071\ Y_m$. This enlarged aperture may be determined by putting $OPD'_p = \frac{1}{16}v_8 Y_m{}^8$, the height of the hump at $Y = 0.7071\ Y_m$, into the equation

$$OPD'_p = v_8 Y_m{}^4 \cdot Y^4 - 2v_8 Y_m{}^2 \cdot Y^6 + v_8 \cdot Y^8 = \frac{1}{16}v_8 Y_m{}^8$$

and solving the quartic equation in Y^2. This gives $Y = 1.10\ Y_m$ as the semi-aperture to which we may go without an increase in the residual maximum OPD. We therefore have the highly important rule for the calculation of systems (almost exclusively microscope objectives exceeding $0.5\ NA$):

'To secure the most perfect compensation of zonal aberration, the trigono-metrical calculation should be carried out for a submarginal ray at $10/11$ of the intended clear aperture, and LA'_m and OPD'_m [by OP(4)] should be brought to zero for this submarginal zone.'

The residual greatest OPD for the full aperture will then be $1/16$ of the Y^8-aberration for the computed submarginal zone. The resulting advantage is far larger than would be thought at first sight; for if the same calculation were carried out for the extreme marginal zone instead of $10/11$ of it, we should have a residual OPD equal to $1/16$ of the Y^8-aberration at the extreme margin, and this would be

$(1.1)^8$ or 2.14 times as large as at the submarginal zone, namely, $8\frac{1}{2}$ wave-lengths for our Rayleigh limit case. When working by the rule, we shall therefore have only $1/(16 \times 2.14)$ or $1/34$ part of the Y^8-aberration at the extreme margin appearing as residual OPD at the combined paraxial and submarginal focus.

Finally we must learn to estimate the magnitude of the maximum residual OPD of systems computed by the above rule. A simple method results from the observation that in accordance with our equation for the wave distortion

$$OPD'_p = v_8 Y_m^4 Y^4 - 2v_8 Y_m^2 Y^6 + v_8 Y^8,$$

the primary aberration expressed by the first term becomes exactly equal to the tertiary aberration expressed by the third term for $Y = Y_m$, that is, for the computed submarginal zone. (The secondary aberration is twice as great with opposite sign.) Hence the maximum residual OPD will also be $1/16$ of the primary aberration for the computed aperture. Now we have already learnt in the preceding section how to compute this primary aberration by $OP(4p)$, using the fictitious large paraxial angles. Hence the second rule:

'To determine the residual OPD, calculate $OP(4p)$ for every surface of the system and form the algebraic sum. If the latter amounts to 4 wave-lengths, the correction will be just on the Rayleigh limit', or

$OT(4)$ In systems corrected for zonal aberration, the primary aberration calculated by $OP(4p)$ may reach 4 wave-lengths.

Owing to the probable presence of still higher aberrations, which in general would tend to give a less favourable correction, every effort should be made to keep within about one half of the theoretical tolerance expressed by $OT(4)$.

An almost absolutely safe method of finding the maximum residual OPD' of a system corrected for $LA'_m = 0$ and $OPD'_m = 0$ would consist in tracing a ray completely through the $0.7071 Y_m$ zone and to determine OPD'_z directly by the trigonometrical equation $OP(4)$. If this came out below $1/4$ wave-length, we could be completely satisfied that the Rayleigh limit was not exceeded. The *whole* calculation for this zonal ray and its OPD'_z could be carried out by four-figure logs. The zonal ray ought to be found to arrive very nearly at the combined paraxial and marginal focus, owing to the maximum or minimum of wave-distortion for the 0.7071 zone; but if, owing to Y^{10} and still higher terms, the ray showed sensible spherical aberration at the final focus, the $\delta OPD'$ correction by $OP(4)$ would have to be applied with this aberration as $-\delta l'$, the minus sign being necessary because LA' and $\delta l'$ are measured in opposite directions from the paraxial focus.

CHAPTER XV

CHROMATIC ABERRATION AS AN *OPD*

[101] THE optical-path method leads to a particularly simple treatment of the problem of establishing achromatism at the axial focal point of a lens system.

In Fig. 133, let O represent any lens system which refracts light from an object-point B towards an image-point B'. Physically considered, point B will be sending out spherical waves of light towards the lens system, and the latter will alter the curvature of these waves so as to direct them towards the image-point B'. In the ideal case, the refracted waves W' would be truly spherical with B' as centre. Actually they will be always more or less distorted owing to the inevitable residuals of spherical aberration, and it is important to notice that we make no restriction as to the amount of this distortion of the refracted waves. The formulae which we shall deduce will therefore remain valid however great the spherical aberration of a system may be.

Chromatic aberration, from our present point of view, consists in this: That the

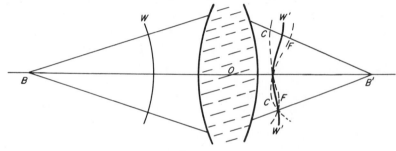

Fig. 133.

refracted waves corresponding to some other colour will have greater or less mean curvature than W' when tangent to it at the optical axis. Perfect achromatism would exist if it were possible to cause waves corresponding to other colours to coincide absolutely with W' when passing through the same point of the axis. As this perfect coincidence is practically unobtainable, we adopt a compromise, by stipulating that the refracted wave of a wave-length slightly differing from that of W' shall, when passing the point where W' cuts the optical axis, simultaneously intersect with W' at the margin of the segment which the lens system allows to pass.

By way of example, let the full-drawn marginal ray in the upper part of Fig. 133 represent the brightest light and W' the emerging wave of that colour. If the lens system were chromatically under-corrected, then red (say C) light would come to a longer focus beyond B' and blue (say F) light to a shorter focus at the left of B'. The emerging waves of these colours would have respectively less and more mean curvature than the wave of brightest light, as indicated by the thinner curves marked C and F. Our new definition of achromatism is illustrated in the lower part of

Fig. 133.   We demand that the waves of other colours shall intersect each other at   [101]
the margin of the clear aperture.   In the absence of the secondary spectrum they
would do so on W'.   If the secondary spectrum is appreciable they will intersect
a little to the left of the margin of W', for we found that the secondary spectrum
lengthens the focus of the more remote colours.   Nevertheless, for light differing
but slightly in wave-length and colour from the brightest light this secondary
departure will be inappreciable, as we found geometrically that the secondary
spectrum only shows itself for relatively large differences of colour.   The two
waves of colours differing from that of brightest light will in general be incapable of
being brought to perfect coincidence throughout the aperture because the spherical
(almost entirely primary) aberration varies for different colours; there is therefore
a gap between them in the intermediate zones, with a maximum close to 0.7071 of
the full aperture.

This physical condition of achromatism assumes a geometrical form suitable for
calculation if we introduce the 'optical paths' passing between the wave-fronts at
W and W'.   By the definition of wave-front these optical paths are exactly equal.
They are easily calculated by multiplying each section of the total path by the
refractive index of the medium through which it passes and thus converting it into
the equivalent optical path.   For these successive sections, we introduce the
symbol d' in the case of the axial thickness and D' in the case of inclined thicknesses
at a distance from the optical axis.   If we extend this nomenclature to the portions
of the total path between W and the first lens surface, and between the last lens
surface and W', we shall have the path along the axis equal to $\Sigma d' \cdot N'$ and that
along the marginal ray equal to $\Sigma D' \cdot N'$; these being equal between the conjugate
waves W and W', we have

(a) $$\Sigma(d' - D') \cdot N' = 0$$

if the sum is extended to all the lens-shaped spaces between W and W'.

The chromatic condition which we have formulated amounts to demanding that
in addition

(b) $$\Sigma(d' - D') \cdot (N' + dN') = 0$$

if dN' is taken as the small increase in the refractive index corresponding to a small
decrease in wave-length.   The difference of the two sums (a) and (b),

OP(5) $$\Sigma(d' - D') \cdot dN' = 0,$$

is the mathematical expression which represents our chromatic condition.

We must, however, pause to examine the validity of the implied assumption that
the part $\Sigma(d' - D') \cdot N'$ contained in (b) is really equal to (a).   Apparently this is
an unsound assumption, as we know that spectral decomposition of the light will
assign slightly different paths to light according to its wave-lengths, so that the
D' would be different for every change of wave-length.   The justification for
neglecting this difference is provided by Fermat's principle of the 'minimum
optical path' proved in section [90].   This principle says that paths between two
given points, differing by small separations and angles of the first order, are equal
to each other within small quantities of the second order.   If we limit our condition

[101] to a small range of wave-lengths, Fermat's principle allows us to claim OP(5) as an exact equation.

The method of computing D' has been given in section [14] of Part I under equations (8) and (9), and our problem is therefore completely solved by the above simple equation OP(5).

As it is usual in all optical calculations to treat air as having an index $N = 1$ exactly and a dispersion $dN = 0$, the sum has to be worked out only for the actual lenses, air-spaces being ignored. The work involved is thus extremely modest in amount, especially so because the equation gives the aberration practically direct and not as the small difference of large quantities, which is what the trigonometrical calculation does. Four-figure logs, or even a good slide rule, are therefore quite sufficient for evaluating the aberration.

The value of dN' for the successive glasses should really be that for a very small change of wave-length from that of the brightest rays for which the main calculation has been made. But as we have here another case (like that of the paraxial calculations by the standard formulae) where *all* the separate items in the total sum will be uniformly multiplied by the same number k, if we use throughout $k \cdot dN'$ instead of dN', we shall prefer to use a number of the order of the actual dispersion for the important section of the spectrum. By long experience, in the case of visual instruments, it has been found convenient to take the dispersion between the C and F lines, which is always given by the glass-makers. For ordinary work, we shall therefore work out equation OP(5) by the direct use of the respective values of $(N'f - N'c)$ instead of the differential dN'.

This $(d' - D')$ method of effecting chromatic correction admits of a particularly simple and perfectly rigorous solution of the problem of determining the value of the last radius of a lens combination *working in air* which will make the system achromatic: up to the time when I worked out the nearly exact solution of this problem given in Part I, it was the only direct solution in existence. It still remains the only one in the case of microscope and photographic lenses of such pronounced curvatures, and consequent non-fulfilment of the assumptions underlying the trigonometrical method, that the latter would break down.

Assume that a system of k surfaces has been calculated up to and including the last-but-one surface and that the $(d' - D')dN'$ sum has also been determined up to that point. The problem then is to determine the value of the last radius r_k which will give the desired chromatic correction. Usually it will be desired that the $(d' - D')dN'$ sum for the whole system shall be zero; but as cases often arise where a certain amount of chromatic over- or under-correction is required, as for instance when a telescope objective is to be adapted to an ordinary non-achromatic eyepiece of the two-lens or the four-lens type, we shall make the solution more general by assuming that it is required to make

OP(5)* $$\sum_{1}^{k-1}(d' - D')dN' = Chr.,$$

where *Chr.* is the symbol for the required under- or over-correction, positive for the former, negative for the latter.

Referring to Fig. 134, it is assumed that a marginal ray has been traced through [101] the first $(k-1)$ surfaces of the system and that the data of the ray emerging from the $(k-1)$st or last-but-one surface are known, and that the axial thickness d'_{k-1} of the last lens has been decided upon. We further assume that $\sum_{1}^{k-2}(d'-D')dN'$ has been calculated for all the lenses to the left of surface $(k-1)$. OP(5)* then yields the equation, simply by writing the last item of the total sum separately,

$$\sum_{1}^{k-2}(d'-D')dN' + (d'_{k-1}-D'_{k-1})\cdot dN'_{k-1} = Chr.,$$

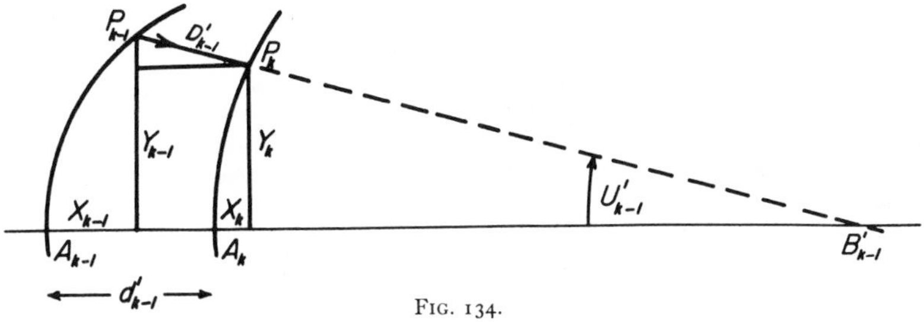

FIG. 134.

in which D'_{k-1} is the only unknown quantity. It is given explicitly by a simple transformation as

$$D'_{k-1} = d'_{k-1} + \left(\sum_{1}^{k-2}(d'-D')dN' - Chr.\right)/dN'_{k-1}.$$

This enables us to calculate the required marginal thickness of the last lens. By (9) we then have

$$D'_{k-1} = (d'_{k-1}+X_k-X_{k-1})\cdot \sec U'_{k-1}.$$

In this equation X_k is the only unknown quantity, and it is found by transposition as

$$X_k = D'_{k-1}\cdot \cos U'_{k-1} + X_{k-1} - d'_{k-1}.$$

From the small triangle below D'_{k-1} in Fig. 134 we then obtain

$$Y_k = Y_{k-1} - D'_{k-1}\cdot \sin U'_{k-1},$$

in which Y_{k-1} is obtained by (8)* (Part I, page 29) as

$$Y_{k-1} = PA\cdot \cos \tfrac{1}{2}(U+I) = r\cdot \sin (U+I),$$

the first form being the more convenient and accurate one when the PA-check has been used.

As X_k and Y_k of the last surface are then known, its radius follows from the spherometer equation

$$r_k = Y_k^2/2X_k + \tfrac{1}{2}X_k$$

[101] and the solution is complete. We shall collect the computing formulae as:

OP(5)**

$$
\begin{cases}
D'_{k-1} = d'_{k-1} + \left(\sum_{1}^{k-2}(d'-D')dN' - Chr. \right) \Big/ dN'_{k-1} \\
X_k = D'_{k-1} \cdot \cos U'_{k-1} + X_{k-1} - d'_{k-1} \\
Y_{k-1} = PA \cdot \cos \tfrac{1}{2}(U+I) = r \cdot \sin (U+I) \\
Y_k = Y_{k-1} - D'_{k-1} \cdot \sin U'_{k-1} \\
r_k = Y_k{}^2 / 2X_k + \tfrac{1}{2}X_k.
\end{cases}
$$

We shall take as a short numerical example of the $(d-D)$ method the telescope objective (Example 2) of Part I for which the complete calculation of a marginal ray, a zonal ray, and the paraxial ray is given in Part I, page 51. For the $(d-D)$ method we only require the data of the extreme marginal ray. We first test the chromatic correction by the direct application of the $(d-D)dN$ sum to the data of the objective that were actually used. The values of X have first to be calculated as $X = \tfrac{1}{2}PA^2/r$. All the data are available from the ray-tracing for the marginal ray and give:

	First surface	Second surface	Third surface
$\log \tfrac{1}{2}$	9.69897	9.69897	9.69897
$+ 2 \log PA$	0.00888	0.00776	9.97768
$+ \text{co-log } r$	9.44977	9.44977n	8.22185n
$\log X$	9.15762	9.15650n	7.89850n

The corresponding X are

$X_1 = \quad 0.14375$	$X_2 - X_1 = \ -0.28713$	$X_3 - X_2 = 0.13546$
$X_2 = -0.14338$	$d'_1 = \ +0.30000$	$d'_2 = 0.20000$
$X_3 = -0.00792$	$d'_1 + X_2 - X_1 = \quad 0.01287$	$d'_2 + X_3 - X_2 = 0.33546$

D'_1 and D'_2 are then found by multiplication into sec U' within each lens. The U' values are $5° \ 39' \ 25''$ and $4° \ 6' \ 48''$ respectively. Hence

	First lens	Second lens
$\log (d'_{k-1} + X_k - X_{k-1})$	8.10958	9.52564
$+ \log \sec U'_{k-1}$	0.00212	0.00112
$\log D'_{k-1}$	8.11170	9.52676
d'_{k-1}	0.30000	0.20000
$- D'_{k-1}$	-0.01293	-0.33633
$d'_{k-1} - D'_{k-1}$	0.28707	-0.13633

By the difference of the indices given in Example 2a (page 53 of Part I) we find

$Nf - Nc = dN'$ to be 0.00817 for the crown and 0.01720 for the flint.   The [101]
$(d - D)$ sum is therefore

$$0.28707 \times 0.00817 \text{ for the crown } = \quad 0.0023454$$
$$\text{and } -0.13633 \times 0.01720 \text{ for the flint } = \quad -0.0023449$$

$$\therefore \ \Sigma(d' - D')dN' = \quad 0.0000005$$

We shall learn that the residue is only one twentieth of the tolerance for chromatic aberration and therefore utterly negligible. We may also note that it would vanish if log dN' for flint were higher by 9 units in the fifth place to make it 8.23562. The corresponding dN' would be 0.0172036; as glass-makers will not guarantee their values of dN within less than 2 or 3 units in the fifth place, we see that the uncertainty from imperfect knowledge of dN amounts to about 8 times the residual chromatic aberration calculated for our objective. The residual is therefore also immaterial from this point of view.

We shall next assume that our objective is being newly designed and that the calculation of the marginal ray in 'brightest light' had been carried through the second—or $(k - 1)$st—surface. We should then want to determine the last radius so as to secure achromatism, given dN'_2 for the flint as 0.01720 and the thickness d'_2 of the last lens as 0.2000; also $U'_{k-1} = 4\text{-}6\text{-}48$, $X_{k-1} = -0.14338$, $(U+I)_{k-1} = -16\text{-}20\text{-}23$, log $PA_{k-1} = 0.00388$ from the preceding calculation and from the calculation on page 51, Part I.

The quantity $\displaystyle\sum_{1}^{k-2}(d' - D')dN'$ represents merely the product for the crown lens and was just found as 0.0023454. *Chr.* is to be zero, as perfect achromatism is aimed at. The calculation by OP(5)** then runs, for the first equation

$$D'_{k-1} = 0.2 + 0.0023454/0.01720 \qquad \log 0.0023454 = \quad 7.37021$$
$$= 0.2 + 0.13636 \qquad\qquad -\log 0.01720 \ = \ -8.23553$$

$$= 0.33636 \qquad\qquad \log \text{ second term} = \quad 9.13468$$

For the second equation we then have

$$\log D'_{k-1} = 9.52680 \qquad \text{First term} = \quad 0.33549$$
$$+\log \cos U'_{k-1} = 9.99888 \qquad\quad +X_{k-1} = -0.14338$$
$$\qquad\qquad\qquad\qquad\qquad\quad -d'_{k-1} = -0.20000$$
$$\log \text{ first term} = 9.52568$$
$$\qquad\qquad\qquad\qquad\qquad\quad X_k = -0.00789$$

The third equation then gives

$$\log PA = 0.00388$$
$$+ \log \cos (8\text{-}10\text{-}11) = 9.99557$$

$$\log Y_{k-1} = 9.99945$$

[101] For the fourth equation we find

$$\log D'_{k-1} = 9.52680 \qquad\qquad Y_{k-1} = \quad 0.99874$$
$$+\log \sin U'_{k-1} = 8.85570 \qquad -\text{sec. term} = -0.02413$$

$$\log \text{second term} = 8.38250 \qquad\qquad Y_k = \quad 0.97461$$

We then find the last radius

$$2\log Y_k = \quad 9.97766 \qquad\qquad \text{First term} = -60.194$$
$$-\log(2X_k) = -8.19811n \qquad\qquad \tfrac{1}{2}X_k = -\ 0.004$$

$$\log \text{first term} = \quad 1.77955n \qquad\qquad r_k = -60.198$$

In Part I $r_3 = -60.000$ was used. It will be noted that the somewhat longer value resulting from the exact solution would tend to correct the positive residual which we found for the actual objective. Nevertheless, it would make no real difference in the quality of the correction if the long radius were changed by the small amount just found.

It is an exceedingly valuable feature of the $(d' - D')$ method of effecting chromatic correction that it requires no separate ray-tracing. It uses the data of the rays of brightest light. Now the present-day glass lists contain many glasses which have practically the same refractive index but a more or less substantial difference in dispersion. Hence, as far as the brightest light is concerned, we may use several different types of glass in a partly computed system without having to recompute for the brightest light; if the refractive index is substantially unchanged, there will be no serious change in the course of the rays of the brightest light. The change in dispersion, however, will alter the chromatic correction very markedly, and the application of the $(d' - D')dN'$ sum will show this with little trouble.

We take advantage of this in the designing of complicated systems such as microscope and photographic objectives. Estimating by approximate formulae the required proportion of crown to flint lens powers so as to make fairly sure that chromatic correction will ultimately be attainable by selecting suitable types of glass, we carry out the calculations for the brightest light alone until satisfactory correction of spherical aberration, coma, and perhaps also of astigmatism and curvature of field has been reached. We then calculate the values of D' for all the component lenses, form the difference $(d' - D')$ for all, and then ring the changes on the available types of glass until a combination has been found which results in a zero value—or a close approach to it—of the $(d' - D')dN'$ sum. With a little practice this simple process nearly always leads to a satisfactory result in a very short time, certainly in a minute fraction of the time that would be consumed in tracing a coloured ray through the whole system for each available combination of types of glass.

In order to secure the best possible prospects of complete success in this game, it is essential to select at least one refractive index which is found in the glass lists associated with various dispersions. $Nd = 1.516$ is one such value, obtainable with dispersions, well graded, from about 0.0080 to 0.0091 or even higher. Another is 1.565 or thereabouts; 1.58 yet another, and there are still more.   In flints we [101] have 1.62 and 1.65 obtainable either with comparatively low dispersion (barium flint) or with decidedly higher dispersion (ordinary flint).   There are thus many opportunities of establishing favourable conditions for the method, and these are greatly added to by the variations between different meltings of any one type.

In this connection the $(d' - D')$ method has another great advantage.   It enables us to estimate closely how nearly it is necessary to fulfil the condition of achromatism so that the residual chromatic aberration may be inappreciable in the use of the instrument.

Apart from the higher aberrations, which are usually small, our method determines the marginal gap between the blue and the red waves emerging from the system, in the same unit of length which we employ for the radii and intersection-length.   In observing with an instrument, we focus for the brightest rays midway between red and blue and thus only half the error (in opposite directions) becomes sensible at the visual focus.   Now any part of a wave converging towards a focus may be allowed to arrive there 1/4 wave-length out of phase with other parts of the wave, in accordance with Lord Rayleigh's rule, without serious detriment to the sharpness of the image.   Hence we may allow a marginal gap of 1/2 wave-length between the red and blue waves when using in our formula a value of the dispersion of the order of that between C and F; in other words, we may put $Chr.$ in formula OP(5)* equal to plus or minus 1/2 wave-length to cover this tolerance for a perfectly achromatic system.

Near the middle of the visible spectrum, 1/2 wave-length is about 0.000 01 inch or 0.000 25 mm., so the well-established Rayleigh limit of the necessary accuracy in the union of rays justifies us in laying down a 'tolerance' in the value of the $(d' - D')dN'$ sum: If it is within 0.000 01 inch or within 0.000 25 mm. of the desired value (either larger or smaller by this amount), the lens system will be nearly as good as it would be if the exact value of the sum were realized.   At first sight the minute lengths of the stated tolerance look as if they were hardly worth playing with.   But it must be remembered that $(d' - D')$ for any one component lens rarely exceeds 0.1 inch, while dN' for the interval from C to F is of the order if 0.01; hence the value of $(d' - D')dN'$ for any one lens is usually not greater (when inch-measure is used) than 0.001 and on this the tolerance of 0.000 01 inch amounts to 1 per cent. Suppose, therefore, that in a given system we had a lens for which $(d' - D')dN'$ came out to 0.001 and that the exact fulfilment of the chromatic condition called for a glass with $Nf - Nc = 0.00867$.   We should then be justified in using any available glass of the required refractive index if its dispersion were within 1 per cent. of the required value, say between 0.00859 and 0.00875, values which really cover the difference between different types of hard crown!   In the first example of the present section, we found the sum to be 0.000 000 5 inch or 1/20 of our tolerance.

When the $(d' - D')$ method is used according to OP(5)** to determine the last radius of a system, so as to establish chromatic correction, the 'tolerance' obviously permits a certain range in the value of the last radius.   Our D' may be the calculated value plus or minus the amount of the tolerance divided by dN'_{k-1}, and as for

[101] the minute quantity we are considering, the factor $\cos U'_{k-1}$ may be put as sufficiently nearly equal to unity, we shall have

$$\text{OT(5)} \begin{cases} \text{Tolerance on } X_k = \pm \dfrac{0.000\ 01 \text{ inch}}{dN'_{k-1}} \quad \text{or} \quad = \pm \dfrac{0.000\ 25 \text{ mm.}}{dN'_{k-1}} \\ \text{Tolerance on } \Sigma(d'-D')dN' = \pm 0.000\ 001 \text{ inch} \quad \text{or} \quad = \pm 0.000\ 25 \text{ mm.} \end{cases}$$

In the second numerical example used above, we find

$$\text{Tolerance on } X_3 = \pm \frac{0.00001}{0.01720} = \pm 0.00058.$$

As the exact value for X_3 was -0.00789, we conclude that the tolerance renders admissible any value of X_3 between -0.00731 and -0.00847 or any value of r_3 between -65.0 and -56.1, which confirms our conclusion that the change from -60.0 to -60.2 following from our solution would make no sensible difference in the chromatic correction.

It should be distinctly understood that it is not suggested that the tolerance which we have proved to exist should be regularly taken advantage of to anything approaching its full extent. That would rob the computer of one of his greatest safeguards, viz., that by calculating for a higher degree of perfection than is really necessary, he leaves a good margin for technical imperfections in the carrying out of his design so that the results will still remain good in spite of these small and mostly unconscious departures from the computed data. But there will frequently be cases, as for instance when a single objective is required at a not excessive cost, when a computer who remembers this chromatic tolerance will be able to produce a perfectly satisfactory result with stock curves which to a less able man would appear too far from the theoretical values to be admissible.

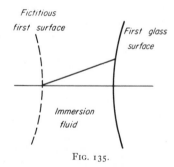

FIG. 135.

One additional point must be mentioned, as it occurs regularly in immersion objectives for the microscope. When the object or image lies in a medium other than air, then the interval between object or image and the first lens surface must also be treated as a lens and its $(d'-D')dN'$ value calculated. This requires no special formula because we can imagine a surface of any curvature we please laid through the object (or image) point; the space filled with the immersion fluid then becomes a lens (Fig. 135) and, as the rays start at the axial point, X for the imaginary first lens surface is necessarily zero.   Therefore we calculate by (9), putting d'_k [101] equal to the distance from object to first glass surface and X for the object side of the imaginary lens equal to zero.

The $(d'-D')$ method is just as applicable to centred systems of non-spherical surfaces of rotation (such as paraboloids, ellipsoids, and hyperboloids of rotation) as it is to the spherical surfaces which we nearly always employ.   The only change in calculating D' is that we must employ the proper formula to find X, instead of that for spherical surfaces.

PARAXIAL FORM OF THE $(d'-D')$ EQUATION

As was pointed out in the early part of this chapter, it is usually not possible to attain perfect chromatic correction for all zones of a given lens system; there is a zonal variation of the chromatic aberration just as there is in the case of spherical aberration and of fulfilment of the sine condition.   The zonal variation of chromatic aberration is accounted for by the fact that the spherical aberration of all ordinary lens systems varies with wave-length; if it is corrected for the brightest light, it is not as a rule also corrected for light of a different colour.   It follows that we shall detect this zonal chromatic aberration most easily in the same way in which we detect the spherical aberration, that is, by calculating the chromatic aberration separately for the full aperture and for the paraxial region and comparing the results.

When using the $(d'-D')$ method we can do this with very little trouble by equation (9p).   We found in Part I, page 38:

(9p)   Paraxial $D'_k = d'_k(1 + \frac{1}{2}u'^2_k) + x_{k+1} - x_k.$

Deducting this from $d'_k = d'_k$ we obtain

Paraxial $(d'_k - D'_k) = x_k - x_{k+1} - \frac{1}{2}d'_k(u'_k)^2.$

The paraxial x are most easily computed by combining (8p) and (8p)* as

$$x = \frac{1}{2}r(u+i)^2$$

and we have the paraxial expression of our chromatic condition

OP(5p) $\begin{cases} \Sigma(d' - D'_p)dN' = 0 \\ \text{Paraxial } (d'_k - D'_k) = \frac{1}{2}r_k(u_k + i_k)^2 - \frac{1}{2}r_{k+1}(u_{k+1} + i_{k+1})^2 - \frac{1}{2}d'_k(u'_k)^2. \end{cases}$

If a lens system can be given such a form as to satisfy OP(5) and OP(5p) simultaneously, that is, so that both sums are zero, then a very high state of achromatism will be realized which corresponds closely to the condition first formulated by Gauss: The spherical aberration shall be corrected for two different colours.

Gauss showed that his condition could be fulfilled in the case of an achromatic object-glass of two lenses by departing widely from the usual type of Fig. 136 (a), and making both lenses of a decided meniscus form: Fig. 136 (b).   In recent years it has been pointed out by American opticians that yet another solution is possible

[101] by introducing a considerable separation between crown and flint, as indicated in Fig. 136 (c). The design of this type of objective was discussed in section [87].

The Gauss form is practically worthless for telescope purposes by reason of the heavy zonal spherical aberration caused by the deep curvatures; the bad effect of this zonal spherical aberration far outweighs the improvement in achromatism. Moreover, the sine condition cannot usually be fulfilled by this type. The American form is free from these drawbacks but suffers, on account of the considerable separation, from a variation in image size with colour. It is therefore unsuitable for direct photography on a plate placed into the principal focal plane, and for visual purposes it calls for compensating eyepieces, which lead to difficulties if micrometric measurements are required. For these reasons the Gauss condition

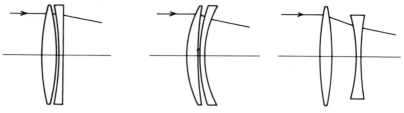

(a) Ordinary object-glass   (b) Gauss object-glass   (c) American object-glass

FIG. 136.

is nearly always ignored in the design of telescope objectives, and types with the two lenses close together and fulfilling the sine condition to a sufficient extent are preferred.

In microscope objectives of high NA and occasionally in photographic lenses, the offence against the Gauss condition can become so serious as to interfere with the utility of the system, and means must then be sought for reducing the offence within tolerable limits. In these cases, the calculation of both the paraxial and the marginal $(d' - D')$ sums represents the easiest method of detecting the defect, and it only remains to determine the maximum difference allowable between the two sums.

We shall discuss this tolerance on the assumption that, in accordance with the arguments on page 641, the marginal $(d' - D')$ sum is always made equal to zero so as to secure equal mean or average curvature of the two different-coloured waves and therefore chromatic correction for a particular intermediate lens zone. The paraxial $(d' - D')$ sum determines the difference in curvature of the two true waves in the paraxial region, and if we compute it with our usual nominal paraxial angles (i.e., first u = first sin U), the paraxial sum measures directly the gap between the two coloured waves which would exist if there were no spherical aberration present: i.e., the paraxial sum measures the distance $C_p F_p$ in Fig. 137. As the true coloured waves are brought to intersection at the marginal zone by making the marginal $(d' - D')$-sum equal to zero, they must be so distorted by spherical aberration as to bring this intersection about. The variation of the *higher* aberrations for the small difference in refractive index between the two colours usually united is quite small

even in boldly curved lens systems. We may therefore attribute the marginal
intersection entirely to the action of a difference of *primary* spherical aberration in
the two colours: on that assumption, the laws of primary aberration will hold and
we can conclude that the gap between the two true, distorted, waves will have a
maximum at 0.7071 of the full aperture and that it will there amount to one
quarter of the marginal gap between the two undistorted waves determined by the
paraxial $(d' - D')$-sum. Furthermore, as the refractive index grows in steady
progression from C through the brightest light to F, we may safely assume that the
wave distortions will show a correspondingly steady progression and therefore that
the true brightest-light wave will fall about midway between the true C and F

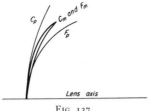

FIG. 137.

waves. If we allow the Rayleigh limit of $1/4$ wave-length for the maximum gap at
0.7071 of the full aperture between the brightest wave and the C and F waves
respectively, the gap between the latter two will have a tolerance of $1/2$ wave-length,
which by the previous argument corresponds to a value of the paraxial $(d' - D')$-
sum of four times this amount, or of two whole wave-lengths. We thus arrive at
the rule for tolerance of zonal variation of the chromatic correction:

OT(5)*   'When the marginal $(d' - D')$-sum has been brought to zero, then the
corresponding paraxial sum may be allowed to reach plus or minus two
wave-lengths, that is 0.00004 inch or 0.001 mm. As an obvious extension
for cases when a residual *Chr.* is left at the margin to cover chromatic
aberration of, say, an eyepiece, the *difference* between the marginal and
the paraxial $(d' - D')$-sums may reach two wave-lengths.'

Again it should be borne in mind that the Rayleigh limit can only be drawn upon
once for any one system; therefore, if we find that we have to take advantage of
OT(5)* to cover zonal variation, we should be especially careful to bring the
marginal $(d' - D')$-sum to zero for the whole system and to have the latter accurately
carried out in glass and brass.

We shall now apply the paraxial $(d' - D')$-sum to the telescope objective in
Part I. We extract from the calculations on p. 51, Part I:

	First surface	Second surface	Third surface
$\log r$	0.55023	0.55023n	1.77815n
$\log u'$	8.98207	8.85235	
$u + i$	0.281689	-0.273583	-0.015949
d'	0.30000	0.20000	
dN'	0.00817	0.01720	

|101| We then calculate the x-values:

$\log \frac{1}{2}$	9.69897	9.69897	9.69897
$+\log r$	0.55023	0.55023n	1.77815n
$+2 \cdot \log (u+i)$	8.89954	8.87418	6.40547
$\log x$	9.14874	9.12338n	7.88259n
x	0.14084	-0.13286	-0.00763

Then we find $(d'-D')_p$ and $(d'-D')_p \cdot dN'$:

	First lens	Second lens
$\log \frac{1}{2}$	9.69897	9.69897
$+\log d'$	9.47712	9.30103
$+2 \log u'$	7.96414	7.70470
$\log \frac{1}{2} d' \cdot u'^2$	7.14023	6.70470
$-\frac{1}{2} d' \cdot u'^2$	-0.00138	-0.00051
$+x$ of first surface	0.14084	-0.13286
$-x$ of second surface	0.13286	0.00763
$(d'-D')_p$	0.27232	-0.12574
$\log (d'-D')_p$	9.43508	9.09947n
$+\log dN'$	7.91222	8.23553
$\log (d'-D')_p dN'$	7.34730	7.33500n

$$(d'-D')_p dN' \text{ of crown } = +0.002\ 225$$
$$(d'-D')_p dN' \text{ of flint } = -0.002\ 163$$

Sum for whole O.G. $+0.000\ 062$

The tolerance by OT(5)* is 0.000 04 inch. The residual zonal chromatic aberration of our object-glass therefore amounts to $1\frac{1}{2}$ times the Rayleigh limit and is objectionable, while we found that the zonal spherical aberration was well below the limit. Telescope object-glasses nearly always break down first with regard to the *chromatic* residuals when the focal length is reduced too much. It should be remembered that we found in Part I, Chapter IV, that this same object-glass also suffers from an amount of the secondary spectrum far in excess of our normal tolerance for that defect.

We can easily determine to what aperture our object-glass would require to be reduced in order to bring the zonal chromatic aberration or, as it is frequently called, the 'offence against the Gauss condition', within the Rayleigh limit. At 2 inches aperture the defect is $1\frac{1}{2}$ times the permissible amount. As the defect is due to primary spherical aberration expressed as *OPD*, it grows (or diminishes) as the fourth power of the aperture. To reduce the defect to 0.67 of its value at 2 inches aperture, the latter therefore requires to be reduced to $2 \times \sqrt[4]{0.67}$ or to

1.81 inches. Our objective, having an equivalent focal length of 8 inches, would   [101]
then work at $f/4.4$, and this ratio would therefore be the highest permissible one
for an object-glass of the type used and of about 8 inches focal length. It should,
however, be noted carefully that this will not be true if we merely cut down the
clear aperture of the original object-glass to 1.81 inches. It will only be true if the
object-glass is recomputed for the new aperture so as to bring the paraxial and the
marginal foci together at the new full aperture and also to make the $(d'-D')dN'$
sum zero for the new, reduced, full aperture. If the original object-glass were
merely cut down in aperture, it would become decidedly under-corrected both
spherically and chromatically and would be very little better than at its original full
aperture. The changes involved in recomputing for the smaller aperture would be
a slight shortening of r_1 and of r_2 and a considerable lengthening of r_3; they would
have to be determined by a complete trigonometrical solution.

One important consideration in connection with all chromatic tolerances requires
to be emphasized. It is that these tolerances are not nearly so definite as those for
spherical aberration. In our numerical example, we worked out the tolerances on
the assumption that the C and F rays should give a good image. As the tolerances
are based on the two $(d'-D')$-sums, they depend on the value chosen for dN', that
is, on the difference between the indices for the two colours which are judged to be
still of sufficient importance to call for good correction. We can therefore, on the
one hand, object that the eye can easily see extreme red and deep blue and violet
parts of the spectrum which lie far outside our adopted $C-F$ range. These extreme
rays would therefore call for a reduced tolerance or permissible variation of, say,
X_3 or r_3 in our object-glass or for a still more reduced f-ratio. But on the other
hand we may claim that the spectrum in the neighbourhood of the C and F lines
is very faint indeed compared with its most brilliant region in the yellow-green and
that therefore it seems unnecessary to call for high correction of even the C and F
regions. A bigger tolerance thus appears permissible from this point of view.
Experience with giant refractors has in fact demonstrated that very good definition
is still obtainable when only the central one-third of the stretch of spectrum
between C and F is brought to focus within the Rayleigh limit; and many of the
productions of rule-of-thumb opticians point in the same direction by giving quite
good images although the chromatic correction is very faulty.

It is impossible to give definite numerical guidance as to how far we may safely
go with the chromatic tolerances. Those given in the above account have been
tested in an extensive practice of many years and have never been found to be too
generous. They may therefore be safely used and may be exceeded up to two or
three times the stated values provided an at least equivalent advantage can be
secured by that means with regard to other aberrations.

Check for the $(d'-D')$ Method

As the d' are successive parts of the axial path of the light from object-point to
image-point, and the D' the corresponding marginal paths, it is clear that our
method depends on differences of these paths and therefore that any other method
of determining these differences must necessarily give the same result and must form

[101] a check on the accuracy of the numerical calculations. In the latter part of Chapter XIII we deduced the trigonometrical equation for differences of optical paths. If we draw a diagram, Fig. 138, based upon the previous Fig. 127 but showing two successive surfaces (1) and (2), and examine the axial and marginal paths d'_1 and D'_1 in the space between the two surfaces, we see, on transferring the marginal path to the optical axis by circles struck from the intersection point

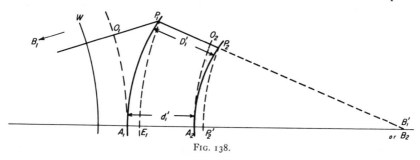

FIG. 138.

B'_1 of the marginal ray P_1P_2, that $E_1P'_2 = D'_1$ and $A_2P'_2 = O_2P_2$. And from the transferred D'_1 we deduce at sight

$$E_1P'_2 = D'_1 = d'_1 + A_2P'_2 - A_1E_1$$

or

$$d'_1 - D'_1 = A_1E_1 - A_2P'_2 = A_1E_1 - O_2P_2.$$

The value of $(d'_1 - D'_1)$ thus appears determined by a difference very similar to that used on page 616 for the determination of OPD'_m, the only difference being that our $(d'_1 - D'_1)$ uses the AE of the preceding surface minus the OP of the following surface instead of both for the same surface.

Now the AE and OP of Chapter XIII are retained in the equations of that chapter down to equation (a) on page 617, and therefore they also reappear in OP(4)Check on page 618, merely multiplied by N' and N respectively. Consequently we can write down our new value of $(d'_1 - D'_1)$ by omitting these factors N' and N in OP(4)Check and putting suffixes (1) and (2) to its two parts. We thus obtain

$$d'_1 - D'_1 = \frac{PA_1 \cdot \sin I'_1}{2 \cos \frac{1}{2}I'_1 \cos \frac{1}{2}U'_1} - \frac{PA_2 \cdot \sin I_2}{2 \cos \frac{1}{2}I_2 \cos \frac{1}{2}U_2}.$$

Obviously this equation can be written down for any space between successive surfaces by putting in the proper suffix numbers. We can include the first space between the spherical entering wave W and surface (1) by noting that because this *wave* surface has its centre of curvature at the object-point B_1, both its OP and its AE are zero and therefore vanish. If we add the factor dN' to form $(d' - D')dN'$, we thus obtain, using plain letters for the wave surface W:

$$(d' - D')dN' = -\frac{PA_1 \cdot \sin I_1}{2 \cos \frac{1}{2}I_1 \cos \frac{1}{2}U_1} \cdot dN'$$

$$(d'_1 - D'_1)dN'_1 = \frac{PA_1 \cdot \sin I'_1}{2 \cos \frac{1}{2}I'_1 \cos \frac{1}{2}U'_1} \cdot dN'_1 - \frac{PA_2 \cdot \sin I_2}{2 \cos \frac{1}{2}I_2 \cos \frac{1}{2}U_2} \cdot dN'_1$$

$$(d'_2 - D'_2)dN'_2 = \frac{PA_2 \cdot \sin I'_2}{2 \cos \frac{1}{2}I'_2 \cos \frac{1}{2}U'_2} \cdot dN'_2$$

$$- \frac{PA_3 \cdot \sin I_3}{2 \cos \frac{1}{2}I_3 \cos \frac{1}{2}U_3} \cdot dN'_2$$

&c.

The items have been put down to show that while the successive horizontal lines give the $(d' - D')dN'$ values for the component lenses, the vertical columns, each of two terms, give the contribution of each *surface* to the total chromatic sum. It is generally more convenient to calculate the check surface-by-surface, and we will therefore record it in the form:

OP(5)Check:

$$\sum_1^{k-1}(d' - D')dN' = \sum_1^k \left(\frac{PA \cdot \sin I'}{2 \cos \frac{1}{2}I' \cos \frac{1}{2}U'} \cdot dN' - \frac{PA \cdot \sin I}{2 \cos \frac{1}{2}I \cos \frac{1}{2}U} \cdot dN \right)$$

For plano: $$= \sum_1^k \left(\frac{Y \sin U}{2 \cos^2 \frac{1}{2}U} dN - \frac{Y \sin U'}{2 \cos^2 \frac{1}{2}U'} dN' \right).$$

In all cases dN_j has been used instead of dN'_{j-1}, with which it is identical, for the sake of symmetry.

As in the case of the $(d' - D')dN'$ sum, terms which involve the dispersion of the air are to be taken as being zero and are therefore omitted.

We shall apply the check to our object-glass, extracting all quantities except the dN from the calculation on page 621 for the Check OPD'_m.

For the first surface (air into crown glass), $dN = 0$ and $dN' =$ that of the crown $= 0.00817$, hence only the first part of the complete expression requires calculation. For the second surface (crown into flint), $dN = 0.00817$ and $dN' = 0.01720$. For the last surface (flint into air), $dN = 0.01720$ and $dN' = 0$, hence only the second part requires calculation.

The calculations then run thus:

	First surface	Second surface	Third surface
log sin I'	9.26890	9.54337n	
+ log PA	0.00444	0.00388	
+ co-log cos $\frac{1}{2}I'$	0.00190	0.00696	
+ co-log cos $\frac{1}{2}U'$	0.00053	0.00028	
+ log $\frac{1}{2}$	9.69897	9.69897	
+ log dN'	7.91222	8.23553	
log first term	6.88696	7.48899n	

	First surface	Second surface	Third surface
$\log \sin I$		$9.57351n$	$8.94410n$
$+ \log PA$		0.00388	9.98884
$+ $ co-log $\cos \frac{1}{2}I$		0.00053	0.00028
$+ $ co-log $\cos \frac{1}{2}U$		0.00805	0.00042
$+ \log \frac{1}{2}$		9.69897	9.69897
$+ \log dN$		7.91222	8.23553
$\log$ second term		$7.19716n$	$6.86814n$
First term	0.0007708	-0.0030831	
$-$ Second term		$+0.0015746$	$+0.0007380$
	0.0007708	-0.0015085	$+0.0007380$
Third surface	0.0007380		
Total positive	0.0015088		
Second surface	-0.0015085		

Therefore chromatic residual $= +0.000\ 000\ 3$ against $0.000\ 000\ 5$ by the $(d'-D')$ method. As the tolerance is $0.000\ 01$ the discrepancy between the two methods amounts to only 2 per cent. of the tolerance and is utterly negligible.

We can easily check the two parts of the original $(d'-D')$ calculation by taking together the terms which have the same dN-factor according to the horizontal lines of the sum from which we obtained the check-formula:

	Crown	Flint
	0.0007708	-0.0030831
	0.0015746	0.0007380
$(d'-D')dN'$	0.0023454	-0.0023451

The results agree absolutely in the case of the crown and within two units in the last place for the flint.

All the figures except dN are taken directly from the calculation for OPD'_m-Check; hence in microscope objectives, where the latter check should not be omitted, there is very little work involved in computing OP(5)Check.

We may make the connection between the two checks still closer by calculating the chromatic sum also in nominal wave-lengths of $0.000\ 02$ inch or 0.0005 mm. We can then take over the whole log-sums of the OPD'-Check and multiply each item by dN/N or dN'/N'. Our check-calculation would then run with:

$$\log dN/N \text{ for crown} = 7.91222 - 0.18087 = 7.73135$$
$$\text{for flint} \quad = 8.23553 - 0.21101 = 8.02452$$

	First surface	Second surface	Third surface	[101]
log first OPD part	3.85458	4.16344n		
$+\log dN'/N'$	7.73135	8.02452		
log first term	1.58593	2.18796n		
log second OPD part		4.16478n	3.54259n	
$+\log dN/N$		7.73135	8.02452	
log second term		1.89613n	1.56711n	
First term	38.54	-154.16		
$-$ Second term		$+\ 78.73$	$+36.91$	
Total	$+38.54$	$-\ 75.43$	$+36.91$	

and we obtain the chromatic residual

$$+75.45 - 75.43\ =\ +0.02 \text{ nominal wave-lengths.}$$

Multiplying by 0.000 02 to turn the result into inches, we find

$$\text{residual} = 0.000\ 000\ 4 \text{ inch,}$$

in close agreement with previous results.

The equations OP(5)Check can be turned into their paraxial form in the usual way, by replacing PA by $y = lu = l'u'$, sines by the corresponding paraxial angles, and cosines by unity. This gives the equations:

$$\text{OP(5p) Check} \quad \sum_{1}^{k-1}(d' - D'_p)dN' = \sum_{1}^{k}(\tfrac{1}{2}y \cdot i' \cdot dN' - \tfrac{1}{2}y \cdot i \cdot dN)$$

$$\text{and for a plane} \quad = \sum_{1}^{k}(\tfrac{1}{2}y \cdot u \cdot dN - \tfrac{1}{2}y \cdot u' \cdot dN').$$

Apart from the value of this paraxial equation as a check on the result obtained by the paraxial $(d' - D')dN'$ sum, it will be found more convenient than the latter for the direct calculation when a check is to be dispensed with.

For the telescope object-glass the figures on pages 51–52 of Part I give:

	First surface	Second surface	Third surface
$\log \tfrac{1}{2}$	9.69897	9.69897	
$\log y$	0.00000	9.98731	
$\log i'$	9.26890	9.53752n	
$\log dN'$	7.91222	8.23553	
	6.88009	7.45933n	

	First surface	Second surface	Third surface
$\log(-\tfrac{1}{2})$		9.69897n	9.69897n
$\log y$		9.98731	9.98090
$\log i$		9.56766n	8.94016n
$\log dN$		7.91222	8.23553
		7.16616	6.85556

Positive terms	0.0007587	
	0.0014661	
	0.0007171	
	0.0029419	
Negative term	−0.0028796	
Final result	+0.0000623	

in exact agreement with the result given by the $(d' - D'_p)$ calculation on p. 652.

INTERPOLATION OF REFRACTIVE INDICES

[102] THE $(d' - D')dN'$ equations are, as was shown, rigorously correct if dN' is the true differential dispersion of the various glasses used in a combination, that is, if dN' is the difference of the refractive indices for two colours differing only very minutely from each other. Such minute differences cannot be determined with sufficient accuracy by direct measurements on a refractometer or spectrometer; they must therefore be deduced mathematically from measurements of a few more widely separated lines of the spectrum. To do this successfully, we must have a sufficiently accurate formula which enables us to calculate the refractive index for any colour or wave-length from the measured indices for a few selected colours or wave-lengths. No absolutely accurate law is known for this purpose, and we must therefore rely on more or less empirical equations which have been proved to represent the connection between wave-length and refractive index with considerable accuracy. The oldest and best-known formula for this purpose was proposed by Cauchy and is as follows:

$$N = N_0 + aw^2 + bw^4 + \ldots,$$

where the symbol w has been introduced for the reciprocal of the wave-length, the latter most conveniently expressed in one-thousandths of a millimetre (microns). This formula, however, converges badly when applied to optical glasses so that four terms have to be used to represent the refractive indices throughout the visible spectrum within one unit in the fifth decimal place.

Another interpolation formula closely resembling Cauchy's, viz.,

$$N = N_0 + aw + bw^4$$

was proposed by Schmidt some time ago, and certainly it approaches more closely to the desired accuracy when used with only the three terms stated. When I went into this question first, I noticed that Schmidt's formula always gave errors of [102] about half the amount of those given by Cauchy's when the latter was also used with only three terms, but that Schmidt's errors were always of the opposite sign, and this caused me, after some trials, to adopt an intermediate formula, viz.,

$$N = N_0 + aw^{9/7} + bw^4.$$

The inconvenient fractional index of the second term later caused me to make a further attempt, and this led me finally to adopt the formula

(1) $$N = N_0 + aw + bw^{3\frac{1}{2}}.$$

When the three constants of this formula are determined by the indices for C, F, and G', the refractive indices for any other colours within the visible spectrum are correctly obtained within one or at most two units in the fifth decimal place. The formula may therefore be safely used to prepare expressions for the quick determination of refractive indices and dispersions by introducing the numerical value of w corresponding to the lines of the spectrum for which makers give dispersion figures. These values of w are:

	wave-length	w	$w^{3\frac{1}{2}}$
A' line (K)	0.7682μ	1.3017	2.5164
C line (H)	0.6563	1.5237	4.3666
D line (Na)	0.5893	1.6969	6.3650
F line (H)	0.4861	2.0572	12.4873
G' line (H)	0.4341	2.3036	18.5534
H line (Ca)	0.3968	2.5202	25.4110

By writing down equation (1) for the C, D, F, and G' lines, and introducing the numerical values for w, four equations are formed from which the three constants N_0, a, and b can be determined in terms of $Ng' - Nf$, $Nf - Nc$, and Nd; in other words, in exactly those numbers which are almost invariably supplied by the glass-maker. My own determination of these constants is as follows:

$$N_0 = Nd - [0.84895](Nf - Nc) + [0.92452](Ng' - Nf)$$
$$a = [0.69108](Nf - Nc) - [0.81775](Ng' - Nf)$$
$$b = -[9.29980](Nf - Nc) + [9.63529](Ng' - Nf),$$

where, in accordance with a well-known astronomical custom, the numbers in square brackets represent the log of the numerical coefficient. By differentiating equation (1) with reference to w, we obtain an expression for the differential dispersion at any point in the spectrum, and by introducing the values of the constants of equation (1), we can prepare formulae for obtaining relative values of the dispersion in any region. We thus obtain the general formula

$$dN/dw = (Nf - Nc)(4.9101 - [9.84387]w^{2\frac{1}{2}}) - (Ng' - Nf)(6.5727 - [0.17936]w^{2\frac{1}{2}}),$$

where any convenient number may be used for the dw, although of course it must be the same for all the glasses used in any particular calculation. Such a factor has been employed in the following three generally useful values of dN so as to

[102] bring the resulting number to about the value of the dispersion between the F and C lines:

$$dNd = [0.1139](Nf - Nc) - [9.7097](Ng' - Nf)$$
$$dNf = [9.4867](Nf - Nc) + [0.0737](Ng' - Nf)$$
$$dNg = -[9.4433](Nf - Nc) + [0.3391](Ng' - Nf)$$

Finally, the general formula for dN enables us to determine the wave-length for which a lens system has minimum focal length when the dispersion between F and C has been used for achromatization, which is the almost universal custom for visual instruments. This is, of course, that value of w for which dN is independent of the dispersion between F and G', or the value of w for which the coefficient of $(Ng' - Nf)$ becomes zero, and this will be found to correspond to a wave-length of 0.5555. We may next determine the refractive index corresponding to this wave-length by using equation (1) with this result:

$$N_{0.5555} = Nd + 0.2157(Nf - Nc) - 0.0474(Ng' - Nf),$$

in which the numerical coefficients themselves are given because the slide rule is the proper means for calculating this.

As the coefficient of $(Ng' - Nf)$ in this formula is small, the third term may be combined with the second by the use of an average proportional number for the dispersion between G' and F and in this way the simple formula

$$N_{0.5555} = Nd + 0.1880(Nf - Nc)$$

is obtained. This formula is sufficiently accurate for all ordinary calculations of achromatic systems, but it should not be used when correction of the secondary spectrum is aimed at.

Ample accuracy in the determination of $N_{0.5555}$ for all purposes will be obtained by using instead of 0.1880 in the last formula a factor interpolated roughly from the following table, which should be entered with the proportional number $(Ng' - Nf)/(Nf - Nc) = Pg'f$ as given in the last column of glass-lists:

$$Pg'f = 0.560 \quad 0.580 \quad 0.600 \quad 0.620$$
$$\text{Factor of} \quad (Nf - Nc) = 0.1892 \quad 0.1882 \quad 0.1873 \quad 0.1863$$

It will be noticed that the factor diminishes approximately by one unit in its fourth place for every two units of the third place by which $Pg'f$ increases.

As to the extent to which the calculations just described have to enter into the ordinary routine, it may be said that for all ordinary visual objectives it is advisable to calculate the spherical aberration for the really brightest rays by determining $N_{0.5555}$ by the simple slide-rule formula

$$N_{0.5555} = Nd + 0.1880(Nf - Nc)$$

and to use the maker's value of $(Nf - Nc)$ for the dispersion in calculating the chromatic aberrations.

For ordinary photographic objectives (i.e., those which have the photographic plate placed at the visual focus) it is generally sufficient to calculate $Nf = Nd +$

[102]

$(Nf-Nd)$ by the maker's figures and to use this for the spherical calculations. Suitable chromatic correction is then obtained by taking $(Ng'-Nd) = (Nf-Nd) + (Ng'-Nf)$ for the dispersion.   It may however be added that the increasing use of orthochromatic plates and of efficient colour screens tends to bring the photographic and the visual foci closer together, so it may even now be considered whether $Ng'-Nc$ would not give a better compromise.

For purely photographic objectives, nearly always astrographic ones, the focus is determined experimentally as so many millimetres within the visual one, and the best spherical correction is demanded for the rays which act most strongly on ordinary gelatine dry plates, which are those in the vicinity of G'.   Such objectives therefore call for the use of $Ng' = Nd + (Nf-Nd) + (Ng'-Nf)$ as the index for the 'brightest' rays and for the interpolation of the dispersion in the neighbourhood of the G' line by the formula already given:

$$dNg' = -[9.4433](Nf-Nc) + [0.3391](Ng'-Nf).$$

The refractive index for the D line and the dispersion in that neighbourhood would probably be nearly the best to use for astronomical objectives specifically intended for the visual observation of the moon and the outer planets on account of the pronouncedly yellow or even reddish colour of these objects.

CHAPTER XVI

THE MATCHING PRINCIPLE AND THE DESIGN OF MICROSCOPE OBJECTIVES

[103] THE methods of dealing with spherical and chromatic aberration which are given in the preceding three chapters find their most important application in the design of microscope objectives. The reason is that deep curvatures and large angles of incidence and obliquity predominate in such systems and lead to unusually large amounts of secondary and tertiary aberrations.

It is most convenient to compute microscope objectives for light travelling backwards with respect to the direction in which it will travel in the instrument as used, that is, from image to object. One reason for this procedure is that we can thus make sure of realizing the desired tube-length; a second and more cogent reason is that the final adjustments are most easily effected by changes in the front lens, which in our chosen computing direction will be the last one of the system, so that it can be freely modified without invalidating the calculations for the other lenses.

The principal types of ordinary microscope objectives are shown in Fig. 139. Those shown at (a), (b), and (c) are simple cemented combinations for low-power work, which can be designed by the methods of Part I. For types (a) and (b) the best results as regards small offence against the sine condition will be obtained by using barium crown of index 1.57 to 1.58 and of high V-value with ordinary dense flint. Type (c), having an additional radius, can be fully corrected for spherical and chromatic aberration *and* for fulfilment of the sine condition; it also admits of a somewhat higher numerical aperture. Light barium crown and ordinary dense flint are suitable glasses. All three types have an unpleasantly curved field which is quite incurable. Type (d), with two widely separated components, is therefore employed even in the lowest powers when a flat field is important; it is of course more expensive.

In all low-power types up to about 0.50 NA, the residual spherical aberration may be treated as being of the ordinary zonal type due to secondary aberration. If it is desired to fix the aperture at the highest value compatible with the Rayleigh limit, the easiest way consists in calculating for the originally adopted aperture the OPD'_p-sum by OP(4p) (page 619) for which the Rayleigh tolerance is 6 wave-lengths (page 634). As the secondary aberration grows with the sixth power of the aperture, we can then determine

Permissible aperture = Original aperture $\times \sqrt[6]{(6 \text{ wave-lengths})/(OPD'_p\text{-sum})}$.

The system must, however, be slightly modified so as to restore strict trigonometrical correction at the new full aperture; it is not permissible merely to cut down or to increase the aperture.

Type (d) represents a generalized form of the first doublet microscope objective ever designed on scientific principles. It was originated in 1830 by J. J. Lister (father of the great Lord Lister) on the strength of an observational result of his.

The form of achromatic lens that was common in his day was plano-convex when [103] considered as a whole, and he noticed that there were two positions of the object for which the lens was free from spherical aberration, and that the coma for these two positions was of the opposite tendency. His original design of type (d) resulted by combining a back lens which realized the first pair of aplanatic (as it was then called) conjugate distances with a front lens which realized the second pair. The combination was thus free from spherical aberration; and as the components had opposite coma-tendencies, the correction for coma could be rendered practically perfect by a suitable ratio of the focal lengths of the components. The originally chosen condition that the components should each be spherically corrected, was soon departed from because it was found that even better results were obtainable by so doing. But the type combines so many virtues, including good flatness of field, that to this day the vast majority of low-power objectives up to 0.30 NA or a little over closely conform to it.

For higher powers, combinations of three achromatic components selected and combined on Lister's principle were at first employed. This form of triplet has not survived, for it has been amply proved that the simpler type (e) of Fig. 139, consisting usually of spherically and chromatically over-corrected back and middle combinations and a simple thick plano-convex front lens, is actually better for objectives having numerical apertures between 0.50 and 0.85. All ordinary 'dry' objectives (that is, those having an air-space between front lens and object) conform to this Amici type.

For the highest powers, almost exclusively immersion objectives, type (f) was worked out by Abbe, apparently on a suggestion by the English microscopist Stevenson. It consists of two over-corrected cemented combinations followed by a *duplex front* composed of a simple meniscus and a thick plano-convex front lens. The latter frequently utilizes more than the complete hemisphere of its convex side and is then said to be hyperhemispherical. In this type, the space between the front lens and the object is filled with water or more usually with cedar-oil, whence the name immersion objective.

In types (d), (e), and (f), modifications are frequently met with, such as a triple-cemented back lens in all three, a front component of Steinheil form in type (d), or a compound meniscus in type (f).

These types, like all other complicated lens systems of bold curvatures, have hitherto always been designed by purely empirical trigonometrical trials. Given a large amount of previous experience and a well-developed optical instinct, good results can be obtained by that method in a reasonable total computing time. But in the absence of such experience and instinct, the purely empirical method becomes a wearisome blind groping for a result which always seems a long way off, a veritable hunt for a pin in a haystack.

The *matching principle* to be expounded presently reduces this hunt to a perfectly systematic process by which anyone reasonably familiar with trigonometrical ray-tracing is bound to find the solution of a problem if there is a solution or is enabled to say definitely that there is no solution if such should be the case. The principle is applicable to almost any centred optical system which is required to produce images of a definitely prescribed state of correction as regards chromatic

[103] and spherical aberration at the axial image-point and offence against the sine condition. Usually the required value is zero for all three. It does *not* work in a straightforward manner in cases like that of ordinary eyepieces, in which an indefinite amount of chromatic and spherical aberration at the axial image-point is tolerated to secure the best compromise for the extra-axial field; for that reason, the special solutions for eyepieces given in Chapter X of Part I are not superseded by the matching principle.

The remarkable feature of the principle is that it involves no mathematical considerations; it is purely a matter of common-sense reasoning. The successive types of microscope objectives will supply progressive examples of its application from which the procedure in other cases of systems of separated components will be easily deduced.

GENERAL STATEMENT OF THE MATCHING PRINCIPLE

Let us assume that the lens system of Fig. 139 (d) were absolutely perfect, so that every ray traced from B_1 would be found to cut the optical axis at B'_6 on emerging from the last surface and so that the sine condition is satisfied by reason

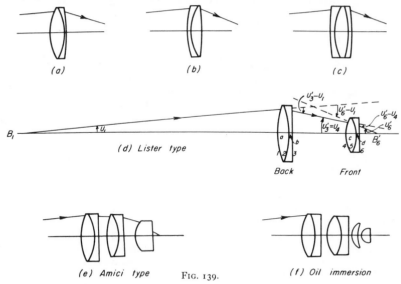

(a)   (b)   (c)

(d) Lister type

Back   Front

(e) Amici type   FIG. 139.   (f) Oil immersion

of a constant value of the ratio $N_1 \sin U_1 / N'_6 \sin U'_6 = M'$. It is at once evident that if we varied the method of computation by tracing a ray from B_1 under any angle U_1 through the back component alone, and another ray from B'_6, right to left, under the angle U'_6 following from the sine condition through the front component alone, then the left-to-right ray emerging from the back component must be absolutely coincident with the right-to-left ray emerging from the front component. [103] This coincidence or perfect *matching* of rays traced in opposite directions is thus a property of any perfectly-corrected system. Furthermore, it is a property *only* of a perfectly-corrected system, for if the system were imperfect, the left-to-right ray on being traced right through would not go through B'_6, or would do so under an angle implying offence against the sine condition, so the right-to-left ray starting from B'_6 under the correct angle U'_6 would not match the left-to-right ray in the space between the two components. A failure to match is thus a sure indication of imperfect correction of the complete system, and this is the main idea of the matching principle.

The practical application of the principle consists in determining by the simple paraxial formulae the focal lengths and apertures of the two components which will produce the desired magnification and numerical aperture, in calculating their separate aberrations for three or sometimes four bendings by the methods of Chapters V and VII in Part I, and then in selecting by a simple graph those bendings which will produce the perfect matching of the rays in the space between the components. These bendings are the solutions of the problem. The great advantage thus secured is that the problem is reduced to one of computing for simple cemented lenses only. Special instructions, to be amplified by individual ingenuity, are required to establish the proper conditions for the final matching in each of the numerous cases which lend themselves to the application of the principle.

DESIGN OF OBJECTIVES OF THE LISTER TYPE

AS a definite problem, let us take this one: [104]
An objective of the Lister type, Fig. 139 (d), is to be designed so as to be free from chromatic and spherical aberration and to fulfil the sine condition. The $NA = \sin U'_6$ shall be 0.25. The magnification $\sin U'_6/\sin U_1$ shall be $-8\frac{1}{3}$ times (as actually used) when B_1 is at a distance $l_1 = -170$ mm. from the nearest lens surface. As $\sin U'_6$ is fixed at 0.25 we get

$$0.25/\sin U_1 = -8\tfrac{1}{3}, \quad \text{or} \quad \sin U_1 = -0.0300.$$

We have thus found the initial data for the ray-tracing.
The glass for the two crown lenses shall be:

Chance's Hard Crown $Nd = 1.5175$, $V = 60.5$, $dN = 0.00856$

and that for the two flint lenses shall be:

Chance's Dense Flint $Nd = 1.6214$, $V = 36.1$, $dN = 0.01722$.

In order to render the problem definite for the application of the matching principle, which is based on the plotting of bendings of the two components, we must now fix upon a suitable focal length for each component. As we do not know the shape of the components which the final solution will lead to, we must sacrifice absolute precision by substituting thin lenses for our components and by working with paraxial thin-lens equations in this preparatory work. We shall have the paraxial $u_1 = \sin U_1 = -0.03$, and as the sine condition is to be fulfilled, we must

[104] aim at $u'_6 = \sin U'_6 = 0.25$. The complete system therefore must produce a change in the direction of the nominal paraxial ray by $u'_6 - u_1 = 0.25 - (-0.03) = 0.28$. One of the chief reasons for using several components in microscope objectives is to subdivide the 'optical work', as it has been loosely called, of producing the total change in direction. The selection of a suitable division of the work between the components is one of the liberties we have in our design, and we restrict ourselves to one set of solutions out of an infinite number of possible sets as soon as our choice has been made.

We shall make the most obvious selection by deciding that each component shall produce half of the total deviation, and we shall carry this particular solution to a finish. But we must remember that if no very favourable result should be forthcoming, we have further opportunities of trying a different subdivision; it may be stated at once that the predominating practice in existing Lister objectives is to produce rather more than half of the total deviation in the back component (the one next to B_1), and we might therefore try subsequently to give 0.6 of the total deviation to this component and the remaining 0.4 to the front component. There is a certain amount of theoretical justification for unequal division, inasmuch as achromatic lenses standing, like the front component, in strongly converging light tend to produce more zonal aberration so that a moderate weakening of their power may reduce the total zonal aberration.

According to our choice, the back component is to produce a deviation of 0.28/2, or $u'_3 - u_1 = 0.14$. Therefore the paraxial ray-slope between the components is $u'_3 = 0.14 + u_1 = 0.14 - 0.03 = 0.11$. We can now fix the requisite focal length of the back component when treated as a thin lens:

By $l_1 u_1 = l'_3 u'_3 = y_1 = y_3$ we have $l'_3 = l_1 \cdot u_1 / u'_3$, or $l'_3 = -170(-0.03)/0.11 = 46.4$ mm., and then by the thin-lens formula $1/l'_3 = 1/f' + 1/l_1$ we have $1/f' = 1/46.4 + 1/170 = 0.02743$ by Barlow's table. The initial $y_1 = -170 (-0.03) = 5.1$ mm. must also be calculated to enable us to fix the necessary thickness of the crown lens.

We can now apply the solution for achromatism. The formula for the total curvature $(1/r_1 - 1/r_2) = c_a$ of the crown component, namely,

TL,Chr. (4)* $$c_a = 1/[f'(V_a - V_b)\delta N_a]$$

then gives $c_a = 0.1313$. It would be a waste of time to calculate c_b also because the third radius is to be determined so as to establish exact achromatism.

We are now ready to select four bendings of the crown lens at equal intervals of $c_1 = 1/r_1$. Taking $c_1 = 0$ as the first value, it is desirable for ease in plotting to select the other three so as to fall on main divisions of the plotting paper. Probably 0.025, 0.05, and 0.075 might be the wisest choice as being likely to include the range within which the solution will be found, but we will cut it rather finely by using 0.02, 0.04, and 0.06. This, with $(c_1 - c_2) = c_a = 0.1313$, gives for the curvature of the second surface the four values $c_2 = -0.1313$, -0.1113, -0.0913, and -0.0713, and taking reciprocals we obtain

$$r_1 = \quad \infty \qquad 50 \qquad\quad 25 \qquad\quad 16.67$$
$$r_2 = -7.62 \quad -8.985 \quad -10.95 \quad -14.03$$

As thicknesses we adopt, on the basis of a drawing, for the crown lens 2.7 mm. and [104]
for the flint lens 1.5 mm., and we can now calculate all four bendings trigonometrically with four-figure logs, solving for the third radius by the $(d'-D')$
method.   For the trigonometrical work we use the indices for the brightest light,
found by $Ny = Nd + 0.188(Nf - Nc)$ to be 1.5191 for the crown and 1.6246 for the
flint.   The results are:

c_1	=	0	0.02	0.04	0.06
r_3	=	-14.576	-20.333	-35.11	-160.92
L'_3	=	41.568	43.229	43.667	42.944
l'_3	=	42.467	42.429	42.568	43.172
LA'_3	=	$+0.899$	-0.800	-1.099	$+0.228$
log sin U'_3	=	9.0920	9.0680	9.0557	9.0540
log u'_3	=	9.0918	9.0792	9.0645	9.0447
OSC'_3	=	$+0.0005$	-0.0260	-0.0205	$+0.0212$

The last line gives the *uncorrected* offence against the sine condition (OSC'),
calculated by the formula

$$OSC' = 1 - (\text{final } u')/(\text{final sin } U').$$

The uncorrected OSC' is used rather than the more familiar form which includes
the correction for the spherical aberration and the exit-pupil position, because we
are actually trying to match ray-slopes in the space between the lenses, and the
uncorrected OSC' is simply a convenient measure of the relation between the
slopes of marginal and paraxial rays.

We must now decide how to proceed in determining the form of the second
component of our microscope objective.   The matching principle is based upon
the simple and obvious fact that in the complete system every ray goes straight
through from the original object-point to the final image-point.   This introduces
no difficulty with regard to the paraxial rays because all paraxial formulae apply to
a whole solid cone of rays within the paraxial region and the paraxial rays are
therefore bound to fit if we bring the intermediate paraxial foci together.   But each
of the marginal rays has a definite angle of inclination, and an intersection-point
which varies with that angle.   Therefore, in order that the marginal ray coming
from *any* first component may be capable of complete coincidence in the space
between the two components with the corresponding ray traced through *any*
second component, we must adopt a definite value of U'_3 for which all the bendings
of both components must be calculated.   Now the log sin U'_3 found for our first
components vary from 9.0920 down to 9.0540 and would defeat the matching
principle.   Obviously they could be made equal by a suitable variation of the
original U_1 used for the various bendings.   But it is not necessary to find these by
laborious trigonometrical trials.   As we only require to plot LA'_3 and OSC'_3, it is
sufficient to determine how these aberrations vary when U'_3 is changed.   Now
both LA' and OSC' vary to a first approximation with the square of any reasonable
measure of the aperture, and therefore with the square of sin U'_3.   If we now
adopt a standard value of U'_3, choosing U'_3 = 6-40-0, with the log sin = 9.0648,
we can convert the LA' and OSC' of our first table into those for the adopted

[104] standard U'_3 by multiplying by the square of the ratio (adopted sin)/(computed sin). For the first vertical column of the table, the log of the ratio will be 9.0648 − 9.0920 = 9.9728 and the log of the squared ratio will be twice this or 9.9456. By adding this to the logs of the LA' and OSC' that we found, we obtain adjusted $LA' = 0.793$ and adjusted $OSC' = 0.0004$. In this way we obtain a new table of LA' and OSC' to be plotted, for the uniform value of $U'_3 = 6$-40-0:

c_1	=	0	0.02	0.04	0.06
Adjusted LA'_3	=	0.793	−0.788	−1.146	+0.240
Adjusted OSC'_3	=	0.0004	−0.0256	−0.0214	+0.0223

We can now calculate all the adopted bendings of the second component by tracing a marginal ray at $U_4 = U'_3 = 6$-40-0 in the left-to-right direction through them and rendering each bending achromatic by solving for its third radius by the $(d'-D')$ method. We shall however require L_4 as well as U_4, and we must assume a value for it. The exact choice is not important because we shall see that we can easily modify the solution for another value of L_4 if the first choice should prove unsuitable. We will choose L_4 as = 25.000, which will give a separation between the two components of about 17.5 mm. considering that the L'_3 and l'_3 average about 42.5 mm.

We now prepare data for the front component by paraxial thin-lens formulae. We have already decided in the first part of the paraxial preparation that $u'_3 = u_4$ is to be 0.11 while $u'_6 = 0.25$ is given. We have decided upon $L = 25.000$. Strictly the paraxial l_4 will differ from this value according to the amount of spherical aberration in the various front components, but as this remains to be found out, we neglect it and calculate the focal length of the front component for $l_4 = 25.000$. By $l_4u_4 = l'_6u'_6$ for thin lenses, we then find $l'_6 = 25 \times 0.11/0.25 = 11$ mm. and then by $1/11 = 1/f' + 1/25$, we find $1/f' = 0.09091 − 0.04 = 0.05091$. This gives by the solution for achromatism $c_c = 0.2437$ as the requisite total curvature of the crown lens.

It is extremely unlikely that the second component should come out with a flat first surface, and we therefore choose the bendings

	$c_4 =$	0.04	0.08	0.12	0.16
giving	$c_5 =$	−0.2037	−0.1637	−0.1237	−0.0837

with the corresponding radii from Barlow's table:

	$r_4 =$	25	12.5	8.333	6.25
	$r_5 =$	−4.909	−6.109	−8.084	−11.95

As thicknesses we adopt 1.8 mm. for the crown and 1.0 mm. for the flint, based on a scale drawing of the crown lens.

We then complete the computing work by tracing a marginal ray at $U_4 = 6$-40-0 and $L_4 = 25.000$ through the four selected bendings, solving in each case for the last radius by the $(d'-D')$ method. Having thus found B'_6 for each bending, we trace a paraxial ray *backwards* from that point in the right-to-left direction, using the sin U'_6 found for the marginal ray as the initial u'_6-value, and the L'_6 as the [104] initial l'_6-value. The l_4 and u_4 at which this paraxial ray leaves the fourth surface (really the third passed by it in the right-to-left direction), can then be compared with the L_4 and sin U_4 of the corresponding marginal ray, and the LA_4 and un-corrected OSC_4 of the four bendings can be found exactly as for the first component. They will be ready for plotting without any correction and the graph can then be discussed and solutions picked off.

The actual calculation was made by four-figure logs for the marginal ray traced through the first bending (r_4 = 25.000, r_5 = −4.909) with U_4 = 6-40-0, log sin U_4 = 9.0648 and L_4 = 25.000. The $(d'-D')$ solution gave r_6 = −10.279 and with this

$$U'_6 = 16\text{-}12\text{-}50; \quad \log \sin U'_6 = 9.4460; \quad L'_6 = 8.996.$$

The tracing of a paraxial ray from the focus so fixed towards the left, with l'_6 = 8.996 and log u'_6 = 9.4460 as starting values, then gives

$$l_4 = 22.639; \quad \log u_4 = 9.0932.$$

Comparing these with the corresponding figures for the marginal ray (its initial figures), namely,

$$L_4 = 25.000; \quad \log \sin U_4 = 9.0648,$$

we obtain, for use in plotting the results against c_4 = 0.04:

$$LA_4 = l_4 - L_4 = -2.361; \quad OSC_4 = 1 - \frac{u_4}{\sin U_4} = -0.0676.$$

On carrying out the same calculation with the other three bendings, the complete results are found to be

c_4	=	0.04	0.08	0.12	0.16
r_6	=	−10.279	−19.121	−405.4	+18.558
LA_4	=	−2.361	−0.103	+0.884	+0.672
OSC_4	=	−0.0676	+0.0173	+0.0421	+0.0194

The results obtained for the selected bendings of the two components can now be plotted side by side, curvatures in the horizontal direction and aberrations in the vertical. A suitable size of the graph will result if for the first component (the 'back combination') one inch is made to represent 0.02 of curvature c_1 and if for the second component one inch represents 0.04 of curvature c_4. The aberrations of the two components must of course be plotted to the same scale, say one inch representing 1 mm. of spherical aberration and 0.05 of offence against the sine condition. To do full justice to the probable precision of the computed bendings, it would be preferable to assign one inch to 0.02 of OSC, but with the method of final adjustment to be described, the scale of one inch to 0.05 of OSC gives sufficiently close readings.

The graph, as drawn by the cubic-parabola method described on page 206 of Part I, or of course by a simple parabola if only three bendings are calculated, will

[104] be similar to Fig. 140. It will be advisable to draw the LA-curves in one colour and the OSC-curves in another to avoid mistakes in picking out solutions.

In accordance with the matching principle, we now have to pick out corresponding bendings of the two components such that $LA'_3 = LA_4$ and simultaneously $OSC'_3 = OSC_4$, for in bendings so coordinated, the coincidence of the marginal B'_{m_3} with B_{m_4} will also imply coincidence of the paraxial B'_3 and B_4 and therefore spherical correction of the complete objective. Furthermore, the equality of OSC'_3 with OSC_4 will assure that the corresponding marginal and paraxial rays will also have the same direction (or u and U values) in the space between the components and will therefore coincide along their whole length. Evidently these conditions are satisfied by those pairs of bendings which allow of the fitting-in of a perfect rectangle with horizontal and vertical sides between the four aberration

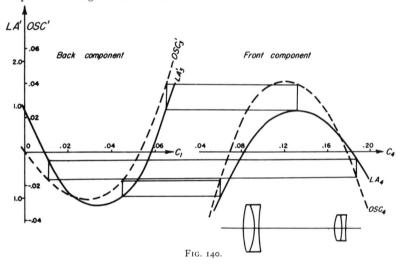

Fig. 140.

curves. In the present case there are three such solutions, as indicated in Fig. 140, but only one, namely, the lowest of the three rectangles shown, falls within the calculated range of bendings. For the highest rectangle, the curves of component (a) have to be extrapolated towards the right and for the very long and narrow intermediate rectangle, the curves of component (b) require similar extrapolation towards the right. The easiest way of finding these rectangles, which we shall call 'matching rectangles', is to slide a ruler over the diagram, always keeping it parallel to the ruled horizontal lines, and to note the vertical distance between the respective LA and OSC curves at each cut of the ruler with, say, the two LA curves. It should not be taken for granted that there are always three solutions; on the contrary, two is the more usual number and there may be not even one useful solution if the chosen glass is very abnormal. When the rectangles have been located and drawn, we read off the coordinated values of c_1 and c_4 corresponding to the respective vertical sides by the horizontal scale of the diagram. A correct [104] diagram thus gives the following three direct solutions:

	c_1	c_4
Top rectangle:	0.0648	0.1332
Long rectangle:	0.0098	0.1876
Lowest rectangle:	0.0447	0.0614

By the total curvature of the respective crown lenses $c_a = 0.1313$ and $c_c = 0.2437$, we then obtain $c_2 = c_1 - c_a$ and $c_5 = c_4 - c_c$ and then the radii by Barlow's tables. For the top solution, which we shall follow up here, we thus find $c_2 = -0.0665$ and $c_5 = -0.1105$ and the prescription:

$$r_1 = \quad 15.43 \qquad\qquad r_4 = \quad 7.508$$
$$2.7 \text{ thick} \qquad\qquad 1.8 \text{ thick}$$
$$r_2 = -15.04 \qquad\qquad r_5 = -9.050$$
$$1.5 \text{ thick} \qquad\qquad 1.0 \text{ thick}$$
$$r_3 \text{ by } (d-D) \qquad\qquad r_6 \text{ by } (d-D)$$

For the purpose of testing and finally correcting this solution, we then trace the usual paraxial and marginal rays, of course with the original initial data $l_1 = L_1 = -170$ and $u_1 = \sin U_1 = -0.0300$. It will nearly always be advisable to carry out these final calculations with five-figure logs and with angles to the nearest second. The figures to be given presently were, however, obtained by careful four-figure work. After tracing the two rays through surfaces (1) and (2), the $(d'-D')$ solution gives $r_3 = -1483$ mm. Using this value, we find for the first component by the check-method:

$$l'_3 = 43.311, \quad L'_3 = 42.521, \quad \log u'_3 = 9.0400, \quad \log \sin U'_3 = 9.0561;$$

and with these $LA'_3 = 0.790$ and $OSC'_3 = 0.0364$, which may profitably be compared (after adjustment to the standard U'_3) with the graph in order to see how near the latter comes to predicting the truth. For the purpose of testing the solution, we however go straight on with the direct results obtained for the back component. The latter having given $L'_3 = 42.521$ and the front components being designed to work with $L_4 = 25.000$, we find the necessary air-space $d'_3 = 42.521 - 25 = 17.521$ mm., and with this, $l_4 = 43.311 - 17.521 = 25.790$ mm., and the starting data for the front combination are therefore

$$l_4 = 25.790, \quad L_4 = 25.000, \quad \log u_4 = 9.0400, \quad \log \sin U_4 = 9.0561.$$

For the actual solution, it is only necessary to trace the marginal ray; but a keen designer will want to increase his stock of experience by finding out how near to perfection is the solution taken directly from the graph, and he will therefore add the slight labour of tracing also the paraxial ray. The $(d'-D')$ solution will give $r_6 = +64.14$, and with this, by the check formulae,

$$l'_6 = 9.034, \quad L'_6 = 9.055, \quad \log u'_6 = 9.4051, \quad \log \sin U'_6 = 9.4054$$

or $LA'_6 = -0.021$ mm. and $OSC'_6 = +0.0007$ (less than one third of the 0.0025 tolerance).

[104]   For LA' the Rayleigh limit tolerance is 0.002 mm$/\sin^2 U' = \pm 0.031$, and the direct solution is therefore also well within this. A self-respecting designer should nevertheless stick to his rule never to draw upon the tolerances when more complete correction means only a little extra work, and he should therefore proceed to the following perfectly systematic method of correcting the graphical solution in one operation.

Having traced the marginal ray through the front component with the results already given, we proceed exactly as in the preparation of the graph by tracing a paraxial ray backwards from the marginal focus already found, that is, right-to-left with

$$l'_6 = \quad 9.055 \qquad \log u'_6 = 9.4054,$$

thus finding

$$l_4 = 25.890 \qquad \log u_4 = 9.0394$$

and by comparison with the corresponding

$$L_4 = 25.000 \quad \log \sin U_4 = 9.0561$$
$$LA_4 = \quad 0.890 \qquad OSC_4 = 0.0377.$$

As the back component of our direct solution gave

$$LA'_3 = \quad 0.790 \qquad OSC'_3 = 0.0364$$

we see that the two components fail to match perfectly and that a slight further bending of the components will be required to effect a perfect match. Since the aberrations plot as smooth curves of comparatively small curvature, the change of any one of them for a small alteration δc of the bending will be closely represented by the first term of Taylor's series:

$$\delta c \cdot \frac{d(LA)}{dc} \quad \text{or} \quad \delta c \cdot \frac{d(OSC)}{dc}$$

and as we seek the values of δc_1 and δc_4 which will lead to equal values of LA and of OSC, we must determine δc_1 and δc_4 from the two equations:

$$LA'_3 + \delta c_1 \cdot d(LA'_3)/dc_1 = LA_4 + \delta c_4 \cdot d(LA_4)/dc_4$$
$$OSC'_3 + \delta c_1 \cdot d(OSC'_3)/dc_1 = OSC_4 + \delta c_4 \cdot d(OSC_4)/dc_4$$

or, slightly transposed,

MP(1) $\begin{cases} \delta c_1 \cdot d(LA'_3)/dc_1 - \delta c_4 \cdot d(LA_4)/dc_4 = LA_4 - LA'_3 \\ \delta c_1 \cdot d(OSC'_3)/dc_1 - \delta c_4 \cdot d(OSC_4)/dc_4 = OSC_4 - OSC'_3. \end{cases}$

The four differential coefficients on which the solution depends can be taken directly from the graph, for they are merely the gradients or slopes of the tangents of the four curves at the respective corners of the matching rectangle, expressed of course with due regard for the plotting scale. For a curve rising towards the right, the differential coefficient will be positive.

For the LA'_3 curve, as an example, the tangent at the upper left-hand corner of the matching rectangle forms a right-angled triangle with the horizontal and [104] vertical rulings of the squared paper such that a vertical side (dLA'_3) equal to 1.45 inches corresponds to a horizontal side (dc_1) of 0.5 inch. Since the vertical scale is 1 inch = 1 mm. of LA'_3 and the horizontal scale for the back component is 1 inch = 0.02 of curvature, we have

$$\frac{dLA'_3}{dc_1} = \frac{1.45}{0.01} = 145.$$

For OSC_4 a downward-sloping tangent results which gives a triangle in which a vertical side of -0.34 inch corresponds to a horizontal side of $+1.0$ inch. Since one vertical inch corresponds to 0.05 of OSC_4 and for the front component one horizontal inch corresponds to 0.04 of curvature c_4, we find

$$\frac{dOSC_4}{dc_4} = \frac{(-0.34)(0.05)}{1 \times 0.04} = \frac{-0.017}{0.04} = -0.4.$$

Working out the other two in the same way, we find $dOSC'_3/dc_1 = 4.0$ and $dLA_4/dc_4 = 0$, the latter corresponding to the horizontal tangent at the pole of the LA_4-curve, on which the corner of the matching rectangle happens to fall. MP(1) therefore gives the two equations

$$145\,\delta c_1 \qquad\qquad = 0.100$$
$$4\,\delta c_1 + 0.4\,\delta c_4 = 0.0013$$

which are easy to work out numerically. In the present case, the first gives directly $\delta c_1 = 0.0007$, and using this in the second we obtain

$$0.4\,\delta c_4 = 0.0013 - 0.0028 = -0.0015$$
or $$\delta c_4 = -0.0038$$

As the direct solution gave for the back combination $c_1 = 0.0648$, we have for the corrected solution $c'_1 = 0.0648 + 0.0007 = 0.0655$ and $c'_2 = 0.0655 - 0.1313 = -0.0658$. Similarly, from the direct solution $c_4 = 0.1332$ and $c'_4 = 0.1332 - 0.0038 = 0.1294$, and with this $c'_5 = 0.1294 - 0.2437 = -0.1143$.

We should therefore secure perfect correction within the possible uncertainty of the numerical calculations with the prescription:

$$r_1 = \quad 15.27$$
$$\qquad\qquad 2.7 \text{ thick}$$
$$r_2 = -15.20$$
$$\qquad\qquad 1.5 \text{ thick}$$
$$r_3 \text{ by } (d' - D')$$

$$r_4 = \quad 7.728$$
$$\qquad\qquad 1.8 \text{ thick}$$
$$r_5 = -8.749$$
$$\qquad\qquad 1.0 \text{ thick}$$
$$r_6 \text{ by } (d' - D')$$

A test of this solution gives $r_3 = +7403$ mm.; with this, $l'_3 = 43.331$, $L'_3 = 42.442$, $\log u'_3 = 9.0392$, $\log \sin U'_3 = 9.0566$, and for the transfer, $d'_3 = 42.442 - 25.000 = 17.442$. Taking both rays straight through the front combination with this separation, we find $r_6 = +95.14$ and $l'_6 = 9.074$, $L'_6 = 9.077$, $\log u'_6 = 9.4061+$, and $\log \sin U'_6 = 9.4063-$. A plus and a minus sign have been added

[104] to the logs of the angles because it happens to be a particularly unfortunate co-incidence that in the final u' and U', the rounded-off figures have nearly the maximum possible value in opposite senses, so that the difference of the logs is practically 0.0001 and not 0.0002 as the rounded-off figures would suggest. Taking the truer value of the difference, namely 0.0001, the final result is

$$LA'_6 = -0.003 \qquad OSC'_6 = +0.0002$$

both of which are within one tenth part of their respective tolerances. If the direct and the final solution had been worked out with five-figure logs, the residuals would undoubtedly have come out correspondingly smaller, provided that proportionately greater care had been taken in determining the differential coefficients in the final correction. The latter will require particular attention when the four tangents happen to approach parallelism because in that case a small failure in matching will lead to large changes in c_1 and c_4, and a second approximation may occasionally be necessary. It should be specially noted that if five-figure accuracy of the final solution is aimed at, the test of the graphical solution must be done with five-figure logs as the differential correction would transfer the full residual uncertainty of a four-figure test of the graph solution to the final results.

Before accepting the solution as correct, it will be wise to guard against a possible mistake in the $(d'-D')$ solutions for r_3 and r_6 by computing OP(5)Check (page 655) for all six surfaces and forming the sum, which should be zero.

The objective which we have obtained by the final solution is highly corrected as regards spherical aberration of the marginal rays, and being chromatically corrected by the $(d'-D')$ method, it also has the best possible achromatism for its full aperture. But before passing it, we must test for zonal residuals of both chromatic and spherical aberrations.

As regards zonal spherical aberration, the least laborious method, and one quite sufficient up to numerical apertures of 0.50 or even 0.60, consists in calculating, from the paraxial data of the trigonometrical calculation, the paraxial OPD'_p-sum and comparing this with the tolerance of 6 wave-lengths. As the present objective has its linear dimensions expressed in millimetres, the formulae will give the result directly in nominal wave-lengths of 0.0005 mm. when calculated for each surface by

OP(4p)   Paraxial $OPD'_p = 250N'y \cdot i'(u-u')(u-i')$

or for a plane: $= 250N'y \cdot u'(u'-u)(u+u')$.

The individual amounts thus found give the sum by simple algebraic addition. The y, if not already used for the check formula, are obtained by the relation $y = l \cdot u = l' \cdot u'$.

For the present objective, the separate values are $+15.56$, -9.78, $+0.95$, $+0.96$, -9.11, and $+4.68$, and the algebraic sum of these gives $+3.26$.

As the tolerance is 6 wave-lengths, the objective would be practically perfect with regard to spherical correction at the calculated aperture. We can decide what would be the safe maximum aperture by the formula given on page 662, which for the present case, when the 'original aperture' $= \sin U'_6 = 0.255$, gives

Permissible $NA = 0.255 \sqrt[6]{6/3.26} = 0.282.$

It is remarkable and may be taken as evidence either of the reliability of our [104] tolerances or of the good judgement of empirical computers that $NA = 0.28$ is just about the aperture given to Lister objectives having the focal length of our specimen.

When it is decided that such an objective is to be raised to the maximum permissible aperture, then it will not be sufficient merely to calculate the increased clear aperture of the lenses and to send the original prescription into the workshop. The test of the graphical solution and the final correction must be repeated for the increased aperture in order to secure perfect correction of spherical and chromatic aberration and zero OSC' at the increased aperture. It will, however, not usually be necessary to recompute the original bendings of the components. The tracing of the rays corresponding to the increased aperture through the original graphical solution will merely give somewhat different values of LA'_3, OSC'_3, LA_4 and OSC_4. These are then reduced to their equivalents at the original aperture by the square-of-the-aperture law, and the reduced values are used for the final correction by the slope of the curves of the graph exactly as has been described.

Zonal variation of the chromatic correction is most conveniently determined by calculating the paraxial $(d' - D')_p$-sum either by OP(5p) (page 649), working by the example (page 651), or by the paraxial form of OP(5)Check (page 657). For our present objective, both methods give the sum, in approximate wave-lengths, as $+1.62$ and therefore within the tolerance of 2 wave-lengths. This residual grows with the fourth power of the aperture and would rise to more than the tolerance at the permissible maximum aperture calculated above from the zonal spherical aberration. But owing to the greater elasticity of chromatic tolerances frequently referred to in these pages, it is not necessary to take a very serious view of moderate transgressions of chromatic tolerances.

These tests for zonal spherical and chromatic aberration should never be omitted in working out a new design; they become more important as the numerical aperture increases.

In addition to the direct solutions obtainable from the graph of Fig. 140 by the matching rectangle, we can obtain by a simple modification all other aplanatic doublets (an infinite number of them!) possible with the selected kinds of glass and the prescribed value of l_1, sin U'_6, and M', in which each component produces half of the total deviation. This huge extension of the possibilities—particularly valuable in the design of photographic objectives by the matching principle—depends on the fact frequently emphasized in preceding chapters that every optical calculation can be carried out on any linear scale without altering the essential properties of the system. In the present case, this means that if the front component were made on a smaller or larger scale, applied uniformly to all linear data like radii, thicknesses, and initial L_4, but with retention of the adopted standard value of U_4, then the LA_4 would come out smaller or larger to scale. Nevertheless OSC_4, being a pure function of the angles, which are ratios or pure numbers and which are therefore independent of the linear scale, would remain absolutely unchanged.

A numerical example will make the significance of this rather abstract proposition clear. A correctly drawn graph like Fig. 140 will show that a back component

[104] with $c_1 = 0.0563$ has $OSC'_3 = +0.0100$ and $LA'_3 = -0.200$ mm. A front component with $c_4 = 0.0748$ also has $OSC_4 = 0.0100$ and will satisfy the matching principle as far as OSC is concerned no matter to what linear scale the front component is made. On the other hand, at $c_4 = 0.0748$ we read $LA_4 = -0.32$ mm., or, to the originally contemplated scale, 1.6 times as large as LA'_3. There would therefore be no correct matching on the original scale. But if we make the front component to 5/8 of the original scale, its LA_4 will also sink to 5/8 of the plotted value and will become equal to LA'_3, thus producing a complete match.

It is now quite simple to make up a prescription for the selected new solution. For the back component, which is not to be altered as to scale, $c_1 = 0.0563$ gives $c_2 = 0.0563 - 0.1313 = -0.0750$, or by taking reciprocals

$$r_1 = \quad 17.76$$
$$\qquad\qquad 2.7 \text{ thick}$$
$$r_2 = -13.33$$
$$\qquad\qquad 1.5 \text{ thick}$$
$$r_3 \text{ by } (d' - D')$$

For the front component, the c_4 taken from the graph is 0.0748 and therefore $c_5 = 0.0748 - 0.2437 = -0.1689$. To the plotted scale the front component would therefore be

$$r_4 = \quad 13.37$$
$$\qquad\qquad 1.8$$
$$r_5 = - \quad 5.921$$
$$\qquad\qquad 1.0$$
$$r_6 \text{ by } (d' - D')$$

and its initial L_4 would be 25.000, as originally chosen. To secure a perfect match, it must however be executed to 5/8 of the original scale: therefore the true prescription is found by dividing all the preceding figures by 1.6 to give

$$r_4 = \quad 8.356$$
$$\qquad\qquad 1.12$$
$$r_5 = -3.701$$
$$\qquad\qquad 0.625$$
$$r_6 \text{ by } (d' - D')$$

and the air-space between the components will have to be adjusted so that $L_4 = 25/1.6 = 15.625$; hence $d'_3 = L'_3 - 15.625$.

When this objective has been trigonometrically tested by the method described, it can be finally corrected as follows:

Suppose (the figures are fictitious, not calculated!) the result were

$$LA'_3 = -0.198 \qquad OSC'_3 = +0.0102$$
$$LA_4 = -0.201 \qquad OSC_4 = +0.0099.$$

We leave the black component untouched, so that the calculations for it are not

disturbed, and seek the additional bending of the front component and the supple- [104]
mentary change of scale which will cause OSC_4 to become equal to OSC'_3 and
LA_4 equal to LA'_3. As the OSC-values are independent of scale, we can determine
the necessary bending at once from the graph, for with an unchanging back component, or $\delta c_1 = 0$, the second of MP(1) gives, to the graph scale,

$$\delta c_4 \cdot d(OSC_4)/dc_4 = OSC'_3 - OSC_4 = 0.0003$$

in our case. By the method already described, the graph gives, at $c_4 = 0.0748$,
$d(OSC_4)/dc_4 = 0.062/0.040 = 1.55$, therefore $\delta c_4 = 0.0003/1.55 = 0.0002$, or the
corrected $c_4 = 0.0750$. This to the graph scale gives $c_5 = 0.0750 - 0.2437 =$
-0.1687 and the full-scale prescription for the adjusted front component becomes

$$
\begin{array}{ll}
r_4 = & 13.333 \\
& \qquad 1.8 \\
r_5 = & -5.928 \\
& \qquad 1.0 \\
r_6 \text{ by } (d' - D')
\end{array}
$$

The bending of the front component by δc_4 will be accompanied by a change in
the spherical aberration by $\delta LA_4 = \delta c_4 \cdot dLA_4/dc_4$. The differential coefficient is
found from the graph to be 45.5, so

$$\delta LA_4 = 0.0002 \times 45.5 = 0.0091.$$

The real front component was made to 5/8 or 0.625 of the plotted scale; δLA_4
would undergo a corresponding reduction to $0.0091 \times 0.625 = 0.0057$, or, as we
only retain three decimal places in millimetres, 0.006. If the bent front component
were made to the 5/8 scale, it would therefore give $LA_4 = -0.201 + 0.006 =$
-0.195, and this would not match the $LA'_3 = -0.198$ of the back component.
As the numerical value of LA_4 requires to be increased, a change of scale is indicated, which evidently must be equal to $-0.198/-0.195$ or, as the scale originally
tried was 0.625, the final front component must be made to the scale $0.625 \times$
$198/195 = 0.634$, and the application of this factor to the above full-scale prescription for the front component gives as the final data

$$
\begin{array}{ll}
r_4 = & 8.453 \\
& \qquad 1.14 \\
r_5 = & -3.758 \\
& \qquad 0.634 \\
r_6 \text{ by } (d' - D')
\end{array}
$$

to be calculated with $L_4 = 25 \times 0.634 = 15.85$. Therefore $d'_3 = L'_3 - 15.85$.
If our assumed figures were real, this amended prescription would give a very close
approximation to complete correction.

It will be noticed that this method of final correction is really simpler than the
first one by the matching rectangle, and may advantageously be substituted for it.
It also renders it easier to make small changes dictated by common sense. As an
example, the back component of our original final solution ($r_1 = 15.27$, $r_2 =$

[104] -15.20, $r_3 = 7403$) would arouse scorn and derision in the glass shop, for it is a great nuisance to have to distinguish the two faces of a crown lens so absurdly close to equiconvexity, and in addition instinct would tell the glass-worker that there could be no serious difficulty in slightly modifying the design so as to replace the excessively long last radius by a plane. A judicious designer would therefore use as back component one with $r_1 = -r_2 = 15.23$ and $r_3 = \infty$, and he would modify the front component by the second method so as to secure quite as good a match as pedantic adhesion to the first method could give.   Any slight residual of chromatic aberration in the back component would of course be transferred to the front one and there corrected.   On the same principle, stock curves close to any radii resulting from the first solution may be utilized.

The choice between the numerous solutions which can be extracted from the graph of Fig. 140 will usually be decided by the magnitude of the zonal spherical and chromatic variation, and it should be borne in mind that still better results may be obtainable by a complete new solution in which, say, 0.6 of the total deviation is produced by the back lens and only 0.4 by the front lens.   If several solutions are found between which there is not much to choose as regards zonal aberration, then the choice should be decided in favour of the form which gives the flattest field.

Good solutions will usually have crown components which do not depart very widely from equiconvexity.   When the object is not, as in our example, to unearth all the possible solutions, both good ones and bad ones, then three bendings of each component will usually be sufficient.   It will be wise to begin with the equiconvex crowns as a first bending, and with deepened counterfaces for a second bending. A preliminary plotting of these will probably suggest whereabout the solution is going to fall, and the third bendings of both components can then be selected accordingly.

DESIGN OF OBJECTIVES OF THE AMICI TYPE

105] THIS type (Fig. 139e) is brought under the matching principle by the simple expedient of selecting a promising fixed form of the thick planoconvex front lens for any one complete solution.   In this way the back and middle components become the only parts subjected to change by bending, and the solution is then obtained from a graph by the matching rectangle.   We must therefore begin by studying the important properties of the thick planoconvex front lens.

As, in our computing direction of the rays, the front lens receives strongly converging light, its curved side could easily be rendered completely aplanatic by making $r = L/(N' + 1)$ and a great increase in convergence could thus be secured without incurring any spherical aberration or coma.   It is, however, nearly always preferable to depart to a moderate extent from the strictly aplanatic curvature. If we make r longer than $L/(N' + 1)$, we shall secure less increase of convergence, but as $(I' - U)$ then becomes negative, the refraction is accompanied by both spherical aberration and astigmatism in the over-corrected sense, and it will be shown subsequently that this has a strong tendency towards flattening the field of view, which is highly desirable.   Hence a departure from the strictly aplanatic curvature in this sense is found in nearly all commercial objectives for everyday use on objects which are usually well within the limit of resolution but which are desired to be in clear view right across the entire field. The opposite departure from the aplanatic radius, making it shorter than $L/(N'+1)$ by a moderate amount, has the advantage of increasing the convergence at a comparatively cheap price in spherical under-correction, and it can be very effective in reducing the zonal residues of spherical and chromatic aberration. The disadvantage is that the field is almost certain to become decidedly curved. This type of front lens is therefore only met with, as a rule, in high-priced objectives in which the highest resolving and defining power is the most important desideratum and is appreciated even when only two thirds of the field are really useful at any one adjustment of the focus.

We shall aim at the generally preferred type; it may be accepted that a useful, but not excessive, amount of the desirable over-correction will result when the radius is so chosen as to yield a value of $I' - U = U' - I$ between $-4°$ and $-8°$ for the marginal ray. That will fix the radius within fairly narrow limits, so that the selection of one definite value does not severely restrict the general utility of the resulting particular solution of a given problem.

If we next consider the flat side of the front lens, we notice at once that on account of the strong convergence of the rays received by the plane surface in the computing direction, very large angles of incidence and refraction will result which may reach and even exceed 60°. Such large angles will lead to extremely heavy spherical aberration, and the only way of keeping this down to a tolerable amount will be to reduce the Y at which the marginal rays pass through the plane. That means a short clear distance to the object-point, and as this is a decidedly undesirable feature, it becomes important to have a criterion by which we may estimate the largest working distance that may safely be attempted at any prescribed numerical aperture of a 'dry' objective.

It was shown in Chapter XIV that in the higher-power dry objectives control can be gained over the secondary spherical aberration, and that then the tolerance for the outstanding tertiary and still higher aberration is of the order of 8 wavelengths. Therefore we must try to ascertain at what working distance the higher aberration reaches this danger point for various numerical apertures. For that purpose, we compute OPD'_m with full five-figure accuracy by OP(4) and by its paraxial form for a suitable range of numerical apertures. The variation for different refractive indices is not very serious, and we therefore take 1.55 as a fair average value. The OPD is strictly proportional to the working distance, so we can find it by proportion for any distance if we calculate for one. Let us select 1 mm. Calculating thus, right-to-left, from an object point at $L' = 1$ mm. and analysing the results for the coefficients of the series

$$OPD'_m = c_4 \cdot NA^4 + c_6 \cdot NA^6 + \&c.$$

the numerical coefficients obtained by successive approximations are found to be, in nominal wave-lengths of 0.0005 mm.,

$$OPD'_m = 145.94 NA^4 + 176.3 NA^6 + 178.8 NA^8 + \&c.$$

[105] The following table results for the assumed free working distance of 1 mm.:

NA	Total OPD'_m	Primary	Secondary	Tertiary and higher
0.1	0.015	0.015	—	—
0.2	0.245	0.235	0.01	—
0.3	1.32	1.18	0.13	0.01
0.4	4.60	3.74	0.72	0.14
0.5	12.79	9.12	2.75	0.92
0.6	31.8	18.9	8.2	4.7
0.7	75.3	35.0	20.7	19.6
0.75	116.7	46.2	31.3	39.2
0.8	184.2	59.8	46.2	78.2
0.85	302.8	76.2	66.5	160.1

It will be seen that the aberrations above the secondary become notable at about $NA = 0.5$ and then rush up to huge amounts at apertures of 0.8 and more, so a working distance of 1 mm. is usually out of the question for numerical apertures above 0.7. Nevertheless, we must remember that the dispersing contact-surfaces of the back and middle combinations will also produce a proportion of tertiary and higher aberration which, being negative, will compensate some of that arising at the front plane. It is therefore not necessarily hopeless to try to obtain correction within the Rayleigh limit when our table would indicate tertiary aberration in excess of our tolerance of 8 wave-lengths for the extreme marginal zone. Bearing in mind the proportional change of OPD' and L', we may put down the following as a fair estimate of working distances which may be attempted with considerable hope of success:

W.D.	NA
1.0 mm.	0.66
0.5	0.72
0.25	0.76
0.125	0.80

Only unusual luck or exceptional endurance in trying modifications will secure correction for the whole aperture within the Rayleigh limit if these limiting working distances are exceeded to any great extent.

The numerous dry lenses of high aperture which give good results although they have working distances or apertures beyond our limits do so only under the usual and time-honoured system of working with an illuminating cone filling at most 3/4 of the full aperture, so that only feeble diffracted light passes through the marginal zone. They all break down when a full illuminating cone is attempted, and nearly all of them are distinctly *improved* when a stop is applied to the back lens which cuts down the NA to 0.8 or less. The least harmful way of computing objectives to meet the persistent demand for impracticably high NA's is to correct as well as possible for 0.8 NA or even a little less, and then to open out the aperture to whatever actual though useless NA people are currently insisting upon! The [105] morality of this procedure is another question.

We shall now develop the systematic method of procedure in the designing of objectives of the Amici type (and of numerous other aplanatic triplets in which the form of one of the three components can be assumed *a priori* on a reasonable basis) and then illustrate the technique by a numerical example. For the latter we will choose the problem:

'Required an objective of the Amici type, of about 8-mm. equivalent focal length, of 0.66 NA, and for use with a tube-length of 170 mm.'

The problem is framed in a way differing from that used for the Lister type to illustrate another and more usual opening of a calculation. The tube-length is always understood as the 'mechanical' one, i.e., the distance from the upper end of the draw-tube, at its normal adjustment, to the shoulder against which the objective mount is screwed (Fig. 141). The objective must form its image in the

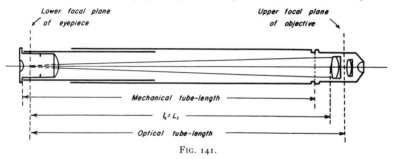

Lower focal plane of eyepiece

Upper focal plane of objective

Mechanical tube-length

$l_1 = L_1$

Optical tube-length

FIG. 141.

lower focal plane of the eyepiece, which we shall assume to lie 10 mm. down the draw-tube—but this figure varies. The linear magnification produced by the objective will, by the general theory of lenses, be found by dividing the distance from the lower focal plane of the eyepiece to the upper focal plane of the objective (the 'optical' tube-length) by the equivalent focal length of the objective. We must therefore estimate next the location of the upper focal plane of the objective with reference to the shoulder of its screw, which corresponds with the lower end of the microscope tube. Probably 20 mm. will not be far wrong and this would make the optical tube-length $170 + 20 - 10 = 180$ mm., and the magnification is thus $-180/8 = -22.5$ times. We shall also require an estimate of the L with which the ray-tracing is to be begun, that is, the distance from the lower focal plane of the eyepiece to the nearest surface of the back lens. The latter is likely to lie about 10 mm. above the upper focal plane of the objective: that would make L 10 mm. less than the optical tube-length, or $L = -170$ mm.

The inevitable uncertainties in this opening have been duly emphasized; but it is not necessary to worry about them as the finished objective can be easily adjusted so as to compensate for the small errors resulting from the inaccurate assumptions.

We have fixed the initial L as -170 mm. and the magnification M as -22.5.

[105] We must next find the initial U. Since our objective, being of comparatively high NA, will have to be finally corrected for the best distribution of the tertiary aberration by the last method of Chapter XIV, it should be computed throughout for the submarginal zone of $10/11$ of the full aperture. Therefore, by the calculation $NA = \sin U'_8 = 0.66 \times 10/11 = 0.60$ and by the sine condition we get $\sin U_1 = \sin U'_8/M = 0.6/(-22.5) = -0.0267$.

In our computing direction, the starting data will therefore be $L_1 = l_1 = -170$ and $u_1 = \sin U_1 = -0.0267$, and for the fixing of a safe thickness of the back crown, we have with sufficient approximation computed $Y = l_1 \cdot u_1 = -170 \times -0.0267 = 4.54$, whence the actual clear aperture is $1.1 \times 9.08 =$ practically 10 mm.

Having thus fixed some of the principal data, we are ready to begin the real designing. We shall name the constituent lenses in the direction from eyepiece towards the object a, b, c, d, and e, and the eight surfaces in the same order from 1 to 8, so that the extreme back lens is 'a' and is bounded by r_1 and r_2 while the front lens is 'e' and is bounded by r_7 and r_8, with $r_8 = \infty$. We begin with the last surface.

A reasonable and usual free working distance for an 8-mm. objective is represented by 1 mm. and this we will adopt: therefore $l'_8 = L'_8 = 1$ mm. As $\sin U'_8 = 0.6$ is stipulated for our objective, we can begin a right-to-left calculation with the data just stated, coupled as usual with $u'_8 = \sin U'_8 = 0.6$ for the paraxial ray. We must however choose the glass first, and we will select Chance's light barium crown No. 4317, for which

$$Nd = 1.5407 \quad Nf - Nc = 0.00910 \quad V = 59.4.$$

These data give for the brightest light $N_y = 1.5424$ or $\log N_y = 0.1882$. With this index we find in a few minutes

$$L_8 = 1.7762 \text{ mm.} \quad U_8 = 22\text{-}53\text{-}40 \quad \log \sin U_8 = 9.5900$$
$$l_8 = 1.5424 \text{ mm.} \quad u_8 = 0.38905 \quad \log u_8 = 9.5900.$$

We must now decide how thick our front lens shall be. The thicker we make it, the longer will be L'_7 and therefore also r_7, as the latter will be determined by a fairly close approach to the condition of aplanatic refraction. We must, however, not make the thickness too great as we must allow for the expansion of the cone of rays in the back lenses, at the last surface of which it is to attain a diameter of 9.08 mm. We will choose $d'_7 = 5$ mm. That will give $l'_7 = 1.5424 + 5 = 6.5424$ mm. and $L'_7 = 1.7762 + 5 = 6.7762$ mm. As at the seventh surface we have $N = 1$ and $N' = 1.5424$, the condition for aplanatic refraction would be realized by $r_7 = 6.776 \times 1.5424/2.5424 = 4.11$ mm. In order to secure the advantages stated above, we will raise this value to $r_7 = 4.40$ mm., and we can now complete the backward tracing of the rays through the front lens with this radius and the data collected above, namely,

$$L'_7 = 6.7762 \quad \log \sin U'_7 = 9.5900 \quad l'_7 = 6.5424 \quad \log u'_7 = 9.5900.$$

For the result we shall find $L_7 = 9.5381, l_7 = 8.8906, U_7 = 16\text{-}6\text{-}50, u_7 = 0.28631$, and for the spherical aberration $LA_7 = l_7 - L_7 = -0.6475$.

It may be noted that $I_7 - U'_7$ is just below $-4°$ and is therefore of the order of [105] the departure from strict aplanatism advised in the general discussion of front lenses. A method of determining r_7 so as to secure a definite value $I_7 - U'_7$ is given in the following section on immersion objectives.

By the matching principle, we can now say that in order to produce with our selected front lens an objective of 0.6 NA, corrected for spherical aberration and the sine condition, we must design the back and middle combinations so as to secure a spherical over-correction $LA'_6 = -0.6475$ mm. and so that the paraxial ray emerges under a nominal angle $u'_6 = 0.28631$ and the marginal ray under an angle $U'_6 = 16\text{-}6\text{-}50$. The problem is therefore now reduced to the simpler one of applying the matching principle to the two cemented combinations exactly as we did in the case of an objective of the Lister type, the only difference being that we have to aim at a prescribed spherical over-correction and offence against the sine condition instead of trying to secure perfect correction in both respects.

We must first decide how we shall secure achromatism in the final solution. Our back and middle lenses will have to be chromatically over-corrected sufficiently to neutralize the chromatic aberration of the front lens. Evidently we must begin by determining the latter; but as we cannot hope to establish a perfect balance of the chromatic aberration for the full aperture, we merely make sure that the solution from the eventual graph will be nearly enough achromatic to allow of final correction by ringing the changes on the $(d' - D')dN'$ sum and by picking types and meltings of glass of the selected indices to secure the required zero value. We therefore use paraxial formulae for the preliminary achromatization and begin by adding a coloured paraxial ray to the two already traced through the front lens. We calculate $N_v = N_y + (Nf - Nc) = 1.5424 + 0.0091 = 1.5515$ in our numerical example, and find $l'_{7_v} = 8.9730$. The paraxial chromatic aberration of the front lens is therefore $l_{7_y} - l_{7_v} = 8.8906 - 8.9730 = -0.0824$ mm. We then know that the back and middle components together must supply this amount of longitudinal chromatic over-correction.

The determination of suitable forms for the back and middle components by paraxial thin-lens equations is now very similar to that employed for the Lister type.

The total deviation to be produced is $u'_6 - u_1$ or, in our case, $0.2863 - (-0.0267) = 0.3130$. There is the possibility of distributing this total in various proportions between the two components; we will again assign one half ($= 0.1565$) to each component. That gives

$$u'_3 = 0.1565 - 0.0267 = 0.1298$$

as the nominal angle of the paraxial ray between the two components.

We must next distribute the required chromatic over-correction of -0.0824 mm. between the two components. It is generally more favourable to assign rather more than half to the back component because the latter produces less zonal aberration. We therefore decide to produce an over-correction of -0.0374 mm. in the middle component and to transfer the remaining -0.045 mm. to the back component. This does *not* mean that the back component must produce -0.045 mm. of longitudinal chromatic aberration at its own focus, for as the aberration

[105] extends along the optical axis, it is depicted or transferred according to the law of *longitudinal* magnification, which means the square of the ordinary magnification. Our middle components will have $u'_6 = 0.2863$ whilst we decided upon $u'_3 = u_4 = 0.1298$. Consequently we have in the direction of the transfer, by the theorem of Lagrange, $M = 0.2863/0.1298$, and applying the square of this to the amount, -0.045 mm., to be transferred, we find: over-correction of back lens $= -0.045 \times (0.2863/0.1298)^2 = -0.218$ mm.

It should be noted that this is an extremely important correction as it leads in the present case to a change of nearly five times.

The result up to this point is:

'The back components must be designed for a chromatic over-correction of -0.218 mm. and for $u_1 = -0.0267$ and $u'_3 = 0.1298$; the middle component must produce a chromatic over-correction of -0.0374 mm. and must change the convergence from $u_4 = 0.1298$ to $u'_6 = 0.2863$.'

The requisite focal length of the back component can now be determined as before by the thin-lens equations, which give, from $l_1 u_1 = l'_3 u'_3 : l'_3 = l_1 u_1/u'_3$. With our values, this means that

$$l'_3 = -170 \times -0.0267/0.1298 = 35.0 \text{ mm.,}$$

and then from $1/l'_3 = 1/f' + 1/l_1$ by Barlow's table: $1/f' = +0.03445$ mm.

With this determined, we can calculate the total curvature c_a of the crown component on choosing the glass. As the final achromatization is to be produced by selecting the right dispersion, we introduce for the index of the crown $Nd = 1.5160$, which is obtainable with dispersions from almost 0.0080 up to more than 0.0090. Crown of very low dispersion is nearly always preferable for microscope objectives, but as experience has shown that, with our rather rough method of allowing for the preliminary chromatic correction, the final test usually discloses a moderate amount of under-correction which must be removed by substituting a glass of lower dispersion, it would be unsafe to choose the lowest obtainable dispersion at the preliminary stage. Therefore we solve for the c_a of the back lens with a crown having $Nd = 1.5160$ and a fictitious dispersion of 0.0082, which give $V_a = 0.5160/0.0082 = 62.9$. For the flint we select Chance 360: $Nd = 1.6225$, $V_b = 36.0$, $Nf - Nc = 0.01729$.

On account of the over-correction of -0.218 to be provided for, we must use TL,Chr.(4),

$$c_a = \frac{1}{f'} \cdot \frac{1}{V_a - V_b} \cdot \frac{1}{\delta N_a} - \frac{R}{\delta N_a} \cdot \frac{V_b}{V_a - V_b},$$

in which

$$R = \frac{\text{required chromatic aberration}}{l'^2} = -\frac{0.218}{35^2} = -0.000178,$$

and find

$$c_a = 0.1562 + 0.0290 = 0.1852.$$

A back combination nearly always comes out with a crown lens which is biconvex

but with the greater curvature on the contact side, and we therefore choose the [105]
three bendings

$c_1 =$		0.05	0.07	0.09

whence $c_2 =$ −0.1352   −0.1152   −0.0952

or, by Barlow, $r_1 =$ 20.00   14.29   11.11

                          $r_2 =$ −7.40   −8.68   −10.50

A scale drawing (it is likely to prove a gross waste of time to omit this!) will
show that an axial thickness of 3 mm. is sufficient for even the most unsymmetrical
first bending.   We therefore decide on this and give to the flint lens a thickness of
1.5 mm.

The trigonometrical calculations which now follow were carried out by Mr. J. S.
Watkins with four-figure logs taken from the six-figure tables.   The brightest-light
indices were found by the usual formula to be 1.5175 and 1.6258 respectively.

As only approximate chromatic correction can be hoped for by the direct solution,
a new method of solution for the last radius was employed, by which r_3 is determined so as to give a prescribed value of U'_3.   This gives in the present case
$\sin^{-1}(0.1298) = 7\text{-}27\text{-}30$ for the emerging marginal ray.   This procedure avoids
the adjustment which had to be applied to the back components of the Lister
objective in order to secure uniform matching with the front components.

The problem is: Given the angle U under which a ray arrives at a surface, also
its L for that surface (as the L' of the previous surface minus the thickness of the
lens which it is traversing), to find the value of r which will lead to a prescribed U',
the indices N and N' on the two sides of the new surface being also given, of course.

U and U' being known, we have by transposition of fundamental equation (3),
$I' - I = U - U'$; hence $I' - I$ is also known.   A transposition of equation (7) of
Part I gives

$$\tan \frac{I'+I}{2} = \frac{N+N'}{N-N'}\cdot\tan \frac{I'-I}{2} = \frac{N+N'}{N-N'}\cdot\tan \frac{U-U'}{2},$$

by which we can calculate $\frac{1}{2}(I'+I)$ and then $I = \frac{1}{2}(I'+I)-\frac{1}{2}(I'-I)$.   From
fundamental equation (1) we then obtain

$$\frac{\sin U}{\sin I} = \frac{r}{L-r}; \quad \frac{\sin U}{\sin U+\sin I} = \frac{r}{L};$$

$$r = L \sin U \Big/ \Big(2 \sin \frac{U+I}{2} \cos \frac{U-I}{2}\Big).$$

The problem is therefore solved by the three equations:

$$\mathrm{MP(2)} \quad \left\{ \begin{array}{c} \tan \dfrac{I'+I}{2} = \dfrac{N+N'}{N-N'}\cdot\tan \dfrac{U-U'}{2} \\[2mm] I = \tfrac{1}{2}(I'+I)-\tfrac{1}{2}(U-U') \\[2mm] r = L \sin U \Big/ \Big(2 \sin \dfrac{U+I}{2} \cos \dfrac{U-I}{2}\Big). \end{array} \right.$$

[105]   There is obviously less work in this solution than in that for perfect achromatism, even when the latter is carried out by the $(d' - D')$ method.

If a sufficiently close choice of dispersions is available, this solution for a prescribed U' may also be advantageously employed in the trigonometrical designing of simple achromatic lenses so as to realize exactly the prescribed equivalent focal length or magnification. It is also available under the same conditions as a simplification of the solution for objectives of the Lister type.

Using this method, the following results are obtained for the three bendings of the back component:

$c_1 =$	0.05	0.07	0.09
$r_3 =$	-23.88	-52.30	$+151.6$
$LA'_3 =$	-2.964	-1.787	$+0.260$
$OSC'_3 =$	-0.0824	-0.0352	$+0.0382$
$L'_3 =$	33.836	32.951	32.122

and the LA' and (uncorrected!) OSC' are ready for plotting by the simple parabola dip-method.

The middle lens can now be dealt with. In order to minimize the difference of magnification for different colours which inevitably results from the combination of chromatically over-corrected back and middle combinations with a non-achromatic front lens, it is desirable to keep the air-spaces down to a minimum. As all the middle lenses must be calculated for the same value of L_4, this means that L_4 must be chosen only a little shorter than the shortest L'_3 found for the bendings of the back component. $L_4 = 31.5$ was therefore decided upon by Mr. Watkins.

The rough determination of the proportions of the middle component by paraxial thin-lens formulae can then be carried out exactly as for the Lister type. As we do not know how much spherical aberration the eventually selected bending of the back component will introduce, we assume $l_4 = L_4 = 31.5$, and as the middle component must change the paraxial convergence from $u_4 = 0.1298$ to $u'_6 = 0.2863$, we find by the thin-lens relation $l_4 u_4 = l'_6 u'_6$, $l'_6 = 31.5 \times 0.1298/0.2863 = 14.3$, and then by $1/l'_6 = 1/f' + 1/l_4$, we find $1/f' = 0.03818$.

This middle combination is to produce a chromatic over-correction of -0.0374 and is therefore calculated by TL,Chr.(4), which gives as the total curvature of the crown component $c_c = 0.1731 + 0.0300 = 0.2031$ for the same kinds of glass as in the back combination.

Middle lenses usually do not depart very far from the equiconvex form of the crown, hence the bendings were selected as

$c_4 =$	0.08	0.10	0.12
giving $c_5 =$	-0.1231	-0.1031	-0.0831
or $r_4 =$	12.50	10.000	8.333
$r_5 =$	-8.123	-9.699	-12.034

The same thicknesses as in the back component, namely, 3.0 for the crown and 1.5 for the flint, were found to be suitable for the middle combination.

Each one of these bendings must now be trigonometrically fitted to our selected [105] front lens. For that purpose, we first trace the marginal ray in the left-to-right direction with $L_4 = 31.5$ and $U_4 = 7\text{-}27\text{-}30$ through surfaces (4) and (5). In order to match the slope of the marginal ray of our front lens, the radius of the sixth surface must now be determined by MP(2) so as to produce $U'_6 = U_7 = 16\text{-}6\text{-}50$, with the results for the three bendings

$$c_4 = \qquad 0.08 \qquad\qquad 0.10 \qquad\qquad 0.12$$
$$r_6 = \quad -53.248 \qquad +132.46 \qquad +26.804$$

and from a trace of the marginal ray,

$$L'_6 = \qquad 11.705 \qquad\qquad 11.340 \qquad\qquad 10.970$$

A right-to-left paraxial ray must then be matched with the paraxial ray previously traced through the front lens, when u_7 was found to be 0.28631 and the longitudinal spherical aberration to be -0.6475. Allowing for this, we find the proper values of l'_6 for tracing the reversed paraxial ray through the bendings to be

$$l'_6 = \qquad 11.0575 \qquad\quad 10.6925 \qquad\quad 10.3225$$
$$= L'_6 - 0.6475$$

and this ray can now be traced with $u'_6 = 0.28631$ to give

$$l_4 = \qquad 30.588 \qquad\qquad 30.640 \qquad\qquad 30.216$$
$$u_4 = \qquad 0.12995 \qquad\quad 0.13081 \qquad\quad 0.13383$$

Comparison of the l_4 with fixed $L_4 = 31.500$ and of the u_4 with the fixed log sin $U_4 = 9.1133$ then gives the LA_4 and OSC_4 for plotting as

$$c_4 = \qquad 0.08 \qquad\qquad 0.10 \qquad\qquad 0.12$$
$$LA_4 = \quad -0.912 \qquad\quad -0.860 \qquad\quad -1.284$$
$$OSC_4 = \quad -0.0012 \qquad\quad -0.0076 \qquad\quad -0.0311$$

On plotting these alongside the corresponding values of LA'_3 and OSC'_3 of the back components exactly as in the case of the Lister type, the matching rectangle (Fig. 142) will be found to fit at $c_1 = 0.08$ and $c_4 = 0.083$, which, with $c_a = 0.1852$ and $c_c = 0.2031$, give the partial prescription for the solution:

$$r_1 = \quad 12.5 \qquad\qquad r_4 = \quad 12.048$$
$$\qquad\qquad\quad 3.0 \qquad\qquad\qquad\qquad\quad 3.0$$
$$r_2 = -9.506 \qquad\qquad r_5 = -8.326$$
$$\qquad\qquad\quad 1.5 \qquad\qquad\qquad\qquad\quad 1.5$$
$$r_3 \text{ by MP(2)} \qquad\qquad r_6 \text{ by MP(2)}$$

This can now be submitted to trigonometrical test. MP(2) gives $r_3 = -150.94$, and with this

$$l'_3 = 31.645 \qquad L'_3 = 32.539.$$

[105] Strictly, the air-space should be such as to make $L_4 = 31.500$ and therefore $d'_6 = 1.039$. The air-space was, however, found by plotting the three values required by the original bendings of the back component and was thus found to be 1.01 mm. The middle component was therefore computed with $l_4 = 30.635$ and $L_4 = 31.529$, and taking both rays straight through, MP(2) gave $r_6 = -67.0$ and

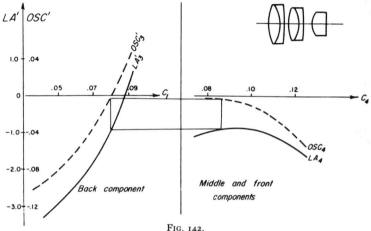

FIG. 142.

$l'_6 = 11.003$, $L'_6 = 11.671$. The air-space should really produce $L_7 = 9.538$ as calculated originally right-to-left through the front, which would give $d'_6 = 11.671 - 9.538 = 2.133$, but the graphical method was again used and gave $d'_6 = 2.09$ and therefore $l_7 = 8.913$ and $L_7 = 9.581$. On taking the rays through the front lens as originally selected ($r_7 = 4.4$, $d'_7 = 5.0$), the final residual aberrations were found to be

$$LA'_8 = -0.0016 \qquad OSC'_8 = +0.0005,$$

again a very close result and one well within our tolerances by the straightforward graphical solution

Although the solution taken directly from the graph is very well corrected, it must be brought to perfect spherical correction before the crucial test for zonal aberration by OPD'_m can be legitimately applied. It is therefore desirable, though more a matter of computer's pride than of practical necessity, also to remove the residual OSC' of one fifth of our tolerance. This could be done by a further slight bending of the corrective combinations by either of the two methods fully described in connection with the Lister type. But in objectives of the Amici type, it is far less laborious to establish the exact correction by modifying the front lens.

For this purpose another very simple and practically direct solution is available which the author has used for many years. Let the system be one of k surfaces (in the above example, $k = 8$); the curved side of the front lens therefore will have the suffix $(k-1)$ and its plane surface will have the suffix k. The marginal and

[105]

paraxial rays are supposed to have been traced through all the surfaces behind the front lens, and the air-space between the latter and the lens next behind it having been decided upon, we shall know l_{k-1}, u_{k-1}, L_{k-1} and U_{k-1} by the usual transfer equations. We shall call the index of the front lens by its simplest name, N_k.

As the front plane is the last surface of the system and as the system must fulfil the sine condition if it is to be at all acceptable, we must have $u'_k = \sin U'_k$. At a plane surface, we have the refraction formula $u_k N_k = u'_k N'_k$ and $\sin U_k \cdot N_k = \sin U'_k \cdot N'_k$, so it follows that, by the universal transfer equations $u_k = u'_{k-1}$ and $U_k = U'_{k-1}$, the system can only fulfil the sine condition if the curved side of the front lens fulfils the condition $u'_{k-1} = \sin U'_{k-1}$.

This argument supplies the first key to the solution: We must find that value of r_{k-1} which makes $\log u'_{k-1}$ equal to $\log \sin U'_{k-1}$. A direct solution has not yet been found; if simple, it would be a highly desirable discovery. But r_{k-1} can be quickly found by a few systematic trials. The graph will always result in a close approximation to the fulfilment of the condition. If OSC' has come out positive, we try a slight lengthening of r_{k-1}, in the reverse case a slight shortening, and if necessary we follow this attempt up by a further change suggested by the result of the first trial. In our example, $r_7 = 4.46$ was in this way found by two trials to give $u'_7 = \sin U'_7$ and therefore to ensure exact fulfilment of the sine condition. The tracing of the paraxial and the marginal rays will also give the values of l'_{k-1} and of L'_{k-1}; in the present example, $l'_7 = 6.5970$ and $L'_7 = 6.8407$. We can then solve directly for the thickness d'_{k-1} of the front lens which will make the spherical aberration zero, in the following way.

The intersection-lengths at the front plane will be $l_k = l'_{k-1} - d'_{k-1}$ and $L_k = L'_{k-1} - d'_{k-1}$. Their difference gives $L_k = l_k + (L'_{k-1} - l'_{k-1})$. L_k and l_k will be changed by the refraction in accordance with Pl(3*) and Pl(3p) into

$$L'_k = [l_k + (L'_{k-1} - l'_{k-1})] \cdot N'_k \cos U'_k / N_k \cos U_k$$
$$l'_k = l_k \cdot N'_k / N_k.$$

As the spherical aberration must be brought to zero, we must have $L'_k = l'_k$, and this gives, on cancelling out N'_k/N_k,

$$l_k = [l_k + (L'_{k-1} - l'_{k-1})] \cos U'_k / \cos U_k$$

or

$$l_k \cos U_k / \cos U'_k - l_k = L'_{k-1} - l'_{k-1},$$

which gives

$$l_k = \frac{L'_{k-1} - l'_{k-1}}{(\cos U_k / \cos U'_k) - 1}$$

and then

$$d'_{k-1} = l'_{k-1} - l_k.$$

If the equation for l_k is to be calculated as given, the cosines must be taken out with the greatest possible precision, especially when U and U' are fairly small. It is preferable to transform the denominator by the formula

$$(\cos U / \cos U') - 1 = (\cos U - \cos U') / \cos U'$$
$$= 2 \sin \tfrac{1}{2}(U' + U) \cdot \sin \tfrac{1}{2}(U' - U) / \cos U',$$

[105] which gives the computing formulae:

$$\text{MP(3)} \begin{cases} l_k = \frac{1}{2}(L'_{k-1} - l'_{k-1}) \cdot \cos U'_k \cdot \operatorname{cosec} \frac{1}{2}(U'_k + U_k) \cdot \operatorname{cosec} \frac{1}{2}(U'_k - U_k) \\ d'_{k-1} = l'_{k-1} - l_k. \end{cases}$$

Although apparently more complicated than the original equation, this last equation really implies less total work because the two sines required for it are also wanted for the *OPD* equation.

A wise computer will of course test the result by calculating through the front plane in the usual way after d'_{k-1} has been found.

In the case of our numerical example, Mr. Watkins' results are: By the first part of the solution, $r_7 = 4.46$. With this, $L'_7 - l'_7 = 0.2437$, $u'_7 = 0.38709$, $U'_7 = 22\text{-}46\text{-}30$. The calculation of MP(3) is then carried out thus:

$$U'_{k-1} = U_k = 22\text{-}46\text{-}30 \quad \longrightarrow \quad \begin{array}{ll} \log \sin U_k & = 9.5878 \\ + \log N_k & = 0.1882 \end{array}$$

$$U'_k = 36\text{-}39\text{-}30 \quad \longleftarrow \quad \log \sin U'_k = 9.7760$$

$$\begin{array}{ll} \frac{1}{2}(U'_k + U_k) = 29\text{-}43\text{-}0 & \longrightarrow \\ \frac{1}{2}(U'_k - U_k) = 6\text{-}56\text{-}30 & \longrightarrow \end{array} \quad \begin{array}{ll} \log \operatorname{cosec} \frac{1}{2}(U'_k + U_k) & = 0.3048 \\ \log \operatorname{cosec} \frac{1}{2}(U'_k - U_k) & = 0.9177 \\ \log \cos U'_k & = 9.9043 \\ \log (L'_{k-1} - l'_{k-1}) & = 9.3869 \\ \log \frac{1}{2} & = 9.6990 \end{array}$$

$$\begin{array}{ll} l'_{k-1} = & 6.5970 \\ -l_k = & -1.6319 \quad \longleftarrow \quad \log l_k \qquad\qquad = 0.2127 \end{array}$$

$$d'_{k-1} = \quad 4.9651$$

This solution then gives

$$\begin{array}{rr} L'_{k-1} = & 6.8407 \\ -d'_{k-1} = & -4.9651 \\ \hline L_k = & 1.8756 \\ \log \cos U_k = & 9.9647 \\ -\log \cos U'_k = & -9.9043 \\ \hline & 0.0604 \\ +\log N_k = & 0.1882 \\ \hline \log (\text{corrected } N_k) = & 0.2486 \end{array}$$

$$\begin{array}{rr} \log L_k = & 0.2731 \\ -\log (\text{corrected } N_k) = & -0.2486 \\ \hline \log L'_k = & 0.0245 \end{array} \qquad \begin{array}{rr} \log l_k = & 0.2127 \\ -\log N_k = & -0.1882 \\ \hline \log l'_k = & 0.0245 \end{array}$$

Hence $L'_k = l'_k$ = Working Distance   = 1.0580 mm. |105|

Sin U'_k = Numerical Aperture = 0.597

(both for the calculated submarginal zone).

The objective is now ready for the application of the crucial tests for secondary aberrations, and here we must always be prepared to find that the first solution for a really new design may fail, sometimes disastrously.

The formula for our objective as amended by the slight modification of the front lens now is as follows:

$$r_1 = \quad 12.5 \qquad r_4 = \quad 12.048 \qquad r_7 = 4.46$$
$$\qquad\qquad 3.0 \qquad\qquad\qquad 3.0 \qquad\qquad\qquad\qquad 4.9651$$
$$r_2 = \quad -9.506 \qquad r_5 = \quad -8.326 \qquad r_8 = \quad \infty$$
$$\qquad\qquad 1.5 \qquad\qquad\qquad 1.5$$
$$r_3 = -150.94 \qquad r_6 = -67.00$$

Air-space   1.01   Air-space   2.09   Working distance 1.058

By the trigonometrical equation for OPD'_m, Mr. Watkins found the following separate amounts for the eight surfaces:

$$-18.707, \quad 31.645, \quad -2.359, \quad -0.137, \quad 41.975, \quad -17.766, \quad 1.888, \quad -32.509,$$

which add up to 4.03 nominal wave-lengths of 0.0005 mm. as the difference of optical path with which the submarginal rays meet the axial ray at the final focus. By the criterion for tertiary aberration, this difference should be zero, and even if we stretched the restrictions imposed in Chapter XIV and discussed the residual as if it were of the ordinary zonal type arising from the presence of secondary aberration alone, there would be a limit of tolerance of 2 wave-lengths for OPD'_m. Evidently our residual is too large.   If we calculate the paraxial OPD'_p-sum, as for the Lister objective, we shall find the value 9.05 nominal wave-lengths, again more than any of our tolerances would countenance.

A new solution must therefore be faced, and we have to decide upon the alteration in the data which is most likely to produce the required large reduction in the residuals of higher aberration.

The residuals in our first solution are of the positive sign, which we found in ordinary telescope objectives and again in the Lister objective, and this indicates that they are due to an excess of negative or 'over-corrected' higher aberrations at the contact surfaces.   Therefore the most promising alteration will be one which diminishes this excess, and as the latter depends primarily on the magnitude of the angles of incidence of the marginal rays at the contact surfaces, our aim must be to diminish these angles.

In the solution hitherto referred to, these angles, always taken as those in the lighter medium, were $I_2 = -34$-24-20 and $I_5 = -38$-22-50, while the final angle of emergence at the front plane was $I'_8 = -36$-39-30.

As a valuable simple criterion, it may be accepted that an objective of the dry

[105] Amici type will certainly yield positive residuals of OPD'_m if I'_8 is numerically smaller than either I_2 or I_5, especially the latter. It will be noticed that in our case I'_8 is nearly $2°$ smaller than I_5, and a considerable remnant of zonal aberration was therefore certain to be found.

I_2 and I_5 can most easily be reduced by using a denser flint glass, and the most promising change is therefore to introduce extra dense flint No. 337, $Nd = 1.6469$, $Nf-Nc = 0.01917$, $V = 33.7$, in place of the ordinary flint, making no change in the front lens or in the distribution of the deviation and chromatic aberration. Repeating the whole solution with this denser flint, Mr. Watkins obtained a prescription which gives an OPD'_m-sum of 0.22 nominal wave-length and a paraxial OPD'_m-sum of 3.20 nominal wave-lengths, the critical angles now being

$$I_2 = -32\text{-}26\text{-}50 \qquad I_5 = -31\text{-}19\text{-}30 \qquad I'_8 = -37\text{-}43\text{-}50.$$

It should be noticed that I'_8 is now more than $6°$ in excess of I_5 and more than $5°$ in excess of I_2.

This second solution would certainly yield objectives of very high spherical correction.

We could not always expect to score a bulls-eye at the second shot, and a third attempt will usually be necessary with an appropriate further change in the density of the flint glass in one or both of the components, more or less proportioned to the change secured by the first modification.

Small residuals of the OPD'_m-sum, say of less than two wave-lengths, can frequently be removed with less labour by substituting a front crown of higher index for that originally used and solving for the radius and thickness of this new front lens by the simple method described above and embodied in MP(3). Such a change of glass in the front lens, or for a smaller correction a mere small change of separation from the middle lens, represents the best method of final correction and renders it quite easy to bring the OPD'_m-sum exactly to zero, in accordance with the assumption in Chapter XIV.

Another method, more costly but the only one available when a still denser flint is not obtainable or cannot safely be employed (flints of Nd greater than 1.72 are distinctly hazardous on account of the risk of tarnishing) consists in making the back combination of the Steinheil triple-cemented form shown in Fig. 139 (c). About six-tenths of the total deviation should be assigned to it and at least two-thirds of the required chromatic over-correction. This modification also has a very beneficial effect upon the zonal chromatic aberration.

Even when the first solution turns out to be imperfect as regards zonal spherical aberration, its chromatic correction should be calculated by the $(d'-D')$ method. In the case of the first solution fully described in this section, it is thus found that the lens would become achromatic with a crown in the middle and back combinations of $Nf-Nc = 0.00810$, or 0.0001 less than the 0.0082 used in the TL preparation. Substitution of denser flint in the second solution will tend to aggravate this discrepancy, and as $Nf-Nc = 0.00809$ is the lowest dispersion obtainable by Chance's glasses, it will be wise to raise the fictitious $(Nf-Nc)$ for the TL work to

0.0083. A highly promising programme for a new solution will therefore be as [105] follows:

The original front lens: unchanged.

Back and middle crown: $Nd = 1.5160$, $Nf-Nc = 0.0083$, $V = 62.2$

Back and middle flint: $Nd = 1.6501$, $Nf-Nc = 0.01936$, $V = 33.6$.

The flint is No. 5093 of Chance's list, a little more dispersive than No. 337.

INTERPRETATION OF *OPD*-RESIDUALS

IN the evolution of a new design, we meet with residuals of the paraxial and the [106] marginal OPD'_m-sums which do not comply with the simplifying assumptions on which the tolerances worked out in Chapter XIV are based. An extension of the interpretation is therefore highly desirable in order to enable us to determine the relation of any residuals to the Rayleigh quarter-wave limit.

We stipulate that the paraxial and marginal or submarginal geometrical foci shall have been brought to accurate coincidence by the simple correcting touches already described. This is necessary so that the ray tracks determined by our calculation in the usual computing direction may be identical with those which would be found by calculating in the direction that the light actually travels; for the *OPD*-values in the latter direction would then be obtained by exactly the same angles and intersection-lengths, and the sums at the final actual focus will therefore be identical with those found in the computing direction and can simply be taken over. We must discuss the wave form at the actually observed focus not only because it is the really important one but because the image-forming rays meet there (that is, on the retina or on the photographic plate) under sufficiently small angles to justify the use of the simple formula OP(4)* for the effect of a small departure from the geometrical focus. At the object-point, the large U-values would necessitate the use of more complicated rigorous expressions.

The distortion of the waves converging upon the final image-point must, if the series is broken off beyond the tertiary term, satisfy the equation

$$OPD' = v_4 Y^4 + v_6 Y^6 + v_8 Y^8.$$

The calculated OPD'_m-sum is the value of this series for the submarginal Y_m, and the calculated paraxial OPD'_p-sum is the value of the first term for $Y = Y_m$. Moreover, as the geometrical spherical aberration of the submarginal ray has been fully corrected, there can be no angular aberration, hence $d(OPD')/dY$ must be zero for $Y = Y_m$. That gives the three equations

$$v_4 \cdot Y_m{}^4 = \Sigma OPD'_p$$
$$v_4 Y_m{}^4 + v_6 Y_m{}^6 + v_8 Y_m{}^8 = \Sigma OPD'_m$$
$$4v_4 Y_m{}^3 + 6v_6 Y_m{}^5 + 8v_8 Y_m{}^7 = 0,$$

the last, on division by $Y_m{}^3$ becoming

$$4v_4 + 6v_6 Y_m{}^2 + 8v_8 Y_m{}^4 = 0.$$

[106] Elimination of v_4 from the second and fourth equations by the use of the first gives

$$v_6 \cdot Y_m{}^6 + v_8 \cdot Y_m{}^8 = \Sigma OPD'_m - \Sigma OPD'_p$$
$$6v_6 \cdot Y_m{}^2 + 8v_8 Y_m{}^4 = -(4\Sigma OPD'_p)/Y_m{}^4$$

and the solution of these two linear equations in v_6 and v_8 together with the original first equation gives the value of the three constants

MP(4)
$$\begin{cases} v_4 = (\Sigma OPD'_p)/Y_m{}^4 \\ v_6 = (4\Sigma OPD'_m)/Y_m{}^6 - (2\Sigma OPD'_p)/Y_m{}^6 \\ v_8 = (-3\Sigma OPD'_m)/Y_m{}^8 + (\Sigma OPD'_p)/Y_m{}^8 \end{cases}$$

We can then calculate the phase-difference for any aperture by the formula

$$OPD' = v_4 Y^4 + v_6 Y^6 + v_8 Y^8$$

and can plot and discuss the wave form. The work can be further simplified, without any sacrifice of validity, by measuring the semi-aperture in terms of the full submarginal value as unit, so that $Y_m = 1$.

Mr. Watkins' first solution for the 8-mm. objective gave $\Sigma OPD'_m = 4.03$ and $\Sigma OPD'_p = 9.05$, or by MP(4) with $Y_m = 1$:

$$v_4 = 9.05; \qquad v_6 = 16.12 - 18.10 = -1.98;$$
$$v_8 = -12.09 + 9.05 = -3.04;$$

or $OPD' = 9.05 Y^4 - 1.98 Y^6 - 3.04 Y^8$.

The second solution gave $\Sigma OPD'_m = 0.22$, $\Sigma OPD'_p = 3.20$, or:

$$v_4 = 3.20; \qquad v_6 = -5.52; \qquad v_8 = 2.54;$$

or $OPD' = 3.20 Y^4 - 5.52 Y^6 + 2.54 Y^8$.

The most instructive way of discussing wave forms of this description consists in plotting OPD' on squared paper against Y^2 (*not* Y itself!) as abscissa, or mathematically expressed, in changing the variable to $x = Y^2$, so that the formula now is

$$OPD' = v_4 \cdot x^2 + v_6 \cdot x^3 + v_8 \cdot x^4.$$

Selecting $x = 0.2$, 0.4, 0.6, 0.8, 1.0, and 1.2 as a suitable sequence of values, we obtain for the first solution:

$Y^2 = x =$	0.2	0.4	0.6	0.8	1.0	1.2
$v_4 \cdot x^2 =$	0.36	1.45	3.26	5.79	9.05	13.03
$v_6 \cdot x^3 =$	−0.02	−0.13	−0.43	−1.01	−1.98	−3.42
$v_8 \cdot x^4 =$	0.00	−0.08	−0.39	−1.25	−3.04	−6.30
$OPD' =$	0.34	1.24	2.44	3.53	4.03	3.31

The second solution gives in the same way

$OPD' =$	0.09	0.22	0.29	0.26	0.22	0.34

and the immense superiority of the second solution at the calculated focus is [106] rendered obvious.

We must however inquire as to the result of a suitable departure from the calculated combined foci of the paraxial and submarginal rays, and here the change of variable proves most valuable. By OP(4)* a change of focus adds a term proportional to Y^2 and therefore to our x, and we can give to this term any positive or negative coefficient we like merely by departing from the calculated focus in one direction or the other.

Using the graphical method, we can study the effect of terms added to the original OPD'-equation by an application of the matching principle. If we plot the original equation correctly as to sign, and the added term or terms separately with reversed sign, then the separation between the two graphs at any ordinate will represent the sum of the original and of the added terms, for $a-(-b) = a+b$.

If we plot the above values of OPD', we obtain two curves like those of Fig. 143.

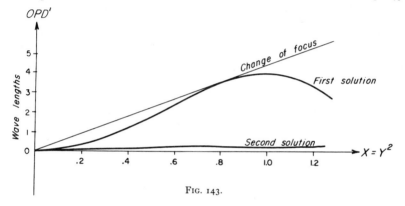

FIG. 143.

A term proportional to x plots as an inclined straight line passing through the origin, and we see at once that such a line touching the curve of the first solution at about $x = 0.8$ fits our curve very much more closely than the original x-axis; in fact, it does not depart from the curve by much more than half a wave-length up to a value of $x = Y^2$ representing a little more than the calculated aperture. Therefore even this objective would be within the doubled Rayleigh limit if made to, say, 0.61 NA provided the best focus is carefully looked for.   In the case of the second solution, an inclined straight line from the origin and touching the curve from below it at a value of $x = Y^2$ that is very nearly unity nowhere departs by more than one sixth of a wave-length up to a little beyond $x = 1.2$ or $Y = \sqrt{x} = 1.1$.   This objective therefore is within two thirds of the Rayleigh limit for an aperture of nearly 0.7 NA and is excellent as far as spherical correction is concerned.

This is not all the information obtainable from Fig. 143.   In critical work the skilled microscopist does not merely seek the best image at the engraved tube-length; he also tries the effect of changing the tube-length.   As will be shown subsequently, a change of tube-length introduces spherical aberration, almost

[106] purely of the primary or Y^4-type. Mathematically this means that we can introduce a term in x^2, with any coefficient that is likely to be beneficial, merely by a change of tube-length. Such a term in x^2 plots on our graph as a parabola with vertical axis and with pole at the origin. A glance at the 'first solution' curve in Fig. 143 will show that up to $x = 0.6$ it is quite indistinguishable from such an ordinary parabola; hence the objective can be rendered practically perfect up to the corresponding $Y = 0.775$ or $NA = 0.6 \times 0.775 = 0.465$ by a suitable change of tube-length. Inasmuch as microscopists rarely use more than a 3/4 illuminating cone, which corresponds to $NA = 3/4 \times 0.66 = 0.495$ for our contemplated full aperture, it becomes apparent that in this way the correction can be made almost absolute for all the strong direct light; the residual OPD' in the outer zone will only affect the much weaker diffracted light and will not greatly affect the definition and 'crispness' of all the coarser features of the image.

The possibilities are not exhausted yet. As change of tube-length and choice of focus always go together, the microscopist really applies a correcting term $a \cdot x + b \cdot x^2$ (a and b being arbitrary constants) to the OPD' formula. Such a term plots on our graph as a vertical parabola passing through the origin but having its pole either at the right or the left of it. Evidently such a parabola can be made to fit the true curve closely for quite a long stretch, in any part of the aperture. That explains the high resolution secured by skilled observers with decidedly imperfect objectives even when oblique or annular illumination has to be employed.

In accordance with an oft-repeated warning, the fact that high resolution may be obtained from a decidedly imperfect lens system by skilful manipulation should not be drawn upon by a self-respecting designer as justification for resting content with residuals in excess of the previously stipulated tolerances unless it absolutely cannot be avoided—as will happen when the practically senseless demand for dry lenses exceeding 0.85 NA has to be met, or a similar demand for immersion objectives in excess of 1.35 NA. In these latter regrettable cases, the best possible correction should be established for about 0.8 NA and 1.3 NA respectively and the excess left to take care of itself and to be made the most of by the skilful use of tube-length adjustment referred to.

It will be noted that in addition to its great value to the designer in estimating the importance of residual higher aberrations, this important discussion also supplies the explanation of many otherwise puzzling peculiarities of microscope technique.

The curves found by plotting OPD' against $x = y^2$ of course do *not* represent the *true* wave form as used in previous diagrams, and they cannot be extended to the left of the vertical zero-axis as x would there be negative and therefore y imaginary.

DESIGN OF IMMERSION OBJECTIVES

[107] OBJECTIVES of the four-component type, Fig. 139 (f), evidently can be brought under the matching principle by adopting fixed data for two of the components and leaving the other two, which should be adjacent to each other, to be modified by bending so as to secure the desired aplanatism. We may therefore either assume fixed forms of the back component and of the front lens and submit the two inner

components to the matching principle, or we may select a fixed prescription for the [107]
front lens and the meniscus and apply the matching principle to the two compound
lenses. The latter procedure is probably the better and will be adopted in the
following instructions.

The most usual equivalent focal length for oil-immersion objectives (water-
immersion is practically, and glycerin-immersion totally, obsolete) is 1.8 mm. or
1/14 inch, but in accordance with an ancient custom, objectives of this power are
called 2 mm. or 1/12 inch. Computers should bear this strange tradition in mind,
for if a true one-twelfth is produced, the average buyer will consider himself
defrauded because his lens does not magnify as much as that of other makers.
Occasionally oil objectives of 1/10 and 1/8 inch focal length are met with, but there
is little demand for them. On the other hand, speculative opticians will offer
short focal lengths like 1/16, 1/20, and even 1/40 inch because a certain type of
amateur will buy them in the hope of seeing new microcosms opening before him.
The real performance is nearly always much worse than that of any respectable
standard 'one-twelfth', simply because the excessively tiny lenses cannot be
produced with sufficient precision. There is no difficulty of any kind in computing
them.

The best immersion oil is distilled from genuine cedar wood and has the pleasant
scent of the latter; it must be exposed to the air (ozone and bright light accelerate
the oxidation) until it becomes decidedly syrupy and it then has the double quali-
fication of agreeing in optical properties with the ordinary cover glass both as
regards Nd, which is about 1.515, and dispersion, which may be assumed to be
about $Nf - Nc = 0.0090$. Designers should keep a watchful eye on the immersion
oil supplied with their productions, for spurious substitutes defective in optical
properties are deplorably prevalent, and as these are cheaper and of course are
claimed to be superior, the purely commercial 'buyer' of a firm is likely to urge
their adoption or even to introduce them surreptitiously. Any notable difference
of the refractive indices of oil and cover glass renders an oil-immersion objective
sensitive to the thickness of the latter and so forfeits one of the great advantages of
the 'homogeneous' immersion principle.

As cover glasses are likely to reach or even slightly to exceed a thickness of
0.20 mm., the least distance from object-point to front plane that can possibly
be tolerated is 0.25 mm., but at least 0.30 mm. should be provided if at all practicable. On the other hand, a clear distance of 0.40 mm. or more is likely to prove a
great nuisance because the oil will run away from the wide space between front
lens and cover glass when the microscope is put into the comfortable inclined
position. Between 0.30 and 0.35 mm. is thus the most generally useful distance
from front plane to object-point.

We shall now deal with actual computing problems. Let us assume that an
oil-immersion objective is required of 1.8-mm. equivalent focal length and a full
numerical aperture of 1.32, to work on a tube-length of 170 mm. Making the
same assumptions concerning optical tube-length (= 180 mm.) and initial L_1
(= −170 mm.) as for the 8-mm. Amici objective, we find $M = -180/1.8 = -100$.
As tertiary spherical aberration must again be reckoned with, we have to calculate
trigonometrically for the 10/11 submarginal zone or for $NA = N'_{10} \sin U'_{10} =$

[107] 1.20, and this gives us the initial sin U_1 as $NA/M = 1.20/(-100) = -0.012$ by the sine condition. For the full aperture of 1.32 NA, the initial $u_1 = \sin U_1$ will be -0.0132 and the clear aperture of the back lens (required for fixing the thickness of the crown component by a scale drawing) will therefore be $2(-170)(-0.0132) = 4.5$ mm. A full, edged, diameter of 5.0 to 5.2 mm. will thus be suitable for the back combination.

SELECTION OF THE DUPLEX FRONT

The most exciting feature in the designing of immersion objectives is the remorseless contraction of the cone of rays (in the computing direction) as the rays are brought round to the tremendous final U' of about 60°. Thicknesses and air-spaces must therefore be kept at the lowest safe figure, and even then the front lens will become surprisingly small. It is possible to achieve a radius of curvature of the front lens equal to half the equivalent focal length of the objective or even a trifle more, but in a first attempt it is better to be satisfied with less. We shall choose $r_9 = 0.800$ mm.

High spherical correction is usually more easily obtained by using, for the front lens and also for the meniscus behind it, glass with an index decidedly higher than that of the immersion oil and cover glass. We shall choose the light barium crown, No. 3463 in Chance's list, already specified for the front of the Amici objective.

If we now decide that the clear distance from the object-point to the front plane shall be 0.30 mm., we are ready to begin a right-to-left ray-tracing. We shall have $l'_{10} = L'_{10} = 0.3000$ and may take the N_y of the immersion oil and cover glass to be $1.517 = [0.1810] = N'_{10}$ and have for the front lens from the Amici objective $N_{10} = [0.1882]$. It is highly advisable to add the extreme marginal ray for $NA = 1.32$ to the paraxial and the marginal ray corresponding to the sub-marginal $NA = 1.20$ in order to discover the full working aperture of the front lens and to be able to fix a safe and yet not excessive thickness for the meniscus lens.

For the submarginal zone, we find $u'_{10} = \sin U'_{10}$ as $NA/N'_{10} = [0.0792 - 0.1810] = [9.8982]$ and for the extreme marginal ray by the same method, $\sin U'_{10} = [9.9394]$. Taking the three rays through the front plane, we obtain

	Paraxial	1.20 NA	1.32 NA
$L_{10} =$	0.30500	0.31318	0.32018
$U_{10} =$	0.77804	51-4-50	58-51-20

Having decided that $r_9 = 0.800$, we must choose the thickness of the front lens. It was shown in the general discussion of front lenses that it is desirable to produce a certain amount of spherical over-correction at the curved surface, which means a positive value of $U'_9 - I_9$, for which we estimated 4° to 8° as a suitable amount. We will choose $U'_9 - I_9 = 5°$ for the submarginal zone, therefore $I_9 = U'_9 - 5° = U_{10} - 5° = 46-4-50$. By the law of refraction we then have $\sin I'_9$ (*within* the front lens) $= \sin I_9/N'$, where N' stands for the glass index, hence $I'_9 = 27-50-20$. We now know r_9, I'_9 and U'_9 and can calculate L'_9 by the second equation of Chr.(2) on page 195 of Part I:

$$r = L \sin U/2 \sin \tfrac{1}{2}(U+I) \cos \tfrac{1}{2}(U-I),$$

which, on transposing for L and writing in 'dashed' symbols on account of the [107] right-to-left calculation, gives

$$L' = 2r \sin \tfrac{1}{2}(U'+I') \cos \tfrac{1}{2}(U'-I') \operatorname{cosec} U'$$

or in the present case, with $r_9 = $ o.8, $U'_9 = $ 51-4-50, $I'_9 = $ 27-50-20, $L'_\text{s} = $ 1.2800 mm. for the 1.20-NA ray. As L_{10} for this ray was 0.31318, we have $d'_9 = $ 0.9668 mm. and the front lens is completely determined. (The fifth decimal has been dropped as being beyond the accuracy of four-figure calculation.) Taking the three rays through the curved surface, we obtain in the previous order

$$
\begin{aligned}
L_9 &= 1.8700 \qquad 1.8617 \qquad 1.8735 \\
U_9 = U'_8 &= 0.52914 \qquad 32\text{-}51\text{-}20 \qquad 36\text{-}47\text{-}0,
\end{aligned}
$$

and on allowing an axial air-space of 0.1 mm., quite as much as is safe in the yet strongly diverging cone,

$$L'_8 = 1.9700 \qquad 1.9617 \qquad 1.9735.$$

It is interesting to note that these display a decided zonal aberration, for the 1.20-NA ray has an intersection-length shorter than that of either the paraxial or the extreme marginal ray. It amounts to more than it appears because small longitudinal aberrations in cones of great convergence correspond to a considerable OPD. The sense of the zonal aberration is in agreement with that produced by cemented combinations with deep contact surfaces and promises a reasonable approach to freedom from zonal aberration of the complete system.

The calculation for the extreme marginal ray gives $(U+I)_9$, the angle at the centre of curvature, as 90-15-10. Our front lens will therefore be filled with light just beyond the hemisphere and thus it must be held in position with cement: it cannot be bezeled in.

The conflicting desiderata usual in bold designs present themselves when we try to choose a radius for the next surface. A deep concavity would give small angles of incidence and therefore small aberration, but it would also reduce the divergence of our ray only a little; it would leave too much work to be done by the following surfaces. A flat or convex surface would bend the rays by a considerable angle but would lead to excessive aberration. Experience shows that for a final NA of about 1.20 an angle of incidence between $-25°$ and $-30°$ is the most promising compromise. We will select $I'_8 = -28°$. That choice enables us to calculate r_8 directly by the second equation of Chr.(2), again written in 'dashed' symbols, with the result that $r_8 = +14.57$. Tracing the three rays through this surface, by the check formula because the radius is decidedly long in comparison with the intersection-lengths, we find

$$
\begin{aligned}
L_8 &= 2.8314 \qquad 3.0186 \qquad 3.0917 \\
U_8 &= 0.36822 \qquad 22\text{-}34\text{-}30 \qquad 25\text{-}13\text{-}10.
\end{aligned}
$$

The only remaining task is to select the thickness and the radius of the convex side of the meniscus lens, and as the divergence of the rays is still great, we must be parsimonious in fixing the thickness in order to prevent the cone of rays from

[107] becoming too large. We therefore base the calculations upon the extreme marginal ray so that we can fix a small but safe thickness for the meniscus at the point where the extreme rays pass through it. That thickness is the D'_7, and we will fix it at 0.2 mm. Referring to Fig. 144, we know the position of the marginal ray within the meniscus and can calculate the ordinate at which this ray leaves the eighth surface as $Y_8 = r_8 \sin (U+I)_8 = 1.4246$ mm. Owing to its slope of $25° \ 13' \ 10''$,

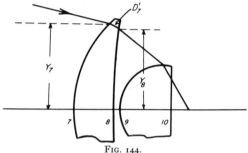

FIG. 144.

the ray will get farther away from the optical axis in traversing the distance $D' = 0.2$ mm. by $\delta y = D' \cdot \sin U_8 = 0.0852$. Therefore $Y_7 = 1.4246 + 0.0852 = 1.5098$.

We must now choose r_7. Evidently we could again choose the value which will secure aplanatism. However, in this case it will be wise to select a shorter radius leading to moderate spherical under-correction to secure a large final diminution of the divergence of the rays. We must therefore ask for a negative value of $U'_7 - I_7$, and choosing it as $-8°$, we obtain $I_7 = U'_7 + 8° = 25$-13-10 + 8-00-00 = 33-13-10, and by the law of refraction find $I'_7 = 20$-48-30. Knowing now $U'_7 = 25$-13-10 and $I'_7 = 20$-48-30, we have $U'_7 + I'_7 = 46$-1-40, and since $Y = r \sin (U+I)$, we can calculate $r_7 = Y_7 \csc 46$-1-40 = 2.098 mm.

We have found r_7 and I'_7 and can calculate L'_7 and then d'_7 exactly as for the front lens.

The transposed second equation of Chr.(2) gives $L'_7 = 3.847$, and as the extreme marginal ray had $L_8 = 3.092$, we obtain by difference $d'_7 = 0.755$ mm. as the thickness of the meniscus. Since it would be unusual to prescribe the thickness within thousandths of a millimetre, the calculation is completed with $d'_7 = 0.75$ mm. and gives

$$l_7 = 5.810 \qquad L_7 = 6.763 \qquad (L_7 = 7.251)$$
$$u_7 = 0.22699 \qquad U_7 = 12\text{-}14\text{-}30 \qquad (U_7 = 12\text{-}51\text{-}0)$$

The figures for full aperture are bracketed because they will no longer be required.

If a coloured paraxial ray with $N_v = 1.526$ for oil and the previous value for the glass is taken through the now complete duplex front, it gives $l_{7_v} = 6.014$. The front system therefore has a longitudinal spherical aberration $l_7 - L_7 = -0.953$ mm. and a longitudinal chromatic aberration $l_{7_y} - l_{7_v} = -0.204$ mm., which have to be allowed for in the back and middle combinations exactly as described in the [107] previous section.

Several solutions worked out by Mr. J. S. Watkins have demonstrated that the rather rough preparation for the chromatic over-correction which works quite well for Amici objectives leads to unpleasantly high chromatic under-correction [positive value of the $(d' - D')dN'$-sum for the complete objective] in the case of immersion lenses. This is due to the convergence of the rays within the crown components of the corrected compound lenses, which causes the paraxial cone of rays to have a decidedly smaller diameter at the contact surface. The difficulty can be sufficiently diminished by a rough second correction of the curvatures of the bendings: Calculate c_a (and c_c for the middle lens) and select three equidistant values of c_1 (or c_4 for the middle) as usual. Calculate also $c_2 = c_1 - c_a$ for each bending and begin the calculation with the bending which has the most unsymmetrical crown lens. Determine a provisional thickness d' by a scale drawing based on $c_2 = c_1 - c_a$. Having traced the paraxial ray through the first surface and found l'_1, calculate a corrected c_2 by

$$\text{corrected } c_2 = (c_1 - c_a) \cdot l'_1 / (l'_1 - d'_1),$$

which obviously shortens r_2 in just the ratio required to secure the same angle of incidence which the infinitely thin lens assumed in the preparatory work would give. Having ascertained that the assumed thickness is sufficient for the shorter r_2 so found, or having made a slight increase of the thickness if found necessary, complete the calculation with the corrected r_2. For middle lenses the procedure is the same, but the corrected c_5 may be worked out as

$$\text{corrected } c_5 = (c_4 - c_c) \cdot L'_4 / (L'_4 - d'_4)$$

simply because the submarginal ray is taken first through middle lenses. This slight modification will be found very effective and may also be tried on Amici objectives.

In the final correction of the graphical solution for immersion objectives, care must be taken that the working distance is maintained within the desirable 0.30 to 0.35 mm. If correction by the front-lens data only causes a sensible transgression of these limits, then r_8 must be slightly changed. A shortening of r_8 will tend to increase the working distance, a lengthening of r_8 will diminish it.

Comparatively small changes of all the data of the duplex front or of the index of the glass will also be found highly effective in changing the value of the residuals of higher aberrations indicated by the OPD-sums; such changes are usually less laborious than a complete new solution.

SOME ADDITIONAL NOTES ON MICROSCOPE OBJECTIVES
EFFECT OF COVER GLASS

IN most applications of the microscope, the object is mounted under a cover glass, [108] a thin glass plate about 0.15 to 0.20 mm. thick, which from the designer's point of view is to be regarded as a plano-parallel plate interposed between the objective and the object; the refractive index of the cover glass may be taken as about 1.518

[108] for 'brightest light'.   If the numerical aperture exceeds 0.25, this plate introduces sensible—and for high NA quite serious—spherical over-correction, which is most conveniently determined by Pl.(5) and (5p) as the difference of the shift of the image produced by the plate for paraxial rays and rays at finite angles respectively.   The equations are

Pl.(5) $B_1 B'_2 = (d'/N)(N - \cos U_1 \sec U'_1)$

Pl.(5p) $(B_1 B'_2)_p = d'(N-1)/N,$

N being the refractive index of the plate, d' its thickness, U_1 the angle of incidence in the air, and U' the corresponding angle of refraction in the glass.   It will well repay the slight trouble to calculate $B_1 B'_2 - (B_1 B'_2)_p$ for $U_1 = 30°, 40°, 50°, 60°$, and $70°$ and to do this for $N = 1.518$ (the usual index of cover glass) and also for $N = 1.56$ and $N = 1.60$, taking d' as unity—say 1 mm.—for convenience.   Four-figure logarithms will be quite sufficiently accurate.   It will be found that the aberration varies only quite moderately for different refractive indices, and as the thickness of the cover glass is never very accurately known, this small effect of the refractive index justifies an extremely simple method of allowing for the use of a cover glass of an assumed thickness (0.17 mm. is a very usual conventional thickness) in designs with plano-convex front lens.

As computed in accordance with our method, dry objectives will be right for use on uncovered objects (chiefly polished metal surfaces in metallurgical work) and for such purposes the computed formula is therefore ready for the workshop.   In the case of objectives intended for covered objects, we diminish the computed thickness of the plano-convex front lens by the adopted standard thickness of cover glass (usually 0.17 mm.), for, inasmuch as a plano-parallel plate produces the same effect wherever it may be placed across a cone of rays and as the refractive index has only a slight effect on the aberration of a plate of given thickness, the cover glass will be a practically perfect substitute for the plate of equal thickness sliced off the front lens and the correction of the aberrations will be correspondingly undisturbed.

In oil-immersion objectives, the indices of oil and cover glass are so nearly alike that no correction of any kind is called for.   The objective as computed may be used on either covered or uncovered objects.   But in designing oil lenses, the probability of cover glasses being used must be borne in mind as the cover glass will fill a corresponding part of the working distance so that the space for oil is diminished.   That is the reason why we should in general not go below 0.3 mm. for the gross working distance used in the designing.   A cover glass of 0.2 mm. will leave only 0.1 mm. for oil, and that is a rather small amount.

The cover glass thus implies no additional computing work, even in dry objectives, when the front lens is of the usual plano-convex form.   This advantage, together with the convenient direct solution for its data, renders the plano-convex form a desirable one from the designer's point of view.   But it also has decided and important advantages from the point of view of the workshop: On account of the huge angles at which the marginal rays emerge in objectives of high NA, the front lens, and especially its outside surface, requires incredibly accurate centring; a slight untruth is invariably most easily detected at a plane surface, and the almost

mathematical accuracy required is therefore realized most closely when the front [108]
lens is plano-convex.

Final Adjustment of Microscope Objectives

When the design has been carefully completed with due regard to the tolerances
specified in Chapters XIV and XV and when the best modern methods are employed
in realizing the prescribed radii, thicknesses, and separations, the results of actual
tests of the first specimen should very closely agree with the predicted properties
of the objective and only very slight departures from the calculated data should be
necessary to compensate for the cumulative effect of residual technical imperfections. This final adjustment can practically always be carried out with ample
approximation by a slight change in the separation between the front lens and the
lens next to it. The preponderating effect of this alteration is to increase or diminish
the free working distance and therefore the spherical aberration arising at the
front plane. Spherical correction can consequently be established by this alteration, and as the effect of small residual technical imperfections on chromatic
aberration and on fulfilment of the sine condition is negligible, the cure is practically
perfect. In oil-immersion objectives it will usually be less troublesome to effect
the correction by a shift of the meniscus lens, bringing this nearer to the front lens
or nearer to the back combination according to the sense of the residual aberration.
Whichever way may be adopted, a decrease of the separation between the front
lens and the lens next behind it (technically known as a 'shortening down' of the
objective) introduces spherical under-correction and therefore removes any over-
correction found in the first test.

In the higher power, the required alteration of the calculated separation should
not exceed a few hundredths of a millimetre. Amounts reaching or exceeding
0.1 mm. would indicate either insufficient precision in the computation or unduly
large workshop errors in radii, thicknesses, and separations.

Change of Tube-length

Generally speaking, the spherical aberration of any lens system varies with the
conjugate distances of the object and image. In the case of thin lenses and systems,
it was shown in section [83] that the spherical aberration of a collective or positive
system attains a minimum and that of a dispersive or negative system a maximum
for some particular distance of object and image. In the case of systems of thick
and more or less widely separated components, the conditions are more complicated.
When an indefinitely small or differential change of the conjugate distances is
considered, it can easily be shown on the principle of equal optical paths that any
lens system would have stationary spherical aberration in the immediate vicinity of
a given pair of conjugate points if it fulfilled the following condition for corresponding first U and final U':

$$N(1 - \cos U) = \text{a constant} \times N'(1 - \cos U'),$$

which can also be written

$$N \sin^2 \tfrac{1}{2}U = \text{a constant} \times N' \sin^2 \tfrac{1}{2}U',$$

[108] and which therefore contradicts the sine condition:

$$N \sin U = \text{a constant} \times N' \sin U'$$

in all cases when the conjugate U and U' differ in magnitude. The proof depends on Fermat's theorem and consequently holds only for a very small change of conjugate distances. The sense of the change in spherical correction following a change of tube-length with a system fulfilling the sine condition is correctly indicated by this theorem: A lengthening of the tube leads to over-correction of the spherical aberration and a shortening to under-correction. But whilst by the theorem the change should prove independent of the construction of the objective, as the proof depends only on the corresponding values of the first U and the last U', experience shows that the sensitiveness to change of tube-length of objectives which are identical in this respect varies considerably indeed; hence we must depend on direct calculation or actual test to determine the true state of affairs.

Change of tube-length plays a very important part in the testing and in the intelligent using of microscope objectives, especially in those of high numerical aperture. To systematize the testing and adjusting of objectives, we determine and record once for all in the case of each type by how many hundredths of a millimetre the lens requires 'shortening down' to compensate for, say, one centimetre departure from the correct tube-length. This can be done either by ray-tracing or by a few carefully conducted actual tests. In going through a batch of newly mounted objectives, we then simply determine the tube-length at which each one gives the best image and find the required 'shortening down' by simple proportion. For considerable variations of tube-length, the required shortening is more nearly proportional to the reciprocal of the tube-length than to the tube-length itself. In using objectives in fixed mounts, changing the tube-length represents the only means of compensating for variations in the thickness of the cover glass. Moreover, it is the only means of neutralizing the varying distance of the object below the lower surface of the cover glass which, unless the object is in air, is equivalent to a corresponding variation in effective cover-glass thickness. Complaints as to imperfect adjustment of objectives are frequently the result of overlooking this disturbing cause.

Adjustment for Different Tube-lengths

Professional users of the microscope generally prefer the handy instrument of about 170 mm. mean mechanical tube-length which has been developed on the continent of Europe. British amateur microscopists, on the other hand, retain a strong affection for the more imposing English type with a mean tube-length of 8 or even 10 inches. The result is that there is a demand for objectives of otherwise similar types adjusted either for the 'Continental' tube-length of about 170 mm. or for the 'English' tube-length, now usually taken as 250 mm.

Although the difference is large, it can be fairly satisfactorily compensated for by merely shortening down in the case of ordinary objectives with moderate separations and with back lenses of a clear aperture not exceeding 8 or at least 10 mm.

This is the usual practice.   If it is to be adopted, the best course for the designer is   [108]
probably to compute for a mean tube-length of about 200 or 210 mm. in order to
keep the variations of aberrations other than the spherical reasonably low for both
tube-lengths.   For objectives of high quality and for those with unusually large
clear aperture of the back lens, it will be preferable or even necessary to calculate
through them for both tube-lengths and separately to determine the form of the
front lens which will give complete correction.   Probably the refractive index and
dispersion of the glass of the front lens will also have to be slightly varied to obtain
the best result at both tube-lengths.

CORRECTION COLLARS

An alternative to changing tube-length for the removal of residual spherical
aberration consists in a convenient mechanical device by which the distance between
the front lens and the remaining parts of the objective can be delicately and continuously varied, so as to produce the required effect on the shortening-down
principle.   In the usual form of a correction-collar mount, the front lens (or
occasionally the two lowest lenses) is attached directly to the rigid part of the mount.
The remaining inner lenses are attached to an accurately cylindrical tube which can
slide with slight friction but without the least suspicion of lateral play in a corresponding bore of the mount.   The latter has a longitudinal slot cut through it to
guide a sliding block attached to the inner tube.   Screw threads cut upon the
projecting face of this sliding block are engaged by spiral grooves on the inside of a
collar that is rotatable on the mount, and a delicate axial movement can thus be
imparted to the inner tube.   One-sided spring pressure on the inner end of the
latter prevents backlash.

The great difficulty with correction collars arises from the obvious fact that it is
almost impossible to secure the same perfection and permanence of centring which
is possible in rigid mounts; this difficulty, together with the costliness of the device,
probably explains the increasing aversion to it.   Nevertheless, the correction
collar does have the advantages that the modification of the spherical aberration is
accompanied by far smaller disturbance of other corrections, that the magnification remains practically unchanged, and that the object can easily be kept under
steady observation while the collar is being turned.

CHROMATIC DIFFERENCE OF MAGNIFICATION

As has been pointed out in Part I, lens systems built up of separated components
which are not individually achromatized usually produce images in different colours
which differ in magnification and therefore cause primary-colour margins to appear
on objects in the outer part of the field of view if ordinary eyepieces are employed.
It has also been shown that compensating eyepieces can be designed to neutralize
this chromatic difference of magnification.

As all microscope objectives of higher power than that obtainable from the Lister
type have a non-achromatic front lens followed by chromatically over-corrected

[108] back lenses, they all suffer more or less from this defect and really call for a certain amount of the compensating effect in the eyepieces to be used with them. If the objectives are designed with lenses close together and if highly-dispersive glass is avoided in the front lenses, however, the chromatic difference of magnification is small enough to admit of satisfactory results with ordinary Huygenian eyepieces, especially if the latter are slightly 'over-corrected' in accordance with a fairly general practice.

CHAPTER XVII

PRIMARY ABERRATIONS OF OBLIQUE PENCILS

IN Chapter XIII we deduced a general equation OP(2) for the differences of optical [109] path with which light arrives at an extra-axial image-point. We showed that the first two, second-order, terms of this equation vanish when the image is received upon a plane at right angles to the optical axis and passing through the paraxial image-point, and when its distance H' from the optical axis is determined by the theorem of Lagrange or its exact equivalent, the paraxial form of the optical sine condition. We also discussed those higher terms which are independent of H and H' and which therefore affect even the axial image-point, and recognized these as representing spherical aberration of successive orders. The discussion showed that the first, primary, or fourth-order term of the spherical aberration is the only one which leads to reasonably simple expressions for a complete lens system, while even the secondary or sixth-order spherical aberration introduces vast complications. The latter would obviously become still more troublesome in the treatment of oblique pencils, and we shall therefore for the present limit the discussion of the aberrations of oblique pencils entirely to the primary or fourth-order terms of our equation OP(2) as given in Chapter XIII, page 606, as

OP(2)

$$OPD' = \tfrac{1}{2}PA^2\left[N'\left(\frac{1}{r}-\frac{1}{l'}\right)-N\left(\frac{1}{r}-\frac{1}{l}\right)\right] + Y\left[\frac{H'N'}{l'} - \frac{HN}{l}\right]$$

Seidel
aberrations

$$\begin{cases} + \tfrac{1}{8}PA^4\left[8k_4(N'-N)+\frac{N'}{l'}\left(\frac{1}{r}-\frac{1}{l'}\right)^2 - \frac{N}{l}\left(\frac{1}{r}-\frac{1}{l}\right)^2\right] \\[2mm] + \tfrac{1}{2}PA^2Y\left[\frac{H'N'}{l'^2}\left(\frac{1}{r}-\frac{1}{l'}\right) - \frac{HN}{l^2}\left(\frac{1}{r}-\frac{1}{l}\right)\right] \\[2mm] + \tfrac{1}{4}PA^2\left[\frac{H^2N}{l^2}\left(\frac{1}{r}-\frac{1}{l}\right) - \frac{H'^2N'}{l'^2}\left(\frac{1}{r}-\frac{1}{l'}\right)\right] \\[2mm] + \tfrac{1}{2}Y^2\left[\frac{H'^2N'}{l'^3} - \frac{H^2N}{l^3}\right] \\[2mm] + \tfrac{1}{2}Y\left[\frac{H^3N}{l^3} - \frac{H'^3N'}{l'^3}\right] \end{cases}$$

We shall now transform this equation so as to render it more suitable for discussion and for actual calculation. First of all, we stipulate that we will adopt the paraxial focal plane as our reference plane in which we shall study the aberrations and that H' shall be taken at the value following from the theorem of Lagrange. That stipulation causes the two terms in the first line to vanish unconditionally and gives the two conditions by which l' and H' are to be determined:

$$(a) \quad N'\left(\frac{1}{r}-\frac{1}{l'}\right) = N\left(\frac{1}{r}-\frac{1}{l}\right) \qquad (b) \quad \frac{H'N'}{l'} = \frac{HN}{l}$$

[109] They have been assigned reference letters as they will be extensively used in our transformations. We must remember that (a) and (b) simply mean that all image-points are to be located by the ordinary paraxial formulae for intersection-lengths and magnification, or according to the general theory of centred lens systems.

Next, it is desirable to put the Seidel aberrations, which now alone remain, back into terms of the rectangular coordinates Y and Z of the point of incidence instead of PA. The latter is defined by

$$PA^2 = (Y^2 + Z^2) + X^2,$$

and since

$$X = \frac{PA^2}{2r} + k_4 PA^4,$$

it is evident that on squaring X and again introducing the explicit value of PA^2, only terms in the square and higher even powers of $(Y^2 + Z^2)$ would result and we should find

$$PA^2 = (Y^2 + Z^2) + \text{a constant} \times (Y^2 + Z^2)^2 + \&c.,$$

and as all our Seidel aberrations are already of the fourth order, the substitution of the complete equivalent of PA^2 would produce fourth-order terms in which PA^2 is simply exchanged for $(Y^2 + Z^2)$ and the sixth-order, eighth-order, &c. terms. This procedure would merely modify the secondary, tertiary, &c., aberrations and would therefore have no influence on our present strictly fourth-order approximation.   Hence, we are justified in writing $(Y^2 + Z^2)$ instead of PA^2 wherever the latter factor occurs, and in the first term $(Y^2 + Z^2)^2$ instead of PA^4.

The further transformations are then as follows:

In the term in PA^4 we use the form of its factor deduced on page 609 and write the term

$$\tfrac{1}{4}(Y^2 + Z^2)^2 \left[4k_4(N' - N) + \tfrac{1}{2}N^2\left(\frac{1}{r} - \frac{1}{l}\right)^2\left(\frac{1}{N'l'} - \frac{1}{Nl}\right) \right].$$

In the term in $PA^2 \cdot Y$ we extract $HN/l = H'N'/l'$ as a common factor by (b), and also $N(1/r - 1/l) = N'(1/r - 1/l')$ by (a), and the term becomes

$$(Y^2 + Z^2)Y \cdot \frac{HN}{l} \cdot \tfrac{1}{2}N\left(\frac{1}{r} - \frac{1}{l}\right) \cdot \left(\frac{1}{N'l'} - \frac{1}{Nl}\right).$$

The following terms in PA^2—now $(Y^2 + Z^2)$—and in Y^2 we rearrange by taking out of the third term those items which contain r, making of these a separate term, and combining the remainder of the third term with the original fourth term, producing

$$\left[\tfrac{1}{4}(Y^2 + Z^2) + \tfrac{1}{2}Y^2\right] \cdot \left[\frac{H'^2N'}{l'^3} - \frac{H^2N}{l^3}\right] + \tfrac{1}{4}(Y^2 + Z^2)\left[\frac{H^2N}{l^2r} - \frac{H'^2N'}{l'^2r}\right].$$

In both these modified terms we take out $H^2N^2/l^2 = H'^2N'^2/l'^2$ as a common factor by (b) and, on contracting the first factor of the first term and splitting its [109] numerical factor $\frac{1}{4}$ into $\frac{1}{2} \cdot \frac{1}{2}$, the terms become:

$$\frac{1}{2}(3\,Y^2+Z^2) \cdot \frac{H^2N^2}{l^2} \cdot \frac{1}{2}\left(\frac{1}{N'l'}-\frac{1}{Nl}\right) + \frac{1}{4}(Y^2+Z^2) \cdot \frac{H^2N^2}{l^2} \cdot \left(\frac{1}{Nr}-\frac{1}{N'r}\right),$$

the second of which may be further modified into

$$\frac{1}{4}(Y^2+Z^2) \cdot \frac{H^2N^2}{l^2} \cdot \frac{N'-N}{NN'r}.$$

In the last term of OP(2) we merely take out $H^3N^3/l^3 = H'^3N'^3/l'^3$ as a common factor, producing

$$\frac{1}{2}Y \cdot \frac{H^3N^3}{l^3} \cdot \left(\frac{1}{N^2}-\frac{1}{N'^2}\right).$$

Collecting the modified terms of OP(2), we obtain our fundamental equation for the primary aberrations of an oblique pencil at a single refracting surface as

$$\text{OP(2)}^* \qquad OPD' = \frac{1}{4}(Y^2+Z^2)^2\left[4k_4(N'-N)+\frac{1}{2}N^2\left(\frac{1}{r}-\frac{1}{l}\right)^2 \cdot \left(\frac{1}{N'l'}-\frac{1}{Nl}\right)\right]$$

$$+ Y(Y^2+Z^2) \cdot \frac{HN}{l} \cdot \frac{1}{2}N\left(\frac{1}{r}-\frac{1}{l}\right)\left(\frac{1}{N'l'}-\frac{1}{Nl}\right)$$

$$+ \frac{1}{2}(3\,Y^2+Z^2) \cdot \frac{H^2N^2}{l^2} \cdot \frac{1}{2}\left(\frac{1}{N'l'}-\frac{1}{Nl}\right)$$

$$+ \frac{1}{4}(Y^2+Z^2) \cdot \frac{H^2N^2}{l^2} \cdot \frac{N'-N}{NN'r}$$

$$+ \frac{1}{2}Y \cdot \frac{H^3N^3}{l^3} \cdot \left(\frac{1}{N^2}-\frac{1}{N'^2}\right).$$

The successive lines in this equation represent the five Seidel aberrations in their usual order: spherical aberration, coma, astigmatism, Petzval curvature and distortion.

Equation OP(2)* is a perfectly general solution of our problem, but in its present form it does not lend itself to a treatment of the oblique aberrations for a complete system of lenses; the equation could only be worked out numerically, surface by surface, by first tracing a paraxial pencil right through in order to find the values of l and H at each surface, and then determining also the values of Y and Z for each individual ray at each surface. The equation therefore calls for further transformation in order that these restrictions may be removed.

To effect this transformation, we must first introduce a simplifying assumption and prove that the latter is legitimate as long as we retain the restriction to primary or fourth-order aberrations. Our equation was originally based upon Fig. 119; if we reproduce this diagram with the omission of quantities no longer required, we

[109] obtain Fig. 145, in which B_h represents an extra-axial object-point at a projected distance l from the pole A of the surface, and at distance H—in the Y-direction— from the optical axis AX. Moreover, P, with coordinates X, Y, Z, represents the point of incidence of any ray on the refracting surface indicated by the curve AP. The simplification to be introduced consists in substituting for the true point of incidence P the closely neighbouring point P' at which the incident ray passes through the YZ-plane, and to use the coordinates Y' and Z' of this substituted point of incidence instead of the true Y and Z in working out OP(2)*. It means that we adopt the paraxial fiction of treating the depth of curvature X as negligible as far as its effect on the numerical value of the aberration terms in OP(2)* is concerned. This is a legitimate simplification because the distance PP' is sensibly

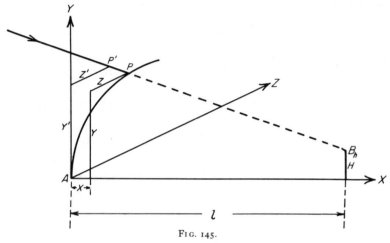

FIG. 145.

equal to X, and therefore it is equal to $(Y^2 + Z^2)/2r$ or a small quantity of the second order. The separation of P from P' by this amount produces an effect on Y' and Z' which is proportional to PP' and to the angle of the incident ray with reference to the XZ- and XY-planes respectively. These angles must necessarily be restricted to small magnitudes of the first order if our first approximation to the aberrations is to be of any value. Therefore the differences $(Y' - Y)$ and $(Z' - Z)$ are small of the third order, or, as Y and Z themselves are small of the first order, the differences are small of the second order compared with the true Y and Z. As all terms of OP(2)* are already small of the fourth order, the correction of Y and Z by a fraction which is small of the second order would only produce terms of the sixth or higher order and would only affect the secondary aberrations, which are neglected anyway.

Adopting this shift of the nominal points of incidence into the YZ-plane, we draw a new diagram, Fig. 146, which will give us the key to the required generalization of equation OP(2)*.

Let A represent the pole of a refracting surface so that our new plane substitute-   [109]
surface coincides with the YZ-plane; let B be an axial and B_h a corresponding
extra-axial object-point.   If the aperture of our surface were limited by a diaphragm
in contact with it, it would admit a cone of rays towards B, and a corresponding
cone with the same base in the refracting surface towards B_h.   In that case, OP(2)*
would be quite convenient because Y and Z would refer to the same point of the
aperture for object-points in any part of the field, In reality, optical systems nearly
always have an aperture-limiting diaphragm which does not coincide with any lens
surface, and then this diaphragm or the image of it which is presented to any one
surface becomes the common base of the cones which are allowed to pass, and the
oblique pencils will pass eccentrically through the surfaces.   We will call the ray
which passes from the centre of the diaphragm or of its image to an extra-axial

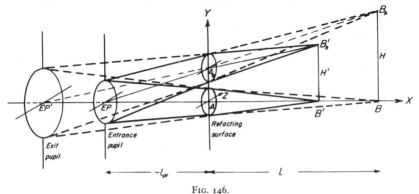

FIG. 146.

object-point the central or 'principal' ray of the corresponding oblique pencil of
rays.   Every image of the actual diaphragm we call a 'pupil', and with reference to
any refracting surface, we call the pupil which defines the entering rays the
'entrance' pupil, and the image of this entrance pupil produced by the refracting
surface, which similarly defines the emerging rays, we call the 'exit' pupil for the
surface in question.   The exit pupil of any one surface is identical with the
entrance pupil of the next following surface.   By tracing an inclined principal ray
from the centre of the actual diaphragm right through the system, the centres of the
pupils are found as the successive intersection-points of this ray with the optical
axis.   We have made use of this process in Part I and have called the intersection-
lengths l_{pr} if the ray was traced by paraxial formulae and L_{pr} if it was traced by
trigonometrical formulae at finite angles.   The diameters of the various pupils
corresponding to a given opening of the actual diaphragm can be determined with
sufficient accuracy respectively by the theorem of Lagrange or by the optical sine
condition.

If we refer again to Fig. 146 and assume that the entrance pupil for our
surface has been located at EP, whose distance is $A\text{-}EP = l_{pr}$ from A, then the
axial object-point will be the apex of a cone of rays which will cut the surface in a

[109] circle having A as its centre, and the extra-axial object-point B_h will receive an oblique cone of rays with the same base—namely, the entrance-pupil—as the axial cone. As the substitution of the YZ-plane for the actual refracting surface makes the latter a plane lying parallel to that of the entrance pupil and also to the plane containing the object-points, it at once follows that cones of rays of any obliquity, but having their apices in the object plane, cut the YZ-plane in figures which are perfectly congruent among themselves and are similar to the aperture of the entrance pupil; hence the usual circular aperture of the pupil will cause all incident cones of rays to cut the YZ-plane in equal perfect circles, and rays from any one point of the pupil aperture to any object point will cut through these circles in precisely similar positions. With reference to the two cones shown in the diagram, with apices at B and B_h respectively and with intersection-circles having centres at A and A_h respectively, we may therefore conclude that if a ray passing from a particular point of the entrance pupil towards the axial object-point B cuts the YZ-plane in a point with coordinates Y and Z, then a ray from the same point of the entrance pupil to B_h will cut the YZ-plane in a point with coordinates $Y^* = Y + A\text{-}A_h$ and the same Z. As our equation OP(2)* was originally built up for coordinates of the point of incidence referred to the three axes AX, AY, and AZ, we have now determined the proper value of these coordinates for any object point in the XY-plane, namely, $Y + A\text{-}A_h$ and Z. The next problem is to find a convenient expression for $A\text{-}A_h$.

Figure 146 shows by the similar triangles $EP\text{-}A\text{-}A_h$ and $EP\text{-}B\text{-}B_h$ that, for a given position of the entrance pupil and of the object-plane, $A\text{-}A_h$ is in a fixed proportion to H, namely, as $EP\text{-}A$ is to $EP\text{-}B$. We may therefore put $A\text{-}A_h = E \cdot H$, where E is a constant simple ratio or a pure number and equal to $(EP\text{-}A)/(EP\text{-}B)$. Bearing in mind that in our diagram $A\text{-}EP = l_{pr}$ is negative while $AB = l$ is positive, the last ratio gives the definition of E as

(c) $$E = \frac{-l_{pr}}{-l_{pr} + l} = \frac{l_{pr}}{l_{pr} - l}$$

and E can be calculated by this equation for every surface as soon as the l and l_{pr} have been determined for the system. E is a very important constant; we shall call it the 'eccentricity constant' for the surface to which it refers. Equation (c) gives its value referred to the data of the incident pencil. If the latter is refracted so that the refracted rays correspond to an exit pupil at a distance l'_{pr} and so that their reference focus at B'_h is defined by l' and H', we shall still have the coordinates of the point of incidence and emergence of a ray defined by $Y^* = Y + A\text{-}A_h$ and by Z; and by the reasoning employed above, we may conclude that the eccentricity constant E' for the emerging pencil is defined by $A\text{-}A_h = E'H'$, and that E' has the value

(d) $$E' = \frac{l'_{pr}}{l'_{pr} - l'}$$

and as $E'H' = EH =$ the fixed $A\text{-}A_h$, we have from (b) the further relation

(e) $$E' = E \cdot H/H' = E \cdot N' \cdot l/N \cdot l'.$$

We are now ready to transform OP(2)* so as to make it directly applicable to [109] oblique pencils of any eccentricity by writing $Y*$ in place of the original general Y and then replacing $Y*$ by $(Y+EH)$.

Merely to abridge the expressions, we shall temporarily introduce short and easily-remembered symbols for the coefficients which are independent of $Y*$, Z, and H by putting OP(2)* into the form

$$\text{OP}(2)* \quad OPD' = \tfrac{1}{4}(Y*^2+Z^2)^2 \cdot Sph + Y*(Y*^2+Z^2)H \cdot Cm$$
$$+ \tfrac{1}{2}(3Y*^2+Z^2)H^2 \cdot Ast + \tfrac{1}{4}(Y*^2+Z^2)H^2 \cdot Ptz + Y*H^3 \cdot Dist,$$

with the definitions, by comparison with the explicit form:

$$Sph = 4k_4(N'-N) + \tfrac{1}{2}N^2\left(\frac{1}{r}-\frac{1}{l}\right)^2\left(\frac{1}{N'l'}-\frac{1}{Nl}\right)$$

$$Cm = \tfrac{1}{2}\frac{N}{l}N\left(\frac{1}{r}-\frac{1}{l}\right)\left(\frac{1}{N'l'}-\frac{1}{Nl}\right)$$

$$Ast = \tfrac{1}{2}\left(\frac{N}{l}\right)^2\left(\frac{1}{N'l'}-\frac{1}{Nl}\right)$$

$$Ptz = \left(\frac{N}{l}\right)^2\left(\frac{N'-N}{NN'r}\right)$$

$$Dist = \tfrac{1}{2}\left(\frac{N}{l}\right)^3\left(\frac{1}{N^2}-\frac{1}{N'^2}\right).$$

In the abridged equation we now introduce the Y of the axial pencil by using instead of the original $Y*$ its equivalent $(Y+EH)$; hence

for $Y*^2+Z^2$ we have $(Y+EH)^2+Z^2 = (Y^2+Z^2)+2\cdot E\cdot Y\cdot H+E^2H^2$,

for $3Y*^2+Z^2$ we have $3(Y+EH)^2+Z^2 = (3Y^2+Z^2)+6\cdot E\cdot Y\cdot H+3E^2H^2$,

for $(Y*^2+Z^2)^2$ we have $(Y^2+Z^2)^2+4\cdot E\cdot Y\cdot H(Y^2+Z^2)+2E^2H^2(Y^2+Z^2)$
$$+4\cdot E^2Y^2H^2+4\cdot E^3Y\cdot H^3+E^4H^4,$$

for $Y*(Y*^2+Z^2)$ we have $Y(Y^2+Z^2)+EH(Y^2+Z^2)+2\cdot E^2Y\cdot H^2$
$$+2\cdot E\cdot Y^2H+E^2Y\cdot H^2+E^3H^3.$$

Equation OP(2)* then becomes

$OPD' =$

$$Sph[\tfrac{1}{4}(Y^2+Z^2)^2+EYH(Y^2+Z^2)+\tfrac{1}{2}E^2H^2(3Y^2+Z^2)+ E^3YH^3+\tfrac{1}{4}E^4H^4]$$
$$+ Cm[YH(Y^2+Z^2)+ EH^2(3Y^2+Z^2)+3E^2YH^3+ E^3H^4]$$
$$+ Ast[\tfrac{1}{2}H^2(3Y^2+Z^2)+ 3EYH^3+\tfrac{3}{2}E^2H^4]$$
$$+ Ptz[\tfrac{1}{4}H^2(Y^2+Z^2)+ \tfrac{1}{2}EYH^3+\tfrac{1}{4}E^2H^4]$$
$$+ Dist[YH^3+ EH^4].$$

[109] If we rearrange the terms according to the successive vertical columns which are seen to have coefficients in the variables H, Y and Z identical with those of the original equation, we obtain

(f)
$$OPD' = \tfrac{1}{4}(Y^2+Z^2)^2 \cdot Sph$$
$$+ Y(Y^2+Z^2)H[Cm+E\cdot Sph]$$
$$+ \tfrac{1}{2}(3Y^2+Z^2)H^2[Ast+2E\cdot Cm+E^2\cdot Sph]$$
$$+ \tfrac{1}{4}(Y^2+Z^2)H^2 \cdot Ptz$$
$$+ YH^3[Dist+\tfrac{1}{2}E\cdot Ptz+3E\cdot Ast+3E^2\cdot Cm+E^3\cdot Sph]$$
$$\{+H^4[\tfrac{1}{4}E^4\cdot Sph+E^3\cdot Cm+\tfrac{3}{2}E^2\cdot Ast+\tfrac{1}{4}E^2\cdot Ptz+E\cdot Dist]\}.$$

A sixth term in H^4 (in brackets) has come into the equation and apparently amounts to a complication by adding a sixth aberration to the five which we previously had. But this term has a simple significance. Our equation in its last form still refers to the OPD' with which a ray from any point of the clear aperture of the refracting surface meets the ray from A at the point B'_h. The sixth line of this equation represents this value of OPD' for the ray from A_h, the centre of the oblique pencil; for if we put the coordinates of A_h, namely, $Y^* = EH$ and $Z = 0$, into the general equation OP(2)* on page 713, we obtain exactly the terms in the sixth line. Hence the omission of this line amounts to deducting from the OPD' of any ray the path difference with which ray A_h-B_h meets ray A-B_h and therefore transforms our equation so that it gives directly the difference of optical path with which any ray of an oblique pencil meets the central ray of that pencil at our adopted reference point B'_h. As this is obviously a more reasonable basis of comparison, the omission of the sixth line is not only justifiable but is highly desirable because it leads to a more direct evaluation of the differences of those optical paths which really come into effective operation at any one image-point. The previous standard ray from A would quite frequently not be included at all in the more oblique pencils of a system with fairly distant entrance and exit pupils.

The only remaining problem is to form the sum of the terms in (f) for any number of centred surfaces at any finite separation so as to obtain a single equation for the aberrations of the complete system. Owing to the introduction of the eccentricity constant E, Y and Z in (f) now have the same significance for all pencils, including of course the axial one; they measure the location in rectangular coordinates of any ray with reference to the central or principal ray of the pencil of which it is a part. It is obvious that the changes in Y and Z at successive surfaces will be most easily and conveniently obtained from the axial pencil. We therefore draw a diagram, Fig. 147, showing an axial pencil passing from a surface (1) to a following surface (2), and we see at once that by similar triangles $Y_1/Y_2 = Z_1/Z_2 = A_1P_1/A_2P_2 = l'_1/l_2$. Both coordinates of a point of incidence therefore change in the same constant ratio, which is easily deduced from the paraxial ray-tracing. A repetition of this reasoning will evidently lead to the result that both coordinates of the point of incidence at any one surface of any ray traced through the system will be in some fixed ratio to its corresponding coordinates at any other surface. Hence we may select one surface (usually the first or the last) as the one whose

coordinates Y_0 and Z_0 of the point of incidence will be used in evaluating the [109]
aberrations, and then we put $Y_1 = p_1 \cdot Y_0$, $Y_2 = p_2 \cdot Y_0$, &c., where $p_1, p_2 \ldots$ are
constants directly obtainable from the paraxial ray-tracing.   These constants p

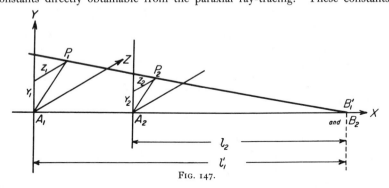

FIG. 147.

will be available equally for functions of Y and Z, for the corresponding Y and Z
have the same p-ratio.   We must also have:

$$Y_1{}^2 + Z_1{}^2 = p_1{}^2(Y_0{}^2 + Z_0{}^2)$$
and
$$(3 Y_1{}^2 + Z_1{}^2) = p_1{}^2(3 Y_0{}^2 + Z_0{}^2).$$

These constant ratios will enable us to extract common factors in terms of Y_0
and Z_0 from the aberrations at the separate surfaces.   But we can deal similarly
with the H-values of the separate surfaces: By (b) we have $H_1 N_1 / l_1 = H'_1 N'_1 / l'_1$.
As H'_1 is the distance from the optical axis to the image of an extra-axial object-
point produced by surface (1) and as this image forms the object for surface (2),
we have $H'_1 = H_2$.   Merely by our nomenclature we also have $N'_1 = N_2$, hence
the last equation may be written

$$H_1 N_1 / l_1 = H_2 N_2 / l'_1.$$

Since by Fig. 147 we have $l'_1 = l_2 \cdot Y_1 / Y_2$, we can further transform the previous
equation into

$$Y_1 H_1 N_1 / l_1 = Y_2 H_2 N_2 / l_2.$$

Inasmuch as the same suffix applies to all terms on each side of this equation, it
follows at once that $Y \cdot H \cdot N / l$ is an invariant for any ray traced right through the
system, and hence it has the same value at any surface as at the surface (o) selected
as the one to which the aberrations of the complete system are to be referred.
Therefore if, in accordance with the symbolism of (f), we return to plain letters
Y, Z, H &c. for *any* surface, we have

$$Y \cdot H \cdot N / l = Y_0 \cdot H_0 \cdot N_0 / l_0.$$

It will be noticed that the invariant YHN/l represents a generalized form of the
Lagrange invariant introduced for strictly paraxial pencils in Chapter I of Part I.
As the invariant property depends upon the assumption that Y, H, N, and l all

[109] refer to the same surface, it is of the utmost importance to remember that *all that follows rests on this assumption.* This means that the N_0, l_0, and H_0 introduced by the invariant must be the values at that surface to which the Y and Z at any other surface are referred by the relations $Y = p \cdot Y_0$ and $Z = p \cdot Z_0$.

We can now modify (f) so as to introduce the data of the selected reference surface. In the first term we use the p-ratio and write the factor of Sph: $\frac{1}{4}(Y_0{}^2 + Z_0{}^2)^2 \cdot p^4$. In the second term we extend the variable factor into $(Y^2 + Z^2)Y \cdot H(N/l)(l/N)$ and then put $Y^2 + Z^2 = (Y_0{}^2 + Z_0{}^2) \cdot p^2$ and $Y \cdot H \cdot N/l = Y_0 H_0 N_0 / l_0$, making the factor $Y_0(Y_0{}^2 + Z_0{}^2)(H_0 N_0/l_0)(p^2 \cdot l/N)$. To modify the third and fourth terms we must first note that, as we also have $l'_1 = l_2 \cdot Z_1/Z_2$, the invariant relation also holds for

$$Z \cdot H \cdot N/l = Z_0 H_0 N_0 / l_0,$$

and on squaring both this and the invariant in Y, we easily find that we may use

$$(3Y^2 + Z^2)H^2 N^2 / l^2 = (3Y_0{}^2 + Z_0{}^2)H_0{}^2 N_0{}^2 / l_0{}^2$$

and similarly $(Y^2 + Z^2)H^2 N^2 / l^2 = (Y_0{}^2 + Z_0{}^2)H_0{}^2 N_0{}^2 / l_0{}^2.$

Therefore, on extending the variable factors of the two terms by $(N^2/l^2)(l^2/N^2)$, they become

Factor of third term: $\frac{1}{2}(3Y_0{}^2 + Z_0{}^2)H_0{}^2(N_0/l_0)^2(l/N)^2$

Factor of fourth term: $\frac{1}{4}(Y_0{}^2 + Z_0{}^2)H_0{}^2(N_0/l_0)^2(l/N)^2.$

In the variable factor of the fifth term, which is the last, we write

$$\begin{aligned}
Y \cdot H^3 &= \frac{Y^3 H^3 N^3}{l^3} \cdot \frac{1}{Y^2} \cdot \frac{l^3}{N^3} = \frac{Y_0{}^3 H_0{}^3 N_0{}^3}{l_0{}^3} \cdot \frac{1}{Y_0{}^2 p^2} \cdot \frac{l^3}{N^3} \\
&= Y_0 H_0{}^3 \left(\frac{N_0}{l_0}\right)^3 \cdot \frac{1}{p^2}\left(\frac{l}{N}\right)^3.
\end{aligned}$$

To form the sum which represents the total primary aberrations for a system of any number of centred surfaces separated by any distances, we have the simplest possible task, for as our aberrations are in terms of differences of optical path and therefore they state the lead of the light travelling along any ray as compared with that travelling along the central or principal ray of the corresponding pencil, the total lead will be the simple algebraic sum of the separate contributions of the surfaces and a perfectly straightforward addition is therefore all that is required. Terms with the suffix (o) are common to all surfaces and go outside as a common factor, whilst terms without suffix apply to individual surfaces and come within the summation sign.   Equation OP(2)* then takes the form

$$\begin{aligned}
OPD' = & \tfrac{1}{4}(Y_0{}^2 + Z_0{}^2)^2 \cdot \Sigma(p^4 \cdot Sph) \\
& + Y_0(Y_0{}^2 + Z_0{}^2)H_0(N_0/l_0) \cdot \Sigma[p^2(l/N)(Cm + E \cdot Sph)] \\
& + \tfrac{1}{2}(3Y_0{}^2 + Z_0{}^2)H_0{}^2(N_0/l_0)^2 \cdot \Sigma[(l/N)^2(Ast + 2E \cdot Cm + E^2 \cdot Sph)] \\
& + \tfrac{1}{4}(Y_0{}^2 + Z_0{}^2)H_0{}^2(N_0/l_0)^2 \cdot \Sigma[(l/N)^2 \cdot Ptz] \\
& + Y_0 H_0{}^3(N_0/l_0)^3 \cdot \Sigma[(1/p)^2 \cdot (l/N)^3(Dist + \tfrac{1}{2}E \cdot Ptz + 3E \cdot Ast + 3E^2 \cdot Cm \\
& \hspace{8.5cm} + E^3 \cdot Sph)],
\end{aligned}$$

in which $E = l_{pr}/(l_{pr} - l)$ and *Sph. Cm*, &c. have the values specified below OP(2)* on page 713. The square brackets of the sums contain (l/N) as a factor; therefore they could not be evaluated if at any surface l became infinitely large. To meet this objection, we take the factor (l/N), its square or its cube, into the following factor by introducing:

$$Cm^* = Cm \cdot l/N; \quad E^* = E \cdot l/N; \quad Ast^* = Ast(l/N)^2;$$
$$Ptz^* = Ptz(l/N)^2; \quad Dist^* = Dist(l/N)^3.$$

The explicit values of these are given below under item (4). It is easily seen that these substitutions completely dispose of the l/N factor in every case. At the same time, the new starred constants are simpler than the original ones. For E^* we find

$$E^* = l_{pr} \cdot l/N(l_{pr} - l) = 1/\left(\frac{N}{l} - \frac{N}{l_{pr}}\right).$$

We now collect the formulae under the heading:

PRIMARY ABERRATION FOR ANY POINT OF FIELD
Equation OP(6)

(1) Trace a paraxial pencil right through the system from object-point to final image-point, calculating also for each surface $y = lu = l'u'$. Having selected the standard reference surface (o) for the total aberrations, tabulate for each surface N, N', l, l', and $p = y/y_0$.

(2) Trace a paraxial pencil from the centre of the diaphragm right through the system (if the diaphragm lies between surfaces, then from its centre to right and left!) finding l_{pr} for the medium to the left of each surface. Calculate for each surface $E^* = 1/\left(\dfrac{N}{l} - \dfrac{N}{l_{pr}}\right)$ and tabulate l_{pr} and E^*.

(3) If any figured surfaces are to be used, the k_4 value as defined in Section [91] by the formula

$$X = (PA^2/2r) + k_4 PA^4$$

must be decided upon. For the usual spherical surfaces, $k_4 = 0$.

(4) Calculate and tabulate for each surface

$$Sph = 4k_4(N' - N) + \tfrac{1}{2}N^2\left(\frac{1}{r} - \frac{1}{l}\right)^2\left(\frac{1}{N'l'} - \frac{1}{Nl}\right)$$

$$Cm^* = \tfrac{1}{2}N\left(\frac{1}{r} - \frac{1}{l}\right)\left(\frac{1}{N'l'} - \frac{1}{Nl}\right)$$

$$Ast^* = \tfrac{1}{2}\left(\frac{1}{N'l'} - \frac{1}{Nl}\right)$$

$$Ptz^* = \frac{N' - N}{NN'r}$$

$$Dist^* = \tfrac{1}{2}\left(\frac{1}{N^2} - \frac{1}{N'^2}\right).$$

[109]   (5) The primary aberrations of the complete system for any point of the field will then be obtained by

OP(6)

$$
\begin{aligned}
OPD' = {} & \tfrac{1}{4}(Y_0{}^2 + Z_0{}^2)^2 \cdot \Sigma(p^4 \cdot Sph) \\
& + Y_0(Y_0{}^2 + Z_0{}^2)H_0(N_0/l_0) \cdot \Sigma[p^2(Cm^* + E^*Sph)] \\
& + \tfrac{1}{2}(3\,Y_0{}^2 + Z_0{}^2)H_0{}^2(N_0/l_0)^2 \Sigma[Ast^* + 2E^*Cm^* + E^{*2}Sph] \\
& + \tfrac{1}{4}(Y_0{}^2 + Z_0{}^2)H_0{}^2(N_0/l_0)^2 \Sigma Ptz^* \\
& + Y_0 H_0{}^3(N_0/l_0)^3 \Sigma \left[\frac{1}{p^2}\,(Dist^* + \tfrac{1}{2}E^*Ptz^* + 3E^*Ast^* + 3E^{*2}Cm^* \right. \\
& \hspace{8cm} \left. + E^{*3}Sph) \right].
\end{aligned}
$$

In accordance with (b), $H_0 N_0/l_0$ may be replaced by $H'_0 N'_0/l'_0$ when it is desired to refer the aberrations to an image surface, as will usually be the case. It is not even necessary that the reference surface (o) be an actual refracting surface of the lens system; it may be any plane at right angles to the optical axis, such as one of the pupils or one of the principal planes of the system, for inasmuch as the in-variant property of $Y \cdot H \cdot N/l$ holds for any refracting surface, in spite of the bending of the rays which takes place there, it must certainly hold for a similar surface at which there is no change of medium. If such an arbitrary plane is to be used, it is only necessary to determine correctly its distance l_0 from the image-point associated with the medium in which the plane is located, and of course to use the index of that medium as the N_0 of the general equation. In addition, we must determine Y_0 as the diameter of the axial pencil in the selected plane and use that for the determination of all the p-values.

The quantity Ptz^* is the well-known Petzval sum; it was discussed on page 288 of Part I and put into other forms which sometimes are more convenient.

There are no exceptions to the validity of OP(6) barring the case when an image falls into one of the refracting surfaces. In this one special case, the series-development of a square root, on which the whole work is based, is inadmissible because the series is then necessarily not convergent for extra-axial object-points. Usually an independent consideration of such a surface will show that it is not a source of aberrations and that it is permissible to skip over it in forming the sums. The case when l_0 or l'_0 in the outside factor is infinite at first sight appears to be another exception; but for any finite angle of field, H_0 or H'_0 must then also be infinite and H_0 and l_0 can be combined because H_0/l_0 (or H'_0/l'_0) then represents $\tan O_0$ (or $\tan O'_0$) if O_0 (or O'_0) represents the angle of subtense of the distant object or image. From the purely mathematical point of view E^* would become infinite if both l and l_{pr} happened to become infinite at the same surface. That is optically impossible, however, because the corresponding l' and l'_{pr} would then be equal and all other corresponding finite l- and l_{pr}-values of the whole system would also be equal. Therefore the final exit pupil would coincide with the final image-plane and the particular diaphragm erroneously introduced into the calculation would be merely a field stop and not an aperture stop.

The numerical calculation of OP(6) is less laborious than it may appear to be because all the quantities entering into it are derived from a few simple items, and [109] as it is only a primary approximation and gives the aberrations directly and not as small differences of large quantities four-figure logs or even a well-handled slide rule will do it ample justice.   A reliable calculating machine would probably be the best of all means for the calculation.   Nevertheless, we shall not employ OP(6) to any great extent for numerical calculations.   As a rule we shall find simplified methods to be deduced from it more expeditious, and in other cases adaptations of our trigonometrical methods and of the matching principle will give us a better approximation, frequently in a shorter total time.

The chief value of the equation lies in its theoretical significance.   It proves definitely that the extra-axial image-points of any centred lens system may be afflicted with five and only five primary aberrations represented by the five lines of OP(6).   We must therefore discuss these to acquire a sound knowledge of the nature of the primary defects of lens systems.

DISCUSSION OF THE SEIDEL ABERRATIONS

IN the final equation, OP(6), which represents a complete and explicit statement of [110] all the primary aberrations of any centred lens system for objects at a finite distance l_1 from the first surface and for an entrance pupil at distance lpr_1 from the first surface, Y_0, Z_0, and H_0 are the only quantities which vary with the aperture of the system and with the distance of the image- or object-point from the optical axis. The sums are constants for the system.   For purposes of discussion we may therefore introduce simple symbols a_1 to a_5 for those constant sums, and the equation for a complete lens system then takes the form

OP(6)*
$$OPD' = \tfrac{1}{4}(Y_0{}^2 + Z_0{}^2)^2 \cdot a_1$$
$$+ Y_0(Y_0{}^2 + Z_0{}^2)H'_0(N'_0/l'_0) \cdot a_2$$
$$+ \tfrac{1}{2}(3Y_0{}^2 + Z_0{}^2)H'_0{}^2(N'_0/l'_0)^2 \cdot a_3$$
$$+ \tfrac{1}{4}(Y_0{}^2 + Z_0{}^2)H'_0{}^2(N'_0/l'_0)^2 \cdot a_4$$
$$+ Y_0 \cdot H'_0{}^3(N'_0/l'_0)^3 \cdot a_5.$$

Here Y_0 and Z_0 are the rectangular coordinates of the point of incidence in the selected reference surface referred to the point of incidence of the central or principal ray of any complete pencil, H'_0 is the distance of the ideal image-point from the optical axis, taken in the Y-direction and measured in the paraxial image-surface, and l'_0 is the distance from the selected reference surface to the paraxial image-point.   We have a perfectly free choice as to the reference surface to which the quantities with suffix (o) shall be referred.   We will select the final exit pupil as that surface for the purposes of our discussion.   Then l'_0 represents the distance from that pupil to the final paraxial image-point and H'_0 the distance of the final ideal extra-axial image-point from the optical axis.

From the complicated explicit expressions for the constant factors a_1 to a_5, it may be expected that they will have positive or negative values covering a wide range and that in a system of even only a few surfaces they will be more or less independent of one another, some being positive and some negative, and perchance one or two

[110] may be zero. Obviously zero values of all five constant coefficients is the ideal as far as primary aberrations are concerned, for OPD' would then be unconditionally zero for any part of the aperture and field, and a perfect union of all light at each ideal image-point would result. It will become apparent subsequently why numerical factors of $\frac{1}{2}$ or $\frac{1}{4}$ have been retained in some of the terms instead of including them in the a-values as would at first sight appear to be a more reasonable procedure.

The discussion of the five aberrations depends almost entirely on the determination of the displacement of the focus which corresponds to them. In Chapter XIII, page 623, we deduced a formula for the change in OPD' which corresponds to a departure of the focus from the point for which the OPD' is known, and with the omission of a corrective term which was shown to be negligible, we arrived at the formulae:

OP(4)* $\delta(OPD') = \frac{1}{2}N' \cdot \delta l'(Y^2 + Z^2)/l'^2$ for longitudinal shifts, and

OP(4)** $\delta(OPD') = N' \cdot \delta H' \cdot Y/l'$ for transverse shifts.

The term corresponding to a shift of focus in the Z-direction has been omitted.

DISTORTION

We shall begin the discussion of OP(6)* with the last term:

$$OPD' = \text{Other aberrations} + Y_0 \cdot H'_0{}^3 \cdot \left(\frac{N'_0}{l'_0}\right)^3 \cdot a_5.$$

The term depends on the aperture in the same way as the $\delta(OPD')$ of OP(4)**, that is, its magnitude is proportional to Y. If we consider a point at a distance $\delta H'_0$ from the ideal focal point B'_h, the term will become

$$Y_0 H'_0{}^3 \cdot \left(\frac{N'_0}{l'_0}\right)^3 \cdot a_5 + \frac{N'_0}{l'_0} \cdot Y_0 \cdot \delta H'_0 = Y_0 \frac{N'_0}{l'_0}\left[H'_0{}^3 \cdot \left(\frac{N'_0}{l'_0}\right)^2 \cdot a_5 + \delta H'_0\right].$$

This will become zero if the expression in the square bracket vanishes. Therefore the equation for the total OPD' of the system will be reduced to its first four terms if we refer it, not to the point B'_h determined by the theorem of Lagrange, but to a point shifted in the Y-direction, that is, towards or away from the optical axis, by an amount $\delta H'_0$ determined by the condition

$$\delta H'_0 = -a_5\left(\frac{N'_0}{l'_0}\right)^2 \cdot H'_0{}^3.$$

The fifth term is thus proved to represent a radial (with reference to the optical axis) displacement of the image which grows as the cube of its distance from the optical axis and is negative or downwards if H'_0 and a_5 are both positive. The fifth term vanishes completely for any aperture when the image is received in the position defined by $\delta H'_0$ in our last equation; the term therefore implies no defect in the sharpness of the image but only affects its location. Our ideal image-point B'_h was fixed by applying the paraxial magnification resulting from the theorem of

Lagrange to the whole field.   Hence this ideal image B'_h corresponds to constant [110] magnification in every part of the paraxial focal plane and therefore to an undistorted image, or an image which has the same perfect perspective as a pinhole image and renders every straight line in the object as a straight line in the image.   If we imagine a focusing screen placed in the paraxial focal plane and the image inspected in the usual way, we should see the axial and extra-axial image points B' and B'_h in their ideal positions if no aberrations were present.   If only our fifth term were of finite magnitude and a_5 were positive, then the extra-axial image point would be displaced towards B' by the $\delta H'_0$ which we have just determined and found to be proportional to the cube of $H'_0 = B'B'_h$.   Let $B'_h{}^*$ be the displaced image, shown in Fig. 148 where, as is unavoidable, $\delta H'_0$ is grossly exaggerated.   Now suppose that B'_h is the central point of a straight horizontal line in the complete ideal image and B'_{h_1} is another point of this line.   As our lens system is supposed to be centred and therefore perfectly symmetrical with reference to its optical axis, we

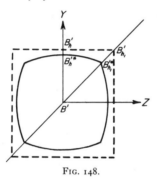

can turn our system of coordinates around the X-axis so as to bring the point B'_{h_1} into the XY-plane; our equations for $\delta H'_0$ will then hold for the point, but the displacement of $B'_h{}^*$ will be proportional to the cube of the new $H'_0 = B'B'_{h_1}$ and will be much larger than that of the point B'_h.   It is easily shown, by treating the displacements as very small, that the straight line of the ideal image becomes a parabolic curve in the displaced image, and as this amounts to a violation of the laws of correct perspective, the displacement of the extra-axial image-points corresponding to a finite value of a_5 is called *distortion*.   It is evident that if we extended the above reasoning to the four lines forming a perfect square with B' as centre in the ideal image, the corresponding distorted image would have four curved sides and would have some resemblance to a perspective view of a barrel. Hence the distortion corresponding to a *positive* value of a_5 is known as 'barrel' distortion.   It will be accepted by analogy that a negative a_5 will throw the dislocation into the opposite direction and will lead to an image of a perfect square which resembles a pincushion; hence the distortion resulting from a negative a_5 is usually called 'pincushion' distortion.   Straight lines passing in any direction through the axial point B' are the only ones which are rendered as straight in a distorted image;

[110] but if equidistant points were marked on the object-line, these points would be rendered at gradually decreasing distances towards the outer margin of the field in barrel distortion and at increasing distances in pincushion distortion.

As distortion equally affects all the rays of an oblique pencil, it is most easily determined trigonometrically by tracing the central or principal ray to intersection with the paraxial focal plane and comparing its true H'_0-value so calculated with the ideal H'_0-value corresponding to the paraxial magnification. Trigonometrical distortion is defined as $\delta H'_0/H'_0$, for which our definition of $\delta H'_0$ gives a value proportional to $H'_0{}^2$.

THE PETZVAL TERM

We have learnt that the fifth term of OP(6)* disappears if we substitute the displaced image-point $B'_h{}^*$ for the ideal one B'_h. We therefore adopt for the next stage of our discussion the reference point $B'_h{}^*$ and so reduce our equation to the first four terms. The equation then becomes

$$OPD' = \text{Other aberrations} + \tfrac{1}{4}(Y_0{}^2 + Z_0{}^2)H'_0{}^2\left(\frac{N'_0}{l'_0}\right)^2 \cdot a_4,$$

and by comparison with the explicit statement of our coefficients on pages 718 and 719, we will note that

$$a_4 = \sum \frac{N' - N}{N \cdot N' \cdot r},$$

the sum to be taken over all surfaces of the lens system.

A comparison of the form of the fourth term given above with OP(4)* shows that both vary with $(Y^2 + Z^2)$, and we may conclude that this term can be compensated by a suitable longitudinal shift $\delta l'$ of the focus. Adding the expression for such a shift to that of our fourth term, we obtain

$$\tfrac{1}{4}(Y_0{}^2 + Z_0{}^2)H'_0{}^2\left(\frac{N'_0}{l'_0}\right)^2 \cdot a_4 + \tfrac{1}{2}(Y_0{}^2 + Z_0{}^2) \cdot \delta l'_0 \cdot N'_0/l'_0{}^2,$$

which may be rearranged into

$$\tfrac{1}{2}(Y_0{}^2 + Z_0{}^2)\frac{N'_0}{l'_0{}^2}[\tfrac{1}{2}N'_0 \cdot H'_0{}^2 \cdot a_4 + \delta l'_0].$$

Arguing as in the case of distortion, we conclude that the Petzval term will completely disappear for the whole aperture if the expression in the square bracket is brought to zero by choosing

$$\delta l'_0 = -\tfrac{1}{2}N'_0 \cdot H'_0{}^2 \cdot a_4.$$

This proves that the fourth term shares the peculiarity of the fifth in not affecting the sharpness of the image points at all but merely displacing them. Whilst distortion leads to a transverse displacement towards or away from the optical axis, the Petzval term shifts the focus longitudinally towards or away from the lens system and by an amount which is proportional to $H'_0{}^2$. This displacement is therefore zero in the optical axis but grows rapidly towards the outer part of the |110| field.

We can easily prove that the Petzval term leads to *curvature of the field.* As we deduced the general equation OP(4)* for a shift of focus along the central ray of a pencil, the $\delta l'_0$ by which we compensated the Petzval term must be taken in the direction from B'_h* towards or away from the centre of the final exit-pupil of the system (see Fig. 149). But as in our primary approximation all angles must be assumed to be small of the first order, the $\delta l'_0$ measured along the principal ray will be sensibly equal to its projection upon the optical axis. Similarly, the distance of the image-point from the axis will differ only by very small (third-order) quantities from the original H'_0. We may therefore regard $\delta l'_0$ as the abscissa of

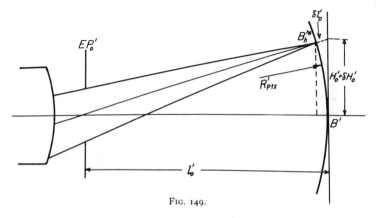

FIG. 149.

the shifted image-point referred to the paraxial image-point B', and H'_0 as the ordinate. Temporarily using the familiar x and y for these coordinates, our equation for $\delta l'_0$ becomes

$$x = -\tfrac{1}{2}(N'_0 \cdot a_4)y^2 \quad \text{or} \quad y^2 = -2\,\frac{1}{N'_0 \cdot a_4}\cdot x,$$

which is the equation of an ordinary parabola with its axis coinciding with AX and its pole at B'; its parameter or radius of curvature at the pole is $-1/N'_0 \cdot a_4$. We thus arrive at the result that the Petzval term represents a curvature of the field given by

$$\frac{1}{R'_{ptz}} = -N'_0 \cdot a_4$$

if we introduce R'_{ptz} as the symbol for the radius of curvature of the final image surface at the axis. We shall in the future call this radius the Petzval radius and the corresponding surface the Petzval surface, the latter necessarily being a surface of rotation with the optical axis as its axis on account of the symmetry of centred lens systems. Our primary approximation does not enable us to determine more

[110] than the curvature of the Petzval surface at its pole or the radius of its osculating sphere; we cannot decide whether it will take spherical, elliptical, hyperbolic, parabolic, or some intermediate form in its remoter parts; that could only be decided by developing the sixth-, eighth-, &c. order terms of the general equation OP(2)—a task of hopeless difficulty—or by trigonometrical calculations, which are more manageable.   In a few highly specialized cases, it is easily shown that the Petzval surface is very nearly a sphere; in the case of most lens systems, an ordinary paraboloid of rotation comes nearer to representing its form for a considerable angular extent.   We have discussed the Petzval curvature in section [55]; for the present we shall merely note that the fourth and fifth terms of the equation OP(6)* vanish if we introduce the displacements of extra-axial image-points by the $\delta H'_0$ and $\delta l'_0$ which we have determined and that the image on the Petzval surface will therefore be affected only by the first three terms of OP(6)*.   We shall now discuss these terms and show that each of them signifies a distinctive defect in the sharpness of the image obtained, coupled with further displacements.

ASTIGMATISM

For the image-points on the Petzval surface, we have brought down equation OP(6)* to the form

$$OPD' = \text{First two terms} + \tfrac{1}{2}(3Y_0{}^2 + Z_0{}^2)H'_0{}^2(N'_0/l'_0)^2 \cdot a_3.$$

The coefficient of the third term, containing Y_0 and Z_0, does not have the same form as the corresponding coefficient of either OP(4)* or OP(4)**, and we may take this as evidence that no shifting of the focus will cause this term to disappear for the whole extent of a circular cone of rays.   If a_3 is not zero, there will be differences of optical path at any focus which we may choose, and the image of an extra-axial object-point will be defective wherever we may examine it.   There are, however, very characteristic special cases.   We can firstly write $3Y_0{}^2 + Z_0{}^2 = 2Y_0{}^2 + (Y_0{}^2 + Z_0{}^2)$ and the second part of this agrees with the Y,Z coefficient of OP(4)* and therefore allows of compensation by a longitudinal shift of the focus. The OPD' resulting from the astigmatic term will then be given by:

$$OPD' = -Y_0{}^2 H'_0{}^2 (N'_0/l'_0)^2 \cdot a_3 + [\tfrac{1}{2}N'_0(Y_0{}^2 + Z_0{}^2)/l'_0{}^2](\delta l'_0 + N'_0 H'_0{}^2 \cdot a_3),$$

and the second part of this will vanish if the focal shift is determined by

$$\delta l'_0 = -N'_0 \cdot H'_0 \cdot a_3.$$

At the focus so located, only the first term, in $Y_0{}^2$, will remain finite; evidently this will disappear for $Y_0 = o$ and will be very small (on account of the squaring) for small values of Y_0.   Hence at our first selected focus there will be a practically perfect union of all light from a rectangular slit-like aperture having the Z-axis as its long centre-line (Fig. 150).   This focus we call the sagittal focus of the oblique pencil.   We will distinguish it by suffix S and will write the last equation which defines its distance from the Petzval surface as

$$\delta l'_{0_s} = -N'_0 \cdot H'_0{}^2 \cdot a_3.$$

We can locate a second equally characteristic partial focus of our oblique pencil [110]
by writing $3Y_0{}^2+Z_0{}^2 = 3(Y_0{}^2+Z_0{}^2)-2Z_0{}^2$, and again introducing a longitudinal
shift of the focus.   We thus obtain

$$OPD' = -Z_0{}^2H'_0{}^2(N'_0/l'_0)^2 \cdot a_3 + [\tfrac{1}{2}N'_0(Y_0{}^2+Z_0{}^2)/l'_0{}^2](\delta l'_0+3N'_0\cdot H'_0{}^2\cdot a_3),$$

and the second term of this will vanish if

$$\delta l'_0 = -3N'_0\cdot H'_0{}^2\cdot a_3.$$

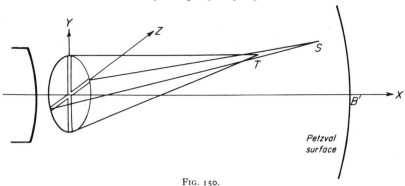

Fig. 150.

The first term in $Z_0{}^2$ which is then left will disappear for $Z_0 = 0$ and will be very
small for small values of Z_0; hence we can claim our second selected focus as giving
practically perfect union of all light from a slit-like aperture in the Y-direction, i.e.,
the rays in the XY-plane.   We call this focus the tangential focus of the oblique
astigmatic pencil, with suffix T, and have

$$\delta l'_{0_t} = -3N'_0\cdot H'_0{}^2\cdot a_3.$$

A comparison of the displacement of the S- and T-foci with reference to the
Petzval surface at once shows that they are in the exact and fixed ratio of one to
three.   This means that if one is known the other automatically follows; that they
can only coincide if a_3 is zero; and, of course, seeing that both have the same sign,
that they always lie on the same side of the Petzval surface.   This extremely
valuable and important property of the astigmatic foci has been strangely ignored.

The distinctive properties of the tangential and sagittal foci are brought out
clearly by discussing the third term of the Seidel aberrations in a more general
form.   If q represents any number, positive or negative, whole or fractional, we
can write

$$3Y_0{}^2+Z_0{}^2 = q(Y_0{}^2+Z_0{}^2)+(3-q)Y_0{}^2+(1-q)Z_0{}^2,$$

and with the focal-displacement term added, the expression for the astigmatism
becomes

$$\tfrac{1}{2}[(3-q)Y_0{}^2+(1-q)Z_0{}^2]H'_0{}^2(N'_0/l'_0)^2\cdot a_3$$
$$+\tfrac{1}{2}[N'_0(Y_0{}^2+Z_0{}^2)/l'_0{}^2](\delta l'_0+qN'_0H'_0{}^2\cdot a_3).$$

[110] The second part of this vanishes if

$$\delta l'_0 = -qN'_0 H'^2_0 \cdot a_3,$$

and this focus will fall between the S- and T-foci if $1 < q < 3$ and outside these foci if q is either greater than 3 or less than 1.

The first term of the generalized expression for the astigmatic difference of optical path would disappear if the term in the square bracket became zero. This would involve

$$(3-q)Y_0^2 + (1-q)Z_0^2 = 0 \quad \text{or} \quad Y_0^2/Z_0^2 = -(1-q)/(3-q) = (q-1)/(3-q).$$

The condition for zero-value of the astigmatic difference of optical path is therefore

$$Y_0/Z_0 = \pm \sqrt{(q-1)/(3-q)}.$$

The expression under the radical is positive only if $1 < q < 3$, that is, for a focus between S and T, and there will then be a definite ratio of Y_0 to Z_0 for which all differences of optical path vanish. This ratio becomes ± 1 for $q = 2$ or midway between S and T. At that point there will therefore be perfect union of all light from a cross-shaped aperture with the thin arms of the cross at $45°$ to the Y and Z axes (Fig. 151). This is a more symmetrical restriction of the aperture than that

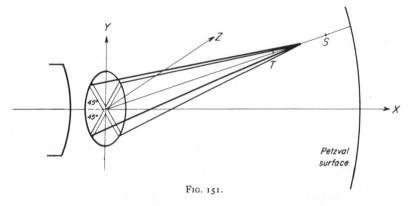

FIG. 151.

for the S- and T-foci and suggests an optimum in the case of an astigmatic pencil of rays. For intermediate foci closer to S, the ratio of Y_0 to Z_0 becomes a proper fraction and represents a cross-shaped aperture having arms at angles less than $45°$ with the Z-axis, and for foci closer to T, the arms of the cross approach the Y-axis. We may say that for the S-focus the arms of our cross become superposed in the Z-axis and for the T-focus they become superposed in the Y-axis, whilst at the midway focus the arms are at their widest separation.

For q greater than 3 or less than 1, the square root to which Y_0/Z_0 was found to be equal becomes imaginary; hence there is no kind of restricted aperture which will give perfect union of the light at points outside the S-to-T range. The S- and

T-foci thus mark the extreme limits of the range within which there exists the possibility of a good image with restricted slit-shaped apertures.

Another remarkable property of an unmutilated astigmatic pencil from a round aperture results from the discussion of the outstanding difference of optical path in our generalized expression, namely,

$$\text{Residual } OPD' = \tfrac{1}{2}[(3-q)Y_0{}^2 + (1-q)Z_0{}^2]H'_0{}^2(N'_0/l'_0)^2 \cdot a_3.$$

Evidently, for any selected value of q, the first term in the square bracket will reach a maximum numerical value for the extreme top and bottom of the clear circular aperture, where $Y_0 = SA$ if SA is taken as the symbol for the full semi-aperture, and the second term similarly reaches its highest numerical value for $Z_0 = SA$. As we have already found that the S- and T-foci mark the limits for tolerable definition, we shall restrict ourselves to that range or to q between 1 and 3. Then $(3-q)$ is necessarily positive and $(1-q)$ necessarily negative. Consequently the largest *difference* of optical path for the complete pencil will be the algebraic difference between the first and second terms in the square bracket when Y_0 and Z_0 are each put equal to SA. Hence

$$\text{Maximum } OPD' = SA[(3-q)-(1-q)]H'_0{}^2(N'_0/l'_0)^2 \cdot a_3$$
$$= 2 \cdot SA \cdot H'_0{}^2(N'_0/l'_0)^2 \cdot a_3.$$

Since this expression does not contain q at all (excepting in so far as we restricted its range between 1 and 3!) it proves that the maximum difference of optical path in an astigmatic pencil is perfectly constant throughout the range between the tangential and sagittal foci. Physically considered, from the Rayleigh-limit point of view, all foci within this range are therefore equally good or bad; deterioration only sets in outside the S- and T-foci.

We must finally discuss the curvature of the field corresponding to the S- and T-foci. We found that the Petzval surface to which they are referred lies at a distance from the paraxial focal plane defined by

$$\delta l'_{0_{ptz}} = -\tfrac{1}{2}N'_0 \cdot H'_0{}^2 \cdot a_4.$$

We have now found that the S- and T-foci lie at a distance from the Petzval surface given by

$$\delta l'_{0_s} = -N'_0 H'_0{}^2 \cdot a_3$$

and

$$\delta l'_{0_t} = -3N'_0 H'_0{}^2 \cdot a_3.$$

Consequently the S- and T-foci lie at a distance from the paraxial focal plane

$$\delta l'_{0_s} + \delta l'_{0_{ptz}} = -\tfrac{1}{2}H'_0{}^2 N'_0(a_4+2a_3)$$
$$\delta l'_{0_t} + \delta l'_{0_{ptz}} = -\tfrac{1}{2}H'_0{}^2 N'_0(a_4+6a_3).$$

We can now determine the corresponding curvatures of field exactly as we did in the case of the Petzval curvature, and we get

$$\frac{1}{R'_s} = -N'_0(a_4+2a_3), \qquad \frac{1}{R'_t} = -N'_0(a_4+6a_3).$$

[110] If a_3 is zero, these two curvatures become equal to each other and to the Petzval curvature; the latter therefore remains as an incurable curvature of the sharp image-surface if the first three Seidel aberrations are zero. The only way to eliminate this image curvature is to make a_4 vanish, as discussed in the preceding section.

COMA

The second term of our equation OP(6)* represents an unsymmetrical deformation of the image called Coma. We shall first discuss it on the assumption that the first and third terms, that is, spherical aberration and astigmatism, are zero. We shall thus obtain the pure coma effect on the Petzval surface.

The second term has the form

$$Y_0(Y_0{}^2 + Z_0{}^2)H'_0(N'_0/l'_0) \cdot a_2.$$

We shall discuss it for a narrow concentric zone of the circular aperture, of radius $SA = \sqrt{Y_0{}^2 + Z_0{}^2}$. The term now has the form

$$Y_0(N'_0/l'_0)SA^2 \cdot H'_0 \cdot a_2.$$

and agrees in its variable coefficient Y_0 with OP(4)**. Hence it can be compensated by a transverse shift (towards or away from the optical axis) of the image-point. The combined effect will be

$$Y_0(N'_0/l'_0)[\delta H'_0 + SA^2 \cdot H'_0 \cdot a_2]$$

and the value will be zero if

$$\delta H'_0 = -SA^2 \cdot H'_0 \cdot a_2.$$

This is a focal displacement of the same kind as that which we found for distortion, but the latter displacement was proportional to the cube of H'_0 and identical for the whole aperture, so that it did not affect the perfect definition of the image in any way. The coma displacement is proportional to H'_0 itself, grows with the square of the aperture, and is therefore zero for the light from the central part of the aperture. As for a zone of given radius SA, the displacement is strictly proportional to H'_0—the distance of the ideal image-point from the optical axis— it follows that with a positive a_2 the image of an extended object produced by a given zone is to a smaller scale than that produced by the central rays. This is equivalent to saying that the lateral magnification is correspondingly changed. This variation of the magnification produced by different zones and its growth with the square of the aperture agrees exactly with the effects which result from non-fulfilment of the optical sine condition and we find in fact that the latter is a measure of coma, but in a more general and precise way than our present primary approximation.

The nature of the coma effect is shown in Fig. 152, which represents the case of a positive value of a_2. Each zone has a focus at which all the light arrives in the same phase, and the distance of these foci from the focus of the central rays is proportional to the square of the aperture; therefore our favourite zone of 0.7071 of

the full aperture finds its physical focus exactly midway between the central and
the marginal foci. With a negative a_2, the foci of the outer zones would fall
farther from the optical axis than the central focus instead of nearer.

The light from each zone reaches its respective physical focus on the Petzval

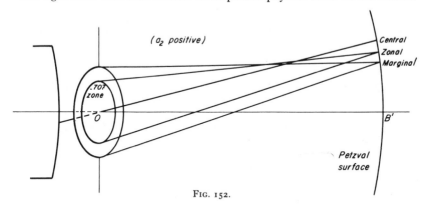

FIG. 152.

surface in exactly the same phase of vibration (provided that spherical aberration
and astigmatism are absent!). The reason is that our general expression for the
coma effect,

$$Y_0(N'_0/l'_0)[\delta H'_0 + SA^2 \cdot H'_0 \cdot a_2],$$

which gives the differences of optical path for any possible values of $\delta H'_0$ and SA,
becomes unconditionally zero for $Y_0 = 0$, that is, for a slit-aperture along the
Z-axis. As this slit runs across all the zones, the light from the nearest and the
farthest points of each zone is proved to arrive at the Petzval surface in unvarying
phase for any possible value of $\delta H'_0$, and as each zone is free from *differences* of
optical path at its own particular value of $\delta H'_0$, the equality of phase is established.
This is a fact of great importance in the integration of the complete light distribution at the coma-focus of a circular aperture.

In the absence of spherical aberration and astigmatism, the coma-focus in the
Petzval surface is the best obtainable one; for if we made a longitudinal shift in
either direction, the optical paths of the rays from the outer zones would be changed
by a smaller amount than those from the central zone and additional differences of
phase would come in, which must lead to a more diffused image of reduced
maximum brightness. Coma alone, therefore, has no influence on the Petzval
curvature of the image surface; it only affects the *distinctness* of the images in the
extra-axial region of the field.

SPHERICAL ABERRATION

The first term in OP(6)* may be disposed of by considering only one zone of
the aperture of radius SA, following the procedure of the previous section, and

[110] combining it with a longitudinal focus shift by OP(4)*. In the absence of all other aberrations this gives us

$$OPD' = \tfrac{1}{4}SA^2(Y_0{}^2 + Z_0{}^2)\cdot a_1 + \tfrac{1}{2}N'_0\delta l'_0(Y_0{}^2 + Z_0{}^2)/l'_0{}^2$$
$$= \tfrac{1}{2}(Y_0{}^2 + Z_0{}^2)(N'_0/l'_0{}^2)[\tfrac{1}{2}SA^2(l'_0{}^2/N'_0)\cdot a_1 + \delta l'_0].$$

Thus any one zone of the pupil will give a perfect image at a position on the lens axis distant from the paraxial focus by

$$\delta l'_0 = -\tfrac{1}{2}SA^2\cdot l'_0{}^2\cdot a_1/N'_0.$$

The properties of spherical aberration were discussed exhaustively in Chapters XIII and XIV. It was shown that this aberration leads to a longitudinal displacement of the best obtainable image and that in our present primary approximation, the best image is located midway between the paraxial and marginal geometrical foci. This image therefore coincides with the geometrical focus of the 0.7071 zone and also with the point at which the extreme marginal light meets the central light without any difference of phase.

Several Aberrations Present Simultaneously

In our present discussion of all the primary aberrations for any point in the field, the most noteworthy fact is that the primary spherical aberration, being the first term of the series of five aberrations and being totally independent of H'_0, affects every point in the field equally. Its magnitude as determined by the convenient formulae for the axial image-point is therefore immediately applicable to every point of the field and we need not determine it separately. In the absence of coma and astigmatism, it will displace the images in every part of the field to precisely the same extent and the images will lie on a curved surface equidistant from the Petzval surface. Since, in our primary approximation, the displacement must be regarded as a small quantity, the image-surface will have practically the Petzval curvature. Therefore spherical aberration, in the absence of coma and astigmatism, will have no sensible effect on the Petzval curvature of the image, which once more proves to be a fixed and unvarying defect of the image surface.

When spherical aberration and coma are present simultaneously, but astigmatism is absent, the coma will displace the physical foci of successive zones of the aperture towards or away from the optical axis (unless the object-point is exactly on the axis) and spherical aberration will displace them towards or away from the lens system. Inasmuch as both displacements grow with the square of the aperture, the zonal foci are arranged on an inclined straight line, as shown by Fig. 153. The foci of the marginal zone will be drawn towards the lens system (for a positive a_1) to the same extent in every part of the field because the spherical aberration is constant for the whole field. There is therefore no reasonable ground for doubting that the Petzval curvature will be maintained by the best compromise images obtainable.

The curvature of the field in the presence of all five aberrations can be discussed by applying the effects of longitudinal and transverse shifts of the focus defined by

OP(4)* and OP(4)** to the complete OPD' equation on page 719. Associating the [110] longitudinal displacement with the three symmetrical aberrations and the transverse displacement with the two unsymmetrical aberrations, we obtain

$$OPD' = \begin{cases} \tfrac{1}{4}(Y_0{}^2+Z_0{}^2)^2 \cdot a_1 + \tfrac{1}{2}(3 Y_0{}^2+Z_0{}^2)H'_0{}^2(N'_0/l'_0)^2 \cdot a_3 \\ \quad + \tfrac{1}{4}(Y_0{}^2+Z_0{}^2)H'_0{}^2(N'_0/l'_0)^2 \cdot a_4 + \tfrac{1}{2}N'_0(Y_0{}^2+Z_0{}^2)\delta l'_0/l'_0{}^2 \\ + Y_0(Y_0{}^2+Z_0{}^2)H'_0(N'_0/l'_0) \cdot a_2 + Y_0 H'_0{}^3(N'_0/l'_0)^3 \cdot a_5 \\ \qquad\qquad\qquad\qquad\qquad\qquad + N'_0 Y_0 \delta H'_0/l'_0. \end{cases}$$

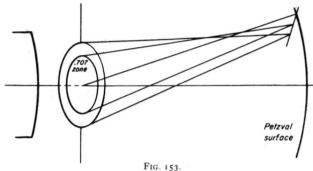

FIG. 153.

We have already shown in the discussion of coma that the terms in the second line can be brought to zero for a zone of any given aperture by the introduction of a suitable value for the transverse displacement $\delta H'_0$, and we need not therefore include these terms in the discussion of curvature of field.   The terms in the first line will be brought to zero by

$$\delta l'_0 = -\tfrac{1}{2}(Y_0{}^2+Z_0{}^2)(l'_0{}^2/N'_0) \cdot a_1 - \frac{3 Y_0{}^2+Z_0{}^2}{Y_0{}^2+Z_0{}^2} H'_0{}^2 N'_0 \cdot a_3 - \tfrac{1}{2}H'_0{}^2 N'_0 \cdot a_4,$$

and this gives the longitudinal displacement of the focus at which the light from particular parts of the aperture will meet in equal phase.   The equation holds for any part of the field and naturally also for the axial point; for the latter H'_0 is zero and gives

$$\text{Axial } \delta l'_0 = -\tfrac{1}{2}(Y_0{}^2+Z_0{}^2)(l'_0{}^2/N'_0) \cdot a_1.$$

At any given aperture of the exit pupil, the longitudinal focal displacement of an extra-axial image-point with reference to the axial one will be the difference of (extra-axial $\delta l'_0$) minus (axial $\delta l'_0$) or

$$\text{Net } \delta l'_0 = -\frac{3 Y_0{}^2+Z_0{}^2}{Y_0{}^2+Z_0{}^2} H'_0{}^2 N'_0 \cdot a_3 - \tfrac{1}{2}H'_0{}^2 N'_0 \cdot a_4.$$

As this is proportional to $H'_0{}^2$, it indicates curvature of field, which can be evaluated exactly like the Petzval curvature on page 722 and the S and T curvatures on page

[110] 724 and gives, if R'_I is used as a general symbol for the radius of curvature of an image surface,

$$\frac{1}{R'_I} = -N'_0 \cdot a_4 - 2\frac{3Y_0{}^2 + Z_0{}^2}{Y_0{}^2 + Z_0{}^2}N'_0 \cdot a_3.$$

The fraction $(3Y_0{}^2 + Z_0{}^2)/(Y_0{}^2 + Z_0{}^2)$ reaches a maximum value of 3 for $Z_0 = 0$ and a minimum value of 1 for $Y_0 = 0$, that is, for T and S rays respectively, for which it gives

$$\frac{1}{R'_S} = -N'_0 \cdot a_4 - 2N'_0 \cdot a_3 = -N'_0 \cdot (a_4 + 2a_3)$$

$$\frac{1}{R'_T} = -N'_0 \cdot a_4 - 6N'_0 \cdot a_3 = -N'_0 \cdot (a_4 + 6a_3).$$

These are precisely the values arrived at on page 727 on the restricting assumption that astigmatism and the Petzval term were alone present. We have therefore now proved that the presence of spherical aberration, coma, and distortion, either simultaneously or in any combination, have no effect on the curvature of the image; hence the latter is invariably fixed by the constants a_3 and a_4 alone, quite regardless of the values a_1, a_2, and a_5. This is an extremely important fact; failure to realize it was the gravest defect in the time-honoured interpretation of the Seidel aberrations. We must, however, remember that the present investigations are strictly limited to primary aberrations. At large apertures and in the outer parts of a large field, certain secondary aberrations reach considerable magnitude and bend the field away from the simple curvatures determined by the above equations. Nevertheless, these equations hold without exception at moderate apertures for the central part of the field, and in photographic lenses and eyepieces usually with close approximation up to fields of 20° to 30° in diameter.

Summary

All primary aberrations of pencils passing through a centred system of refracting surfaces are included in the equation

OP(6)* $OPD' = \frac{1}{4}(Y_0{}^2 + Z_0{}^2)^2 \cdot a_1 + Y_0(Y_0{}^2 + Z_0{}^2)H'_0(N'_0/l'_0) \cdot a_2$

$\qquad + \frac{1}{4}(3Y_0{}^2 + Z_0{}^2)H'_0{}^2(N'_0/l'_0)^2 \cdot a_3$

$\qquad + \frac{1}{4}(Y_0{}^2 + Z_0{}^2)H'_0{}^2(N'_0/l'_0)^2 \cdot a_4 + Y_0H'_0{}^3(N'_0/l'_0)^3 \cdot a_5.$

The five terms represent (1) Spherical aberration, (2) Coma, (3) Astigmatism, (4) Petzval term, and (5) Distortion. Their coefficients $a_1 \ldots a_5$ are constants of the system, each being a sum to be formed for all the refracting surfaces.

The explicit values of these five sums are (page 718):

$$a_1 = \Sigma(p^4 \cdot Sph); \quad a_2 = \Sigma[p^2(Cm^* + E^* \cdot Sph)];$$
$$a_3 = \Sigma[Ast^* + 2E^*Cm^* + E^{*2}Sph)]; \quad a_4 = \Sigma Ptz^*;$$
$$a_5 = \Sigma[(1/p^2)(Dist^* + \tfrac{1}{2}E^*Ptz^* + 3E^*Ast^* + 3E^{*2}Cm^* + E^{*3}Sph)];$$

$p = y/y_0$; $E^* = 1/(N/l - N/l_{pr})$; $Sph = 4k_4(N' - N) + \frac{1}{2}N^2(1/r - 1/l)^2(1/N'l' - 1/Nl)$; $Cm^* = \frac{1}{2}N(1/r - 1/l)(1/N'l' - 1/Nl)$; $Ast^* = \frac{1}{2}(1/N'l' - 1/Nl)$; $Ptz^* = (N' - N)/NN'r$; $Dist^* = \frac{1}{2}(1/N^2 - 1/N'^2).$

The coefficients $a_1 \ldots a_5$ can be calculated by these formulae, but they are [110] in many cases obtained more easily by other methods to be dealt with subsequently. The ideal image-point is located in a plane at right angles to the optical axis and containing the paraxial focus at distance l'_0 from the reference surface (usually the exit pupil) in which the semi-aperture $SA = \sqrt{(Y_0^2 + Z_0^2)}$ is measured. The distance H'_0 of each ideal extra-axial image-point from the optical axis is that corresponding to the paraxial magnification of the system. The ideal image to which the aberrations are referred is therefore perfectly flat and free from distortion.

The significance of the five aberrations is as follows:

(1) Spherical aberration, alone, displaces the best obtainable image-points throughout the field equally by an amount

$$\delta l'_0 = -\tfrac{1}{2} l'_0{}^2 \cdot \frac{a_1}{N'_0} SA^2.$$

(2) Coma, alone, displaces physical extra-axial image-points towards or away from the optical axis by an amount

$$\delta H'_0 = -SA^2 \cdot H'_0 \cdot a_2.$$

(3) Astigmatism, alone, produces longitudinal changes of focus which vary according to the direction of a narrow diametral strip of the full circular aperture. There is a definite focus for any such strip. For a slit-aperture in the Z-axis the displacement of the 'sagittal' focus is

$$\delta l'_{0_S} = -N'_0 \cdot H'_0{}^2 \cdot a_3$$

and the curvature of field

$$\frac{1}{R'_S} = -N'_0(a_4 + 2a_3).$$

For a slit-aperture in the Y-axis, the displacement of the 'tangential' focus is

$$\delta l'_{0_T} = -3N'_0 H'_0{}^2 \cdot a_3$$

and the curvature of field

$$\frac{1}{R'_T} = -N'_0(a_4 + 6a_3).$$

The curvatures of the sagittal and tangential fields are not seriously affected by the simultaneous presence of spherical aberration or coma or both.

(4) The Petzval term represents, by itself, simply a longitudinal shift of an extra-axial image by $\delta l'_{0_{ptz}} = -\tfrac{1}{2} N'_0 H'_0{}^2 \cdot a_4$ with a corresponding curvature of the field $1/R'_{ptz} = -N'_0 \cdot a_4$. If astigmatism is also present, then both the S- and the T-foci always lie on the same side of the Petzval surface and at distances from it which are in the fixed ratio of $1:3$.

(5) Distortion, by itself, displaces the sharp image-points towards or away from the optical axis by an amount

$$\delta H'_0 = -a_5(N'_0/l'_0)^2 \cdot H'_0{}^3.$$

[110] It causes straight lines in the object which do not pass through the optical axis to be depicted as curved lines in the image.

EFFECT OF 'FIGURED' SURFACES

All our deductions apply to lens systems which include one or more non-spherical or 'figured' surfaces, for the only figuring constant k_4 which can affect the primary aberrations has been retained throughout. It is still widely believed that the figuring of surfaces, if rendered a commercially practicable operation, would be a means of overcoming certain grave difficulties in the design of well-corrected lens systems, more especially those concerned with the attainment of a flat field free from astigmatism. Our discussion shows that there is no foundation for this belief. The chief enemy of a perfectly flat field, namely, the Petzval term, depends only on the axial radius of curvature r of the refracting surfaces and is utterly independent of any figuring of the extra-axial parts of the surfaces. The primary condition for a flat field, $a_4 = 0$, therefore applies to figured systems just as absolutely as it does to systems with strictly spherical surfaces, and the former have no advantage in this respect. Similarly the three-to-one ratio in the distances of the tangential and sagittal foci from the Petzval surface was found to be absolutely immutable for primary astigmatism, and figuring has no effect on it. The only advantages of figured surfaces are that they give us independent control over the spherical aberration and therefore, at times, the possibility of attaining a certain result with fewer constituent lenses or with a more compact system. These advantages are worth bearing in mind, but they must be weighed in each case against the additional cost of figuring, the difficulty of exactly reproducing the prescribed amount on every system made to a given specification, and the difficulty of applying the figuring in zones which are strictly concentric with the optical axis.

An important conclusion can be drawn from our expressions for the constants a_1 to a_5, namely, that the figuring constant enters into all of them with the exception of a_4 if E^* has a sensible value, that is, if the oblique pencils pass eccentrically through a figured surface. In the case of photographic objectives of large size computed for strictly spherical surfaces, it is frequently found that the effects of imperfect annealing and of variation of the refractive index within each lens become apparent chiefly in the form of defective spherical correction while the other aberrations are satisfactorily small. Figuring will then be the easiest way of improving the spherical correction; but if this were done to a surface at a considerable distance from the diaphragm, the other aberrations would be seriously affected. Hence this kind of figuring—a very common necessity—should always be applied to the surface closest to the diaphragm; it will then affect practically no aberration except the spherical one, which will be altered nearly uniformly for the whole field.

GEOMETRICAL INTERPRETATION OF THE SEIDEL ABERRATIONS

[111] THE physical interpretation of the primary aberrations of oblique pencils, as a source of a confusion of phase of the light meeting at any selected focal point, is the simplest possible one because it gives the complete information in a single equation,

and it is the most useful one because it allows of the direct application of the [111]
Rayleigh limit to the results as soon as the coefficients a_1 to a_5 have been determined.
But as these coefficients are in many cases deduced most easily from the tracing of
a few geometrical rays and as the geometrical interpretation is as yet practically the
only one recognized elsewhere, it is desirable to transform our equation OP(6)* so
as to express the departure of the rays, or normals of the wave surface, from their
ideal direction towards the ideal focus in the paraxial image plane and at the value
of H'_0 which is determined by the paraxial magnification.

We can effect this transformation by a simple generalization of the method applied
to spherical aberration in Chapter XIII. The OPD' determined by OP(6)* again
represents the gap between the distorted real wave and the strictly spherical ideal
wave at any point P, and again a positive value of OPD' means that the outer part
of the real wave is located at the right of the ideal wave. But whilst in the case of
pure spherical aberration the real wave was a true surface of rotation so that all its
normals cut the central ray of the pencil, we have now to deal with a more complicated form of real wave whose normals depart from their ideal direction not only
towards or away from the central ray but also sideways, so that the vast majority of
the rays do not cut the central ray at all. The generalization of the process
employed on page 614 must therefore consist in determining the departure of the
real ray or normal from the direction of the ideal one in two coordinates, for which
we choose the departures in the Y and Z directions.

In Fig. 154 (a), representing our usual perspective view, let B'_h in the XY-plane
be the ideal extra-axial image-point, let the adopted reference surface (usually the
exit pupil) pass through the origin A, and let AP_i represent an oblique plane section
through the ideal wave with B'_h at its centre of curvature, and AP_r the corresponding
section of the distorted real wave. Then by our definition of the difference of
optical path referred to the central ray AB'_h, we shall have

$$OPD' \text{ by OP(6)*} = N'_0 \cdot P_i P_r.$$

The factor N'_0 on the right is necessary because all our OPD' are obtained by
multiplying the geometrical paths (such as $P_i P_r$) by the index of the medium in
which they lie.

We now lay a plane at right angles to the XY-plane through B'_h and P_i. This
plane will cut the ideal and the real waves in two plane curves which, when projected orthogonally upon the XZ-plane, will appear as shown in Fig. 154 (b). We
also lay a plane at right angles to the XZ-plane through B'_h and P_i and, by projecting this upon the XY-plane, obtain Fig. 154 (c). Remembering that these diagrams
are unavoidably exaggerated to a tremendous extent, that all the curves really
would lie very close to the YZ-plane, that corresponding ideal and real curves
would lie within a few wave-lengths of each other, and that all angles involved
must be small in order to render our primary approximation justifiable, we can
once more introduce a number of simplifying assumptions.

For the section (b), Y_0 will be sensibly constant throughout the extent of the
two curves in which the waves are cut, and for the section (c), Z_0 will similarly be
constant. Consequently, in analogy with the deductions on pages 614–616 of
Chapter XIII, the partial differential coefficient $\partial P_i P_r / \partial Z_0$ will give the inclination

[111] with reference to each other of the tangents to the wave-curves at points P_i and P_r in Fig. 154 (b). Similarly the partial differential coefficient $\partial P_i P_r / \partial Y_0$ will give the corresponding inclination for Fig. 154 (c). As the normals of the waves at points P_i and P_r must be at right angles to the tangents, the two inclinations of

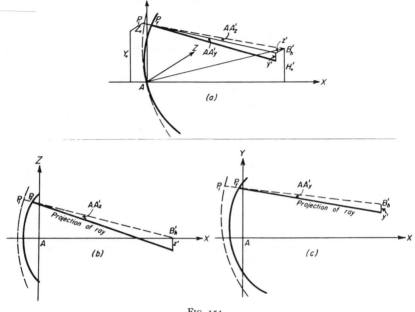

FIG. 154.

corresponding tangents determined by the partial differential coefficients also determine the angles between the projections of the ideal and the real rays in Fig. 154 (b) and (c) respectively, which are the components of the angular aberration in the Z and Y directions respectively.

In carrying out these partial differentiations, we shall add to OP(6)* the two terms $\delta(OPD')$ representing the effects of a shift of the focus from the ideal position B'_h by $\delta l'_0$ and $\delta H'_0$ as previously determined in equations OP(4)* and (4)**. We thus obtain the general expression for $P_i P_r = OPD'/N'_0$:

$$P_i P_r = \tfrac{1}{4}(Y_0^2 + Z_0^2)^2 \cdot \frac{a_1}{N'_0} + Y_0(Y_0^2 + Z_0^2)H'_0 \left(\frac{N'_0}{l'_0}\right)\frac{a_2}{N'_0}$$

$$+ \tfrac{1}{2}(3Y_0^2 + Z_0^2)H'_0{}^2 \left(\frac{N'_0}{l'_0}\right)^2 \frac{a_3}{N'_0} + \tfrac{1}{4}(Y_0^2 + Z_0^2)H'_0{}^2 \left(\frac{N'_0}{l'_0}\right)^2 \frac{a_4}{N'_0}$$

$$+ Y_0 \cdot H'_0{}^3 \left(\frac{N'_0}{l'_0}\right)^3 \frac{a_5}{N'_0} + \tfrac{1}{2}\delta l'_0 (Y_0^2 + Z_0^2)/l'_0{}^2 + \delta H'_0 \cdot Y_0/l'_0.$$

As we have shown that the two partial differential coefficients represent the [111] components of the angular aberration in the corresponding directions, we shall introduce the symbols $AA'_y = \partial(P_iP_r)/\partial Y_0$ and $AA'_z = \partial(P_iP_r)/\partial Z_0$ for these components. Treating Y_0 as the only variable in the first partial differentiation and Z_0 as the only variable in the second, we easily find

$$AA'_y = Y_0(Y_0{}^2+Z_0{}^2)\frac{a_1}{N'_0} + (3Y_0{}^2+Z_0{}^2)H'_0\left(\frac{N'_0}{l'_0}\right)\frac{a_2}{N'_0} + 3Y_0H'_0{}^2\left(\frac{N'_0}{l'_0}\right)^2\frac{a_3}{N'_0}$$

$$+ \tfrac{1}{2}Y_0H'_0{}^2\left(\frac{N'_0}{l'_0}\right)^2\frac{a_4}{N'_0} + H'_0{}^3\left(\frac{N'_0}{l'_0}\right)^3\frac{a_5}{N'_0} + \delta l'_0 \cdot Y_0/l'_0{}^2 + \delta H'_0/l'_0$$

$$AA'_z = Z_0(Y_0{}^2+Z_0{}^2)\frac{a_1}{N'_0} + 2Y_0Z_0H'_0\left(\frac{N'_0}{l'_0}\right)\frac{a_2}{N'_0} + Z_0H'_0{}^2\left(\frac{N'_0}{l'_0}\right)^2\frac{a_3}{N'_0}$$

$$+ \tfrac{1}{2}Z_0H'_0{}^2\left(\frac{N'_0}{l'_0}\right)^2\frac{a_4}{N'_0} + \delta l'_0 \cdot Z_0/l'_0{}^2.$$

These become more useful for discussion if we turn the angular aberration into a transverse aberration. This can be done by multiplying the angular aberrations by the distance $P_iB'_h$, which is sensibly equal to our l'_0. So multiplied, AA'_y gives the distance y' by which the real ray falls below B'_h on the focal surface or below the shifted focus if $\delta l'_0$ and $\delta H'_0$ are retained, and AA'_z gives the distance z' by which the real ray falls in front of B'_h. As both these components thus fall into the negative direction of our Y- and Z-axis when AA'_y and AA'_z are positive, as in our diagrams, we shall compensate for this by multiplying AA'_y and AA'_z by $-l'_0$. We then find the components of the transverse aberration as

TSA(1)
Transverse
Seidel
Aberrations

$$\left\{\begin{array}{l} y' = -Y_0(Y_0{}^2+Z_0{}^2)(l'_0/N'_0)\cdot a_1 - (3Y_0{}^2+Z_0{}^2)H'_0\cdot a_2 \\ \quad - 3Y_0H'_0{}^2(N'_0/l'_0)\cdot a_3 - \tfrac{1}{2}Y_0H'_0{}^2(N'_0/l'_0)\cdot a_4 \\ \quad - H'_0{}^3(N'_0/l'_0)^2\cdot a_5 - \delta l'_0 \cdot Y_0/l'_0 - \delta H'_0 \\ z' = -Z_0(Y_0{}^2+Z_0{}^2)(l'_0/N'_0)\cdot a_1 - 2Y_0Z_0H'_0\cdot a_2 \\ \quad - Z_0H'_0{}^2(N'_0/l'_0)\cdot a_3 - \tfrac{1}{2}Z_0H'_0{}^2(N'_0/l'_0)\cdot a_4 - \delta l'_0 \cdot Z_0/l'_0. \end{array}\right.$$

The coefficients a_1 to a_5 in these equations retain the exact values specified on page 732. The terms in $\delta l'_0$ and $\delta H'_0$ naturally vanish when the transverse aberrations are referred to the ideal image-point B'_h; they are left in the equations merely to facilitate the discussion of the effects of shifting the focus either lengthways or sideways. Figure 155 shows the meaning of the displacements y' and z'. If a focusing screen were placed in the paraxial image plane and B'_h represents the ideal image-point, then in the presence of Seidel aberrations a ray from point P in the exit pupil would reach the screen at some other point; y' and z' are the co-ordinates of this point. These coordinates are reckoned positive in the standard directions of our Y- and Z-axes; negative values of y' and z' are shown in Fig. 155 because in accordance with TSA(1), they will come out negative if $Y_0Z_0H'_0l'_0$ and $a_1 \ldots a_5$ are all positive.

In many cases it is more convenient to locate the point P in the exit pupil by polar coordinates. One coordinate is the radius vector AP, which we shall call

[111] SA and which we shall consider to be intrinsically positive. The other is the angle O measured from the positive direction of the Y-axis towards P; it may be either measured clockwise to $360°$ as an intrinsically positive angle or to $180°$ clockwise

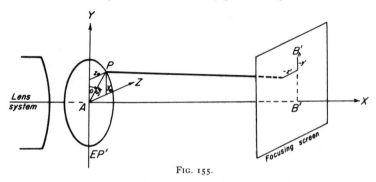

Fig. 155.

and counter-clockwise from AY, in which case it is positive or negative, respectively. Figure 155 shows that

$$Y_0 = SA \cos O; \qquad Z_0 = SA \sin O; \quad \text{and} \quad Y_0{}^2 + Z_0{}^2 = SA^2.$$

By substituting these into TSA(1) we can effect the transformation. The only terms which call for additional explanations are $(3Y_0{}^2 + Z_0{}^2)$ and $2Y_0Z_0$. We put $(3Y_0{}^2 + Z_0{}^2) = (Y_0{}^2 + Z_0{}^2) + 2Y_0{}^2 = SA^2(1 + 2\cos^2 O) = SA^2[1 + (\cos^2 O + \sin^2 O) + (\cos^2 O - \sin^2 O)] = SA^2(2 + \cos 2O)$, and $2Y_0Z_0 = 2SA^2 \cdot \cos O \cdot \sin O = SA^2 \sin 2O$ and obtain the alternative equations:

$$
\text{TSA(1)*}
\begin{cases}
y' = -SA^3 \cos O \cdot (l'_0/N'_0) \cdot a_1 - SA^2(2 + \cos 2O)H'_0 \cdot a_2 \\
\quad - 3SA \cos O \cdot H'_0{}^2(N'_0/l'_0) \cdot a_3 - \tfrac{1}{2}SA \cos O \cdot H'_0{}^2(N'_0/l'_0) \cdot a_4 \\
\quad - H'_0{}^3(N'_0/l'_0)^2 \cdot a_5 - SA \cos O \cdot (\delta l'_0/l'_0) - \delta H'_0 \\
z' = -SA^3 \sin O \cdot (l'_0/N'_0) \cdot a_1 - SA^2 \sin 2O \cdot H'_0 \cdot a_2 \\
\quad - SA \sin O \cdot H'_0{}^2(N'_0/l'_0) \cdot a_3 - \tfrac{1}{2}SA \sin O \cdot H'_0{}^2(N'_0/l'_0) \cdot a_4 \\
\quad - SA \sin O \cdot (\delta l'_0/l'_0).
\end{cases}
$$

Equations TSA(1) and TSA(1)* show that all the transverse Seidel aberrations are small quantities of the third order, for the power indices of SA and H'_0 always add up to three, and the same remark also applies to the angular measurement of these aberrations from which the transverse measure was deduced. For that reason the Seidel aberrations are very frequently referred to as the 'third-order aberrations' of oblique pencils. Since they become of the fourth order when put into the only absolute measure as differences of optical path, and on the other hand become of the second order when turned into longitudinal aberrations as far as that is possible, we shall avoid confusion by adhering to the description of the Seidel aberrations as the primary aberrations of oblique pencils.

DISCUSSION OF TSA(1)*

The *distortion* term a_5 occurs only in the y-component, hence distortion affects [111] only the distance of the image point from the optical axis and does not displace it in the Z direction. Its magnitude follows even more easily than in the case of OP(6)* by taking the distortion term together with the last term $\delta H'_0$, which shows at once that distortion displaces the image-point by

$$\delta H'_0 = -H'^3_0(N'_0/l'_0)^2 \cdot a_5,$$

in exact agreement with the result derived from OP(6)*.

Taking the *Petzval* term together with the $\delta l'_0$-term, we similarly obtain from both the y' and the z' equations

$$\delta l'_0 = -\tfrac{1}{2}H'^2_0 N'_0 \cdot a_4,$$

again in exact agreement with the previous deductions from OP(6)*, and the Petzval curvature of the image would therefore also come out identical.

We now adopt the reasoning given fully in the discussion of OP(6)*, namely, that the terms in a_4 and a_5 merely represent a shift of focus from the paraxial image-plane into the Petzval surface, the new ideal image-point being the point where the central ray of each oblique pencil penetrates the Petzval surface. We can then discuss the first three aberrations with reference to this new reference-point.

ASTIGMATISM

Picking out the astigmatic term in a_3 together with the term for a longitudinal shift of focus, we obtain

$$y'_{ast} = -3SA \cos O \cdot H'^2_0(N'_0/l'_0) \cdot a_3 - SA \cos O \cdot (\delta l'_0/l'_0)$$
and
$$z'_{ast} = -SA \sin O \cdot H'^2_0(N'_0/l'_0) \cdot a_3 - SA \sin O \cdot (\delta l'_0/l'_0).$$

Taking out as a common factor $SA \cdot \cos O/l'_0$ in the first equation and $SA \sin O/l'_0$ in the second, we conclude that

$$y'_{ast} = 0 \quad \text{if} \quad 3H'^2_0 N'_0 \cdot a_3 + \delta l'_0 = 0 \quad \text{or} \quad \delta l'_0 = -3H'^2_0 N'_0 \cdot a_3$$
and
$$z'_{ast} = 0 \quad \text{if} \quad H'^2_0 N'_0 \cdot a_3 + \delta l'_0 = 0 \quad \text{or} \quad \delta l'_0 = -H'^2_0 N'_0 \cdot a_3.$$

The displacements of the focus away from the Petzval surface agree exactly with those found in the previous discussion for the points T and S respectively. But the new discussion shows that at T, y'_{ast} is unconditionally zero whilst at S, z'_{ast} is unconditionally zero, both for any semi-aperture SA or any value of O, for neither of these appears in the solutions for $\delta l'_0$. As y' measures the distance above or below the adopted focal point on the central ray at which any ray intersects a plane laid through that point, and z' a corresponding distance in front of or behind the focal point, we have found a proof of the existence of focal lines in an astigmatic pencil of geometrical rays (Fig. 156). At T all the rays from the whole of the aperture pass through a line-focus parallel to the YZ-plane and at S the

[111] same rays pass through another line-focus lying in the XY-plane. We obtain a still clearer idea of this peculiarity of an astigmatic pencil of geometrical rays if we deduce from the starting equations for y'_{ast} and z'_{ast} a single equation, which gives

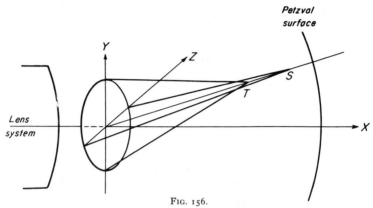

FIG. 156.

the ray distribution in any cross section of the pencil. Transposing those equations so as to have cos O and sin O alone on the right, they become

$$\frac{y'_{ast}}{-(SA/l'_0)(3H'_0{}^2N'_0 \cdot a_3 + \delta l'_0)} = \cos O$$

$$\frac{z'_{ast}}{-(SA/l'_0)(H'_0{}^2N'_0 \cdot a_3 + \delta l'_0)} = \sin O,$$

and if these equations are squared and added together they become

$$\frac{y'_{ast}{}^2}{b^2} + \frac{z'_{ast}{}^2}{c^2} = 1,$$

in which we have put

$$(SA/l'_0)(3H'_0{}^2N'_0 \cdot a_3 + \delta l'_0) = b$$
$$(SA/l'_0)(H'_0{}^2N'_0 \cdot a_3 + \delta l'_0) = c.$$

The equation in $y'_{ast}{}^2$ and $z'_{ast}{}^2$ is that of an ellipse referred to its centre, with semiaxis b in the Y-direction and semiaxis c in the Z-direction. In general, any cross section of an astigmatic pencil from a circular zone of semi-aperture SA is therefore an ellipse whose axes are proportional to SA. The orientation and shape of these ellipses change in a characteristic way at successive sections of the pencil (Fig. 157). In the Petzval surface, $\delta l'_0 = 0$, hence

$$b = 3SA \cdot H'_0{}^2N'_0 \cdot a_3/l'_0$$
$$c = SA \cdot H'_0{}^2N'_0 \cdot a_3/l'_0.$$

The section of the pencil in the Petzval surface is therefore an ellipse with major [111] axis in the Y-direction and minor axis equal to one third of the major.   At point S we know that $\delta l'_0 = -H'_0{}^2N'_0 \cdot a_3$; hence the major axis there is

$$b = 2SA \cdot H'_0{}^2N'_0 \cdot a_3/l'_0$$

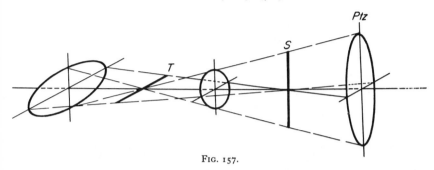

FIG. 157.

and the minor axis is zero.   That is the sagittal focal line, which is thus recognized as an ellipse with an infinitely short minor axis.   Midway between S and T we have $\delta l'_0 = -2H'_0{}^2N'_0 \cdot a_3$; hence

$$b = \quad SA \cdot H'_0{}^2N'_0 \cdot a_3/l'_0$$
$$c = -SA \cdot H'_0{}^2N'_0 \cdot a_3/l'_0.$$

The minus sign of c indicates that the rays have crossed the XY-plane at S but does not affect the interpretation of the curve since only the square of the axes enters into the equation of the curve.   At the midway point the rays are therefore distributed around an ellipse with equal axes, which is only another name for a perfect circle.

At the tangential focus we have $\delta l'_0 = -3H'_0{}^2N'_0 \cdot a_3$ and therefore

$$b = 0 \quad \text{and} \quad c = -2SA \cdot H'_0{}^2N'_0 \cdot a_3/l'_0.$$

This represents the tangential focal line at T, which is therefore of the same length as that at S, but extends in the Z-direction.

If we go another equal step towards the lens system we shall have

$$\delta l'_0 = -4H'_0{}^2N'_0 \cdot a_3$$

and

$$b = -SA \cdot H'_0{}^2N'_0 \cdot a_3/l'_0; \quad c = -3SA \cdot H'_0{}^2N'_0 \cdot a_3/l'_0,$$

which are the parameters of an ellipse identical in size and form with that in the Petzval surface but turned through a right angle.

Observations on lens systems suffering from *heavy* astigmatism closely bear out these geometrical findings.   As in the case of all aberrations affecting definition, this is no longer the case when the amount of astigmatism is so small that it is comparable with the Rayleigh limit.   The focal lines then become ellipses of

[111] moderate eccentricity and of course show ill-defined edges when sufficiently magnified. Geometrically the focal lines might be regarded as the best images because the geometrical concentration of light is there by far the strongest, seeing that all other sections of the pencil have a finite area whilst that of the focal lines is zero.

COMA

For coma, when alone present, TSA(1)* gives

$$y'_{coma} = -SA^2(2 + \cos 2O)H'_0 \cdot a_2$$
$$z'_{coma} = -SA^2 \cdot \sin 2O \cdot H'_0 \cdot a_2$$

or, on multiplying out the right-hand side of the first,

$$y'_{coma} = -2SA^2H'_0 \cdot a_2 - \cos 2O \cdot SA^2H'_0 \cdot a_2$$
$$z'_{coma} = \qquad\qquad -\sin 2O \cdot SA^2H'_0 \cdot a_2.$$

Put into this form, we see at once that for a circular zone of any given value of its semi-aperture SA, and for any one point in the field as defined by H'_0, the displacement y' consists of a constant first term and a second term depending on $\cos 2O$, whilst the displacement in the Z-direction depends purely on $\sin 2O$; moreover, $\cos 2O$ in the y' term and $\sin 2O$ in the z' term have the same coefficient. As both $\sin 2O$ and $\cos 2O$ have the same values as $\sin 2(180° + O)$ and $\cos 2(180° + O)$ respectively, we can draw a first important conclusion, namely, that a ray from any point P of a circular zone of the aperture and the ray from the diametrically opposite point of the aperture (for which O is greater by 180°) have the same displacements y' and z' and therefore meet in the same point of the Petzval surface. That is one geometrical characteristic of pure coma in the absence of other aberrations.

If once more we imagine that we are observing the image on a focusing screen conforming to the Petzval surface, we easily obtain a complete knowledge of the ray distribution. Assuming a_2 and H'_0 to be positive, the constant first term of y'_{coma} is negative; we can therefore draw it downwards from the point where the central ray of the complete pencil pierces the Petzval surface, say to point C. It then becomes evident that a circle drawn around C as centre with $SA^2H'_0a_2$ as radius must contain all the foci of pairs of rays from opposite points of the circular zone of the aperture. For if we consider a radius of this 'coma-circle' at any angle $2O$ from the negative Y-direction, and project this radius upon the Y-axis, the projection will measure $-SA^2H'_0a_2 \cdot \cos 2O$ whilst the perpendicular by which the projection is effected measures $-SA^2H'_0a_2 \cdot \sin 2O$. Every point of the circumference of the coma-circle therefore has the correct coordinates y' and z' of our starting equations. Zones of smaller semi-aperture will of course give similar coma-circles on a scale that is smaller in proportion to SA^2. The totality of these circles will cover the complete figure shown in Fig. 158, which is built up by drawing tangents from the point of penetration of the central ray to the largest coma-circle of the marginal zone. The angle at the apex of the figure is easily shown to be exactly 60°. The total length of the coma figure is $3SA^2H'_0a_2$ and its greatest width is $2SA^2H'_0a_2$. By the physical discussion in section [110] we found that all the light from a narrow zone meets in equal phase at a point distant $SA^2H'_0a_2$ from [111] the central rays.   This displays one of the most remarkable contradictions between physical and geometrical optics yet discovered.   Physically, all the zonal foci are

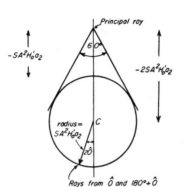

FIG. 158.

concentrated in a line from the apex of the coma-figure to the nearest point of the largest coma-circle; geometrically, the foci are scattered over the entire area of the complete figure!

The right and wrong of this extraordinary contradiction can again be resolved as in the case of astigmatism.   With small and moderate amounts of coma, such as can be tolerated in respectable instruments, the physical measure of coma is always borne out by observation.   A reasonable approach to the geometrical figure is only seen when the coma is of great and utterly intolerable magnitude.†

The fact that the physical interpretation gives a length of the 'coma-patch' which is only one third of that deduced geometrically is of great importance to the lens designer because the sine condition, being based on physical optics, gives the short length whilst the direct trigonometrical tracing of rays a, pr, and b through the centre and upper and lower margins of the aperture gives the geometrical length. The two measures are therefore always found to be approximately in a one-to-three ratio, and as the sine condition represents the more reliable measure, the trigono- metrical method gives an exaggerated indication of the amount of coma present.

When the focusing screen is moved towards or away from the lens system, the two geometrical rays from opposite points of the aperture, which meet on the coma- circle in the Petzval surface, become separated and a double loop takes the place of the simple circle.   These looped curves are of the epicycloidal type and are easily deduced from our equations; but they are not of sufficient interest to be dealt with *in extenso*.   We note them merely as the expression of the fact that in the presence of coma the smallest image of a point is located on the Petzval surface.

The glaring disagreement ˉ between the geometrical coma-patch and the

† More recent integrations have shown that even when the amount of coma is very small, the light distribution closely resembles the geometrical coma pattern, due to the effects of diffraction.

[1111] distribution of the physical foci becomes less pronounced by looking upon the geometrical ray distribution in the presence of coma from a different point of view. The usual coma-patch illustrated in Fig. 158 results when we consider the separate foci of pairs of rays from diametrically opposite points (O and $180° + O$) of the clear aperture.   The rays of a pencil afflicted with coma meet in pairs in yet another way, which in fact appeals more to optical common sense. As the diffused image is necessarily small compared with the aperture of the pencil, it is clear that every ray must *cross the XY-plane* somewhere, at either the left or the right of the usual coma-patch which we have so far considered.   We can locate this intersection-point with the aid of TSA(1) merely by asking for what value of $\delta l'_0$ the displacement z' becomes zero.   Taking the coma term and the $\delta l'_0$ term of the equation for z' together, we obtain

$$z'_{coma} = -2Y_0 Z_0 H'_0 \cdot a_2 - \delta l'_0 Z_0 / l'_0,$$

and this will become zero if

(1) $\delta l'_0 = -2 l'_0 Y_0 H'_0 \cdot a_2.$

The intersection-point with the XY-plane is thus determined in the longitudinal or X-direction.   To locate it completely, we must also determine y'_{coma}, which by the first part of TSA(1) is given as

$$y'_{coma} = -(3Y_0^2 + Z_0^2)H'_0 \cdot a_2 - \delta l'_0 Y_0 / l'_0,$$

and on introducing the value of $\delta l'_0$ already determined by (1),

$$y'_{coma} = -(3Y_0^2 + Z_0^2)H'_0 \cdot a_2 + 2Y_0^2 H'_0 \cdot a_2 = -(Y_0^2 + Z_0^2)H'_0 \cdot a_2$$

or using the semi-aperture $SA^2 = Y_0^2 + Z_0^2$,

(2) $y'_{coma} = -SA^2 H'_0 \cdot a_2.$

By (2) the transverse displacement is identical for all rays from a circular zone of given semi-aperture and is identical with that of the physical coma-focus determined in the preceding section.   It is also identical with that found in the present section for the combined focus of the two rays from the rearmost and foremost points of the aperture zone, which have $Y_0 = 0$, $Z_0 = \pm SA$ and $O = \pm 90°$.

By (1) the longitudinal displacement depends only on Y_0 as far as the aperture is concerned, and it is therefore identical for the two points of a circular zone which lie at the same height above or below the XZ-plane, or at position-angles $+O$ and $-O$ respectively.   The new pairs of rays with a common focus in the XY-plane are therefore of this type, and are placed symmetrically with reference to the XY-plane. A diagram, Fig. 159 (a), renders the state of affairs perfectly clear.   The rays from the front and the rear of the zone of semi-aperture SA, having $Y_0 = 0$, by (1) give $\delta l'_0 = 0$ and therefore meet in the ideal focal plane at the distance below the ideal image-point B'_h determined by (2).   The rays from the top and the bottom of the aperture, having $Y_0 = \pm SA$ (its maximum value), meet at a maximum longitudinal displacement; those from halfway up and down the aperture ($Y_0 = \pm \frac{1}{2}SA$) meet at half the previous distance; and rays with other values of Y_0 fill continuously a longitudinal focal line at the distance y' defined by (2) below the central ray aiming at B'_h.   Aperture zones of other values of SA give corresponding focal lines whose length is proportional to SA by (1) and whose distance below B'_h is pro-

portional to SA^2 by (2).   The whole of these focal lines formed by the rays from   [111]
the complete exit pupil fill a segment of a parabola lying in the XY-plane as shown
in Fig. 159 (b).   These focal lines are special cases of the 'Characteristic Focal
Line' described on page 277 of part I.

It will be noticed that these new coma-foci cover exactly the same range in the
important transverse direction with reference to the central ray of the oblique
pencil which we found in the previous discussion by the OPD'-method.

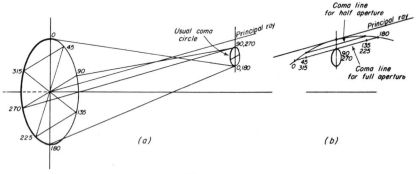

FIG. 159.

The usual coma-circle is shown dotted in Fig. 159 (a) and (b) in order to show its
relation to the new focal lines.

Apart from the close agreement of the new interpretation of geometrical coma
with the more fundamental physical theory, it should be noted that, on account of
the symmetry of oblique pencils with reference to the XY-plane, the pair-foci
used in the present discussion must always be really formed even in the presence
of any other aberrations.   Moreover, it must carry light, for these pairs of rays
meet with absolutely equal optical paths, again on account of the symmetry with
reference to the XY-plane.   The foci which define the time-honoured coma-
circle, on the other hand, hardly ever yield equal optical paths and the two rays
meet even geometrically only in the total absence of any other aberration affecting
the definition of the image.   The very existence of these foci on the coma circle
is thus of a highly problematical nature, with the sole exception of the one which
is formed by the rays from the foremost and rearmost point of the aperture and
which is common to the two different methods of treating the coma pencil.

SPHERICAL ABERRATION

With regard to spherical aberration, we shall take only the first term of either y'
or z' together with the $\delta l'_0$ term to find the location of the point where the marginal
rays of the zone of radius SA cut the central ray.   Either y' or z' gives

$$\delta l'_0 = -SA^2 \cdot \frac{l'_0{}^2}{N'_0} \cdot a_1$$

[111] or exactly twice the amount which we found on page 730 for the point at which the marginal light meets the central light in the same phase, thus confirming the location of the best physical focus midway between the marginal and paraxial geometrical foci.   The $\delta l'_0$ term found from the transverse aberration represents the longitudinal aberration of the marginal rays, but with reversed sign because $\delta l'_0$ is counted from the paraxial towards the marginal focus.   The mathematical expression for the longitudinal aberration is therefore

$$LA' = SA^2 \cdot \frac{l'_0{}^2}{N'_0} \cdot a_1.$$

CHROMATIC VARIATION OF THE ABERRATIONS

[112] THE fundamental equation OP(2) deduced in Chapter XIII and modified in the present chapter is not subject to any restrictive assumptions as to the relations between N, N', l, l', H, and H' and is therefore valid for any values of these data, provided only that the $\sqrt{(1+a)}$ series, by which the equation is obtained, is convergent.   Consequently it is perfectly legitimate to work out the equation for the refractive indices of any colour in the spectrum.   If we now take N and N' as the indices for some colour treated as the most important one, say for what we have usually called 'brightest light', we shall have for some different colour the indices $N+\delta N$ and $N'+\delta N'$, and on introducing these into the equation, we shall obtain a correct statement of the OPD' value of any ray in this new colour.   Moreover, as OP(2) contains either N or N' as a factor in every one of its constituent items, but no powers or reciprocals of N or N', the introduction of $N+\delta N$ and of $N'+\delta N'$ will, on multiplying out, reproduce the original OP(2) in N and N' as a common OPD' for light of any colour and a new series in addition.   In this new series, N is replaced by δN and N' by $\delta N'$, which obviously represents the additional optical path differences which affect the image in any colour other than the 'preferred' or 'brightest' colour, for which N and N' are the true indices.   This supplementary series therefore represents the chromatic variation of all the aberrations contained in the original equation; by simply putting δN in place of N and $\delta N'$ in place of N', we obtain it in the form:

OP(2)Chr. $\delta OPD'_{chr} = \frac{1}{2}PA^2\left[\delta N'\left(\frac{1}{r}-\frac{1}{l'}\right)-\delta N\left(\frac{1}{r}-\frac{1}{l}\right)\right]$

$\qquad + Y[H'\cdot\delta N'/l' - H\cdot\delta N/l]$

$\qquad + \frac{1}{8}PA^4\left[8k_4(\delta N'-\delta N)+\frac{\delta N'}{l'}\left(\frac{1}{r}-\frac{1}{l'}\right)^2-\frac{\delta N}{l}\left(\frac{1}{r}-\frac{1}{l}\right)^2\right]$

$\qquad + \frac{1}{2}PA^2Y\left[\frac{H'\delta N'}{l'^2}\left(\frac{1}{r}-\frac{1}{l'}\right)-\frac{H\delta N}{l^2}\left(\frac{1}{r}-\frac{1}{l}\right)\right]$

$\qquad + \frac{1}{4}PA^2\left[\frac{H^2\cdot\delta N}{l^2}\left(\frac{1}{r}-\frac{1}{l}\right)-\frac{H'^2\delta N'}{l'^2}\left(\frac{1}{r}-\frac{1}{l'}\right)\right]$

$\qquad + \frac{1}{2}Y^2\left[\frac{H'^2\delta N'}{l'^3}-\frac{H^2\delta N}{l^3}\right]+\frac{1}{2}Y\left[\frac{H^3\delta N}{l^3}-\frac{H'^3\delta N'}{l'^3}\right].$

The two first terms are small of the second order as regards aperture and field, [112] while the remaining five terms are, like the Seidel aberrations, small of the fourth order in the same respect.   But inasmuch as these chromatic variations everywhere contain δN instead of N, and as with the usual C-to-F range of colour to be corrected the ratio of δN to N is about as 1 to 180 in crown glass and not less than 1 to 70 even for extremely dense flint glass, the coefficients are very much smaller than those in the original OP(2).   It has therefore been the custom from the earliest days of analytical optics, and has always agreed well with practical experience, to treat the two second-order terms as of the same order of importance as the ordinary Seidel aberrations and to call them the primary chromatic aberrations of lens systems.   We then treat the five fourth-order terms as corresponding in magnitude (or rather in smallness) to the secondary spherical aberrations and therefore as negligible in all investigations in which the Seidel aberrations are deemed to give a sufficient first approximation to the truth for 'brightest light'.   We shall follow this practice and nomenclature and confine our present discussion to the two primary chromatic aberrations after carrying the complete development a little farther.

We must first replace PA^2 by the rectangular coordinates of the point of incidence. As was shown in the transformation of the Seidel aberrations, this implies merely the replacement of PA^2 by $Y^2 + Z^2$ in the fourth-order terms.   But PA^2 also occurs in the first of the second-order terms, and as X is itself small of the second order, the X^2 in $PA^2 = X^2 + Y^2 + Z^2$ will produce a fourth-order term which must be added to the secondary chromatic aberrations.   To a first and sufficient approximation, we have $X = (Y^2 + Z^2)/2r$, and the fourth-order term therefore combines with the third term of OP(2)Chr., which becomes

$$\tfrac{1}{8}(Y^2 + Z^2)^2 \left[8k_4(\delta N' - \delta N) + \frac{\delta N'}{l'}\left(\frac{1}{r} - \frac{1}{l'}\right)^2 - \frac{\delta N}{l}\left(\frac{1}{r} - \frac{1}{l}\right)^2 \right.$$
$$\left. + \frac{\delta N'}{r^2}\left(\frac{1}{r} - \frac{1}{l'}\right) - \frac{\delta N}{r^2}\left(\frac{1}{r} - \frac{1}{l}\right) \right].$$

Next we apply the same rearrangement of the fifth and sixth terms as in the case of the corresponding Seidel aberrations (i.e., astigmatism and Petzval term), and these terms become

$$\tfrac{1}{4}(3Y^2 + Z^2)\left[\frac{H'^2\delta N'}{l'^3} - \frac{H^2\delta N}{l^3}\right] + \tfrac{1}{4}(Y^2 + Z^2)\left[\frac{H^2\delta N}{l^2 r} - \frac{H'^2\delta N'}{l'^2 r}\right].$$

Finally we adopt the paraxial focal plane for light of the preferred or brightest colour as the image surface to which all the optical path differences shall be referred. That reduces the bulk of the aberrations to the Seidel ones dealt with in the preceding sections and subjects l, l', H, and H' to the conditions (a) and (b) of page 707:

$$\text{(a) } N'\left(\frac{1}{r} - \frac{1}{l'}\right) = N\left(\frac{1}{r} - \frac{1}{l}\right) \quad \text{(b) } \frac{H'N'}{l'} = \frac{HN}{l}.$$

We therefore introduce these two equations of condition into our new equation

[112] for the chromatic variation by taking out $N(\mathrm{1}/r - \mathrm{1}/l)$ or HN/l as a common factor, which merely calls for extending all items by N/N, N'/N', or suitable powers of these unit-ratios. We thus obtain the chromatic equation corresponding to, and being additional to, our previous OP(2)* on page 709, in the form

OP(2)*Chr.:

$$\delta OPD'_{chr} = \tfrac{1}{2}(Y^2 + Z^2)\cdot N\left(\frac{\mathrm{1}}{r} - \frac{\mathrm{1}}{l}\right)\left[\frac{\delta N'}{N'} - \frac{\delta N}{N}\right] + \frac{YHN}{l}\left[\frac{\delta N'}{N'} - \frac{\delta N}{N}\right]$$

$$\text{Secondary chromatic aberrations}\begin{cases} + \tfrac{1}{4}(Y^2+Z^2)^2\left[4k_4(\delta N' - \delta N) + \tfrac{1}{2}N^2\left(\frac{\mathrm{1}}{r}-\frac{\mathrm{1}}{l}\right)^2\left(\frac{\delta N'}{N'}\cdot\frac{\mathrm{1}}{N'l} - \frac{\delta N}{N}\cdot\frac{\mathrm{1}}{Nl}\right)\right. \\ \qquad\qquad\qquad\left. + \tfrac{1}{2}N\left(\frac{\mathrm{1}}{r}-\frac{\mathrm{1}}{l}\right)\cdot\frac{\mathrm{1}}{r^2}\left(\frac{\delta N'}{N'} - \frac{\delta N}{N}\right)\right] \\[2mm] + Y(Y^2+Z^2)\frac{HN}{l}\cdot\tfrac{1}{2}N\left(\frac{\mathrm{1}}{r}-\frac{\mathrm{1}}{l}\right)\left(\frac{\delta N'}{N'}\cdot\frac{\mathrm{1}}{N'l} - \frac{\delta N}{N}\cdot\frac{\mathrm{1}}{Nl}\right) \\[2mm] + \tfrac{1}{2}(3Y^2+Z^2)\frac{H^2N^2}{l^2}\cdot\tfrac{1}{2}\left(\frac{\delta N'}{N'}\cdot\frac{\mathrm{1}}{N'l} - \frac{\delta N}{N}\cdot\frac{\mathrm{1}}{Nl}\right) \\[2mm] + \tfrac{1}{4}(Y^2+Z^2)\frac{H^2N^2}{l^2}\cdot\frac{\mathrm{1}}{r}\left(\frac{\delta N}{N^2} - \frac{\delta N'}{N'^2}\right) \\[2mm] + \tfrac{1}{2}Y\frac{H^3N^3}{l^3}\left(\frac{\delta N}{N^3} - \frac{\delta N'}{N'^3}\right). \end{cases}$$

We now limit the further development and discussion to the two primary chromatic aberrations, merely noting that of the secondaries only the first, which obviously represents the chromatic variation of the spherical aberration, ever receives any serious attention. Even this is usually considered only as the source of the zonal variation of chromatic aberration at finite aperture, which we have already studied extensively in connection with the design of telescope and microscope objectives.

In order to adapt our equation to the general case of a centred lens system composed of lenses of any thickness and at any separation, we have to apply to the two primary chromatic aberrations the successive generalizations which were used in the first part of this chapter. This procedure will turn OP(2)* into the final perfectly general OP(6).

Introducing temporary simple symbols for the constant coefficients, we may write the primary part of OP(2*)Chr. as:

$$\delta OPD'_{chr} = \tfrac{1}{2}(Y^2 + Z^2)\cdot Chr_I + (YHN/l)\cdot Chr_{II}.$$

The introduction of $Y^* = Y + EH$ instead of the Y of the general equation, which makes the proper allowance for the eccentric passage of the oblique pencils from an entrance pupil at distance l_{pr} from the refracting surface, then gives

$$\delta OPD'_{chr} = \tfrac{1}{2}(Y^2+Z^2)\cdot Chr_I + EYH\cdot Chr_I + \tfrac{1}{2}E^2H^2\cdot Chr_I$$
$$+ (YHN/l)\cdot Chr_{II} + (EH^2N/l)\cdot Chr_{II}.$$

Here the two terms which involve H only (and not either Y or Y^2+Z^2) have a [112] significance exactly corresponding to similar terms at the same stage of the transformation of the Seidel aberrations; that is, they represent the $\delta OPD'$ of the central ray of the eccentric pencil, and their omission refers the $\delta OPD'$ of any ray to this central ray and so effects a desirable simplification.

Hence the really significant terms are

$$\delta OPD'_{chr} = \tfrac{1}{2}(Y^2+Z^2)\cdot Chr_I + YH\left(\frac{N}{l}\cdot Chr_{II}+E\cdot Chr_I\right),$$

or, on re-introducing the full expressions for Chr_I and Chr_{II} and slightly simplifying,

$$\delta OPD'_{chr} = \tfrac{1}{2}(Y^2+Z^2)N\left(\frac{1}{r}-\frac{1}{l}\right)\left(\frac{\delta N'}{N'}-\frac{\delta N}{N}\right)$$
$$+(Y\cdot H\cdot N/l)\left(\frac{\delta N'}{N'}-\frac{\delta N}{N}\right)\left[1+E\cdot l\left(\frac{1}{r}-\frac{1}{l}\right)\right].$$

In order to form the sum for any number of surfaces at any separations, it is then only necessary to introduce into the first term the constant p which expresses the ratio of Y and Z to the Y_0 and Z_0 at the standard reference surface. The second term has the Lagrange invariant as an outside factor and, being invariant, this can be turned into $Y_0H'_0N'_0/l'_0$ without ceremony.

The terms which have to be added to OP(6) to represent the primary aberrations of any centred lens system for any colour are therefore

OP(6)Chr. $\delta OPD'_{chr} = \tfrac{1}{2}(Y_0{}^2+Z_0{}^2)\sum p^2N\left(\frac{1}{r}-\frac{1}{l}\right)\left(\frac{\delta N'}{N'}-\frac{\delta N}{N}\right)$
$$+\frac{Y_0H'_0N'_0}{l'_0}\sum\left(\frac{\delta N'}{N'}-\frac{\delta N}{N}\right)\left[1+E\cdot l\left(\frac{1}{r}-\frac{1}{l}\right)\right].$$

For finite values of δN and $\delta N'$, the p and E in this equation should really be those for coloured light, which would require the tracing of paraxial pencils from the object point and from the centre of the diaphragm in the colour corresponding to $N+\delta N$ and $N'+\delta N'$. The differences in the l and l_{pr} values, however, will usually be small. Moreover, the chromatic aberrations for an indefinitely small δN or for the immediate vicinity of brightest light are usually the more significant ones and these (or rather fixed multiples of them) will be obtained by using the usual finite δN with the p and E calculated for the 'preferred' or 'brightest' light. The calculation of separate chromatic p and E values is therefore usually unnecessary or even undesirable.

As in the case of the Seidel aberrations, it will be convenient to replace the two sums on which the value of the two chromatic aberrations depends by simple symbols for the purpose of discussion, and we shall choose a_6 and a_7 for these. The equation then takes the simple form

OP(6)Chr. $\delta OPD'_{chr} = \tfrac{1}{2}(Y_0{}^2+Z_0{}^2)\cdot a_6 + (Y_0H'_0N'_0/l'_0)\cdot a_7.$

The two terms are very easily interpreted by the method applied in preceding sections of this chapter to the Seidel aberrations.

[112]   The first term agrees in its aperture coefficient with $OP(4)^*$ and therefore admits of complete compensation by a longitudinal shift of the focus.   Taking the first term and $OP(4)^*$ together we obtain

$$\delta OPD' = \tfrac{1}{2}(Y_0{}^2 + Z_0{}^2)[a_6 + N'_0 \cdot \delta l'_0 / l'_0{}^2],$$

and this will be zero and the $\delta l'_0$ will indicate the sharp focus of the coloured rays if

$$\delta l'_0 = -a_6 \cdot l'_0{}^2 / N'_0 = -(l'_r - l'_v).$$

This $\delta l'_0$ represents the longitudinal chromatic aberration of the lens system with reversed sign; it obviously will become zero if $a_6 = 0$, and this is therefore the condition for paraxial achromatism of our lens.   Unlike the TL,Chr. solutions of Part I, that by a_6 is a strictly correct one, no matter how thick or how widely separated the constituent lenses may be.   Moreover, the sum symbolized by a_6 is simple, so its evaluation represents a convenient method of testing any lens system for chromatic aberration, or of establishing exact paraxial achromatism by determining the last radius so as to bring the sum to zero-value.

As our equation OP(6)Chr. was deduced for an oblique pencil, and as nevertheless the value of $\delta l'_0$ to which it leads is independent of H as well as of the aperture, we have also a valid and binding proof that the primary longitudinal chromatic aberration of any centred lens system is constant throughout the entire field. Moreover, the first of the secondary chromatic aberrations (the spherical variation) is also independent of H, and consequently it is also uniform throughout the field. There can therefore be only very small variations of the longitudinal chromatism in different parts of the field of any centred lens-system; and these variations, if sensible, will be entirely due to the third and fourth of the secondaries, namely, to chromatic variation of astigmatism and of the Petzval curvature.

The second term in OP(6)Chr. agrees in its dependence on aperture with $OP(4)^{**}$, hence it represents a *transverse* displacement of the focus of the coloured light with reference to that of the brightest light.   Taking the two together, we obtain

$$\delta OPD' = \frac{Y_0 N'_0}{l'_0} [H'_0 \cdot a_7 + \delta H'_0],$$

which means that the value of $\delta H'_0$ which leads us from the brightest to the coloured focus is

$$\delta H'_0 = -H'_0 \cdot a_7.$$

As the displacement is proportional to H, it indicates that the coloured light gives an image to a smaller or a larger scale than the brightest light, and we recognize the second chromatic term as the expression for the chromatic difference of magnification; we also learn that $a_7 = 0$ indicates achromatism of magnification.

The achromatism of magnification of a given lens-system can only be disturbed by the second and fifth of the secondary chromatic aberrations, for these are the only two which vary with Y and therefore the only ones which cause additional transverse displacements of the coloured images with reference to the images in brightest light.   These two secondary chromatic aberrations are thus convicted of

being jointly responsible for the difference between the paraxial and the marginal   [112]
g_7-values of the eyepieces which were studied in detail in Chapter X.

We can extract some further information from OP(6)Chr. with reference to 'stability of achromatism'. Achromatism of either kind is considered stable if it remains undisturbed by changes in the distance of object and image, or by shifting the diaphragm. It is easily seen by examining the explicit sums which represent our a_6 and a_7 that it must in general be extremely unlikely that these sums should give the same zero-value whatever the values of l, p, and E may be. Strict achromatism of either kind will therefore as a rule exist only for one particular distance of the object and for one particular position of the diaphragm, and any change in either respect will disturb it. Stability of achromatism is thus recognized as a purely theoretical abstraction; we saw however in the section on thin achromatic lenses (Part I, section [40]) that, with systems built up of such components, an approach to stability of achromatism can be secured, the closeness of which depends chiefly upon the smallness of the unavoidable finite thickness which must be given to the supposedly 'thin' lenses.

NUMERICAL CALCULATION OF THE PRIMARY ABERRATIONS

THE equations developed in the two preceding sections supply the means for   [113]
calculating the contribution made by each surface in the lens to each primary aberration. Since aspheric surfaces are rarely used in actually manufactured systems, they will not be considered here because the formulae will then become particularly simple and convenient.

Spherical aberration. To evaluate this aberration contribution, we refer to the expression on page 746, namely,

$$LA' = SA_0{}^2 \frac{l'_0{}^2}{N'_0} \cdot a_1,$$

and on inserting the explicit value of a_1 from OP(6)* and OP(6), with omission of the Σ-sign, we obtain the spherical contribution term as

$$SC' = SA_0{}^2 \cdot \frac{l'_0{}^2}{N'_0} \cdot p^4 \cdot \tfrac{1}{2} N^2 \left(\frac{1}{r} - \frac{1}{l}\right)^2 \left(\frac{1}{N'l'} - \frac{1}{Nl}\right).$$

Since $p = lu/SA_0$ and $1/r - 1/l = (l-r)/lr = i/lu$, this expression becomes

(a) $SC' = \tfrac{1}{2}(l'u'N'i')^2 \left[\dfrac{1}{N'l'} - \dfrac{1}{Nl}\right] / N'_0 u'_0{}^2.$

The term in square brackets may now be transformed as follows, by replacing u' with the identically equal $(u + i - i')$:

$$\left[\frac{1}{N'l'} - \frac{1}{Nl}\right] = \frac{i'u'}{l'u'N'i'} - \frac{iu}{luNi} = \frac{1}{l'u'N'i'}(i'u' - iu) = \frac{(i-i')(i'-u)}{l'u'N'i'}$$

giving $SC' = \tfrac{1}{2}l'u'N'i'(i-i')(i'-u)/N'_0 u'_0{}^2,$

which agrees perfectly with the geometrically determined expression on page 314 of Part I.

[113] *Coma.* By reference to the discussion of coma on page 742, we find that the sagittal coma of a lens is given by

$$Coma'_s = SA_0{}^2H'_0 \cdot a_2,$$

where

$$a_2 = \Sigma[p^2(Cm^* + E^* \cdot Sph)]$$
$$= \Sigma\left[\tfrac{1}{2}p^2N\left(\frac{1}{r}-\frac{1}{l}\right)\left(\frac{1}{N'l'}-\frac{1}{Nl}\right) + \tfrac{1}{2}p^2E^*N^2\left(\frac{1}{r}-\frac{1}{l}\right)^2\left(\frac{1}{N'l'}-\frac{1}{Nl}\right)\right].$$

On omitting the Σ-sign and taking out some factors, we find the coma-contribution of a single surface in the lens to be given by:

$$CC' = SA_0{}^2H'_0 \cdot \tfrac{1}{2}p^2N\left(\frac{1}{r}-\frac{1}{l}\right)\left(\frac{1}{N'l'}-\frac{1}{Nl}\right)\left[1 + E^* \cdot N\left(\frac{1}{r}-\frac{1}{l}\right)\right]$$

$$= \tfrac{1}{2}H'_0(l'u'N'i')\cdot\left(\frac{1}{N'l'}-\frac{1}{Nl}\right)\left[1+\frac{l}{N}\cdot\frac{l_{pr}}{l_{pr}-l}\cdot N\cdot\frac{l-r}{lr}\right]$$

$$= \tfrac{1}{2}H'_0(l'u'N'i')\cdot\left(\frac{1}{N'l'}-\frac{1}{Nl}\right)\left[\frac{qi}{u}\right],$$

where q was defined on page 260 of Part I as $\left(\frac{r}{l-r}+\frac{l_{pr}}{l_{pr}-l}\right)$. We next remove H'_0 by the Lagrange equation, namely, $H'_0 = HNu/N'_0u'_0$, giving

$$CC' = \tfrac{1}{2}\frac{(qH)Ni}{N'_0u'_0}\cdot l'u'N'i'\cdot\left(\frac{1}{N'l'}-\frac{1}{Nl}\right).$$

Now on page 308 of Part I it was shown that $qH = lu(i'_{pr}/i'')$; hence

$$CC' = \frac{1}{N'_0u'_0}\cdot\left(\frac{i'_{pr}}{i'}\right)\cdot\tfrac{1}{2}(l'u'N'i')^2\left(\frac{1}{N'l'}-\frac{1}{Nl}\right).$$

Since it will always be necessary to calculate SC', it will be convenient to express $(1/N'l' - 1/Nl)$ in terms of SC' by equation (a), namely:

$$\tfrac{1}{2}(l'u'N'i')^2\left(\frac{1}{N'l'}-\frac{1}{Nl}\right) = SC'\cdot N'_0u'_0{}^2.$$

Hence
$$CC' = SC'\cdot\left(\frac{i'_{pr}}{i'}\right)u'_0,$$

in perfect agreement with the formula on page 314 of Part I.

Astigmatism. The sagittal astigmatism, which is the longitudinal distance from the Petzval surface to the sagittal focal line, was found in the previous section to be given by

$$Ast'_s = H'_0{}^2N'_0\cdot a_3,$$

where
$$a_3 = \Sigma[Ast^* + 2E^*\cdot Cm^* + E^{*2}\cdot Sph].$$

On inserting the explicit values of the terms and omitting the Σ-sign, we find the [113] contribution of a single surface to the sagittal astigmatism to be

$$AC' = H'_0{}^2 N'_0\left[\tfrac{1}{2}\left(\frac{1}{N'l'}-\frac{1}{Nl}\right)+2E^*\cdot\tfrac{1}{2}N\left(\frac{1}{r}-\frac{1}{l}\right)\left(\frac{1}{N'l'}-\frac{1}{Nl}\right)\right.$$

$$\left.+E^{*2}\cdot\tfrac{1}{2}N^2\left(\frac{1}{r}-\frac{1}{l}\right)^2\left(\frac{1}{N'l'}-\frac{1}{Nl}\right)\right]$$

$$= \tfrac{1}{2}H'_0{}^2 N'_0\left(\frac{1}{N'l'}-\frac{1}{Nl}\right)\left[1+E^*N\left(\frac{1}{r}-\frac{1}{l}\right)\right]^2$$

$$= \tfrac{1}{2}\left(\frac{H^2 N^2 u^2}{N'_0 u'_0{}^2}\right)\left(\frac{1}{N'l'}-\frac{1}{Nl}\right)\left[\frac{qi}{u}\right]^2$$

$$= \tfrac{1}{2}(qH)^2\frac{N^2 i^2}{N'_0 u'_0{}^2}\left(\frac{1}{N'l'}-\frac{1}{Nl}\right).$$

Since $qH = lu(i'_{pr}/i')$, this becomes

$$AC' = \tfrac{1}{2}\frac{(l'u'N'i')^2}{N'_0 u'_0{}^2}\left(\frac{i'_{pr}}{i'}\right)^2\left(\frac{1}{N'l'}-\frac{1}{Nl}\right) = SC'\cdot\left(\frac{i'_{pr}}{i'}\right)^2,$$

again in perfect agreement with the formula on page 314 of Part I.

Petzval curvature. The sagitta of the Petzval surface was found to be given by

$$X'_{Ptz} = \tfrac{1}{2}H'_0{}^2 N'_0\cdot a_4.$$

$$\therefore\ PC' = \tfrac{1}{2}H'_0{}^2 N'_0\cdot(Ptz^*) = \tfrac{1}{2}H'_0 N'_0\left(\frac{N'-N}{NN'r}\right);$$

which is in agreement with the previous derivation on page 314.

Distortion. The distortion of a lens is given by

$$Dist' = H'_0{}^3\left(\frac{N'_0}{l'_0}\right)^2\cdot a_5$$

and hence the distortion contribution of a surface becomes

$$DC' = H'_0{}^3\left(\frac{N'_0}{l'_0}\right)^2\frac{1}{p^2}[Dist^*+\tfrac{1}{2}E^*Ptz^*+3E^*Ast^*+3E^{*2}Cm^*+E^{*3}Sph^*].$$

The last three terms in this bracket have many common factors, and by inserting the explicit values of Ast^*, Cm^*, and Sph it is easily verified that they add up to

$$\tfrac{1}{2}\frac{q^3 i^2 l}{N u^2}\left(\frac{1}{N'l'}-\frac{1}{Nl}\right)-\tfrac{1}{2}\frac{lu}{Ni}\left(\frac{1}{N'l'}-\frac{1}{Nl}\right).$$

By including the outside factor, which may be reduced to $H^3 N^3 u/N'_0 u'_0 l^2$, the first of these two terms can be completely disposed of by writing it as follows:

$$DCI' = \frac{(qH)^3 N^2 i^2}{N'_0 u'_0 lu}\cdot\tfrac{1}{2}\left(\frac{1}{N'l'}-\frac{1}{Nl}\right) = CC'\left(\frac{i'_{pr}}{i'}\right)^2,$$

[113]   as was shown on page 308. The second of these two terms can be transformed in the following way:

$$-\frac{lu}{2Ni}\left(\frac{1}{N'l'}-\frac{1}{Nl}\right) = \frac{lu}{2Ni}\left[\frac{u}{Nlu}-\frac{u'}{N'l'u'}\right] = \frac{1}{2Ni}\left[\frac{u}{N}-\frac{u'}{N'}\right]$$

$$= \frac{1}{2Ni}\left[\frac{u}{N}-\frac{u+i-i'}{N'}+\frac{i}{N}-\frac{i}{N}\right]$$

$$= \frac{1}{2Ni}\left(\frac{1}{N}-\frac{1}{N'}\right)(u+i)-\frac{1}{2Ni}\left(\frac{iN}{N^2}-\frac{i'N'}{N'^2}\right)$$

$$= \frac{r(u+i)}{2Ni}\left(\frac{N'-N}{NN'r}\right)-\tfrac{1}{2}\left(\frac{1}{N^2}-\frac{1}{N'^2}\right)$$

$$= \frac{lu}{2Ni}\cdot Ptz^* - Dist^*.$$

Hence the remaining terms in DC' can be collected to give:

$$DC\,\mathrm{II}' = \frac{H^3N^3u}{N'_0u'_0l^2}\cdot\frac{Ptz^*}{2}\left[E^* + \frac{lu}{Ni}\right]$$

$$= \frac{H^3N^2u^2}{N'_0u'_0li}\cdot\frac{Ptz^*}{2}\cdot\left(\frac{qi}{u}\right).$$

But $\tfrac{1}{2}Ptz^* = PC'\cdot\dfrac{N'_0u'_0{}^2}{H^2N^2u^2}$ and $qH = lu\left(\dfrac{i'_{pr}}{i'}\right).$

Hence $DC\,\mathrm{II}' = PC'\cdot\left(\dfrac{i'_{pr}}{i'}\right)u'_0,$

as was shown previously on page 308.

Longitudinal chromatic aberration.   In the last section this aberration was shown to be given by

$$Lch' = \frac{l'_0{}^2}{N'_0}\cdot a_6,$$

and omitting the Σ in the expression for a_6 we have

$$LchC' = \frac{l'_0{}^2}{N'_0}p^2N\left(\frac{1}{r}-\frac{1}{l}\right)\left(\frac{\delta N'}{N'}-\frac{\delta N}{N}\right).$$

By repeating the transformations used for the other aberrations, this at once reduces to

$$LchC' = l'u'N'i'\left(\frac{\delta N'}{N'}-\frac{\delta N}{N}\right)\Big/N'_0u'_0{}^2,$$

thus agreeing with the formula derived on page 312 of Part I.

Transverse chromatic aberration.   For this aberration we found that

$$Tch' = H'_0 \cdot a_7.$$

Hence

$$TchC' = H'_0\left(\frac{\delta N'}{N'} - \frac{\delta N}{N}\right)\left[1 + El\left(\frac{1}{r} - \frac{1}{l}\right)\right].$$

The quantity in the square bracket is $\left[1 + E^* \cdot N\left(\frac{1}{r} - \frac{1}{l}\right)\right] = \frac{qi}{u}$ and, as before, we replace H'_0 by $HNu/N'_0 u'_0$.   Then since $qH = lu(i'_{pr}/i')$ we find

$$TchC' = \frac{Nu}{N'_0 u'_0}\left(\frac{\delta N'}{N'} - \frac{\delta N}{N}\right) \cdot \frac{i}{u} \cdot lu\left(\frac{i'_{pr}}{i'}\right)$$

$$= \frac{l'u'N'i'}{N'_0 u'_0}\left(\frac{\delta N'}{N'} - \frac{\delta N}{N}\right) \cdot \frac{i'_{pr}}{i'}.$$

$$\therefore TchC' = LchC' \cdot \left(\frac{i'_{pr}}{i'}\right)u'_0$$

as was proved previously on page 313.

CONVERSION FACTORS FOR THE g AND a COEFFICIENTS

The series of aberration coefficients g_1 to g_7 given on pages 314–315, and the present series a_1 to a_7, are evidently closely related, the actual connection between them being as follows:

$$LA' = SA^2 \cdot g_1, \quad \text{where } g_1 = \left(\frac{l'^2_0}{N'_0}\right) \cdot a_1$$

$$Coma'_s = SA^2 H'_0 \cdot g_2, \qquad g_2 = a_2$$

$$Ast'_s = H'^2_0 \cdot g_3, \qquad g_3 = N'_0 \cdot a_3$$

$$X'_{Ptz} = \tfrac{1}{2}H'^2_0 \cdot g_4, \qquad g_4 = N'_0 \cdot a_4$$

$$Dist' = H'^3_0 \cdot g_5, \qquad g_5 = \left(\frac{N'_0}{l'_0}\right)^2 \cdot a_5$$

$$L'ch = g_6, \qquad g_6 = \left(\frac{l'^2_0}{N'_0}\right) \cdot a_6$$

$$T'ch = H'_0 \cdot g_7, \qquad g_7 = a_7$$

EFFECT OF A LONGITUDINAL SHIFT OF THE EXIT PUPIL
(THE 'SECOND FUNDAMENTAL LAW')

THIS important law was derived on the basis of purely geometrical optics in   [114] Chapter VI of Part I, but the alternative derivation by physical methods to be given in this section has much to recommend it and it will be included to complete the *OPD* study.

[114]   If we combine the general equation OP(6)* for the differences of optical path in an oblique pencil with the supplementary equation OP(6)Chr. for the two chromatic aberrations, we have

$$OPD' = \tfrac{1}{4}(Y_0^2+Z_0^2)^2 a_1 + Y_0(Y_0^2+Z_0^2)H'_0(N'_0/l'_0)a_2$$
$$+ \tfrac{1}{2}(3Y_0^2+Z_0^2)H'^2_0(N'_0/l'_0)^2 a_3 + \tfrac{1}{4}(Y_0^2+Z_0^2)H'^2_0(N'_0/l'_0)^2 a_4$$
$$+ Y_0 H'^3_0(N'_0/l'_0)^3 a_5 + \tfrac{1}{2}(Y_0^2+Z_0^2)a_6 + Y_0 H'_0(N'_0/l'_0)a_7.$$

We have shown that this equation is valid not only if the quantities with suffix (o) refer to the data of the ray at any refracting surface, but even when these quantities are measured in and from *any* given plane at right angles to the optical axis. We shall now adopt the exit pupil for which the coefficients a_1 to a_7 are known as the plane in which Y_0 and Z_0 are to be measured, and we shall take H'_0 to be the height of the final image and l'_0 to be the corresponding distance from the exit pupil to the final paraxial image-point, whereupon N'_0 will be the index of the medium in which the exit pupil and the final image lie. The index N'_0 will therefore practically always be unity, for air, but we shall retain it in our equations for the sake of maintaining absolute generality of the solution.

We now ask how the coefficients a_1 to a_7 are changed when the aperture-limiting diaphragm of the lens-system is shifted along the axis so as to produce a new exit pupil at axial distance $l'pr_0$ from the original exit pupil, it being understood that the aperture of the shifted diaphragm is adjusted so as to admit an axial pencil of the same angular extent or marginal U'-value as before.

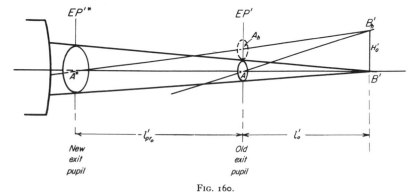

FIG. 160.

If we now refer to Fig. 160, in which EP' marks the given exit pupil for which the coefficients a_1 to a_7 are known, and $EP'*$ the shifted position for which these coefficients are to be determined, and compare it with Fig. 146 on page 711, we see a complete analogy. An oblique pencil, which for the original exit pupil had AB'_h as its central ray, will on shifting the pupil to $A*$ have $A*B'_h$ as its central ray. This ray will pass eccentrically through the plane of the original pupil, say at A_h. As before, we see that AA_h is in a fixed proportion to $B'B'_h = H'_0$ and that we may put $AA_h = W'_0 \cdot H'_0$, in which W'_0 is an easily determined numerical [114] constant, as before. Then, by similar triangles, on allowing for $l'pr_0$ being negative in Fig. 160, the value of W'_0 is found to be

$$W'_0 = -l'pr_0/(-l'pr_0 + l'_0) = l'pr_0/(l'pr_0 - l'_0).$$

(This is the value of the quantity W' on page 344 of Part I, when the original principal ray crosses the axis at the 'last' surface making l'_{pr} equal to zero in the expression for W'.)

To a marginal ray travelling towards B'_h which had the coordinates Y_0 and Z_0 in the original exit pupil, there will correspond a ray from the shifted exit pupil which cuts the original one at $Y^* = Y_0 + W'_0 \cdot H'_0$ and at the same Z_0. All these quantities correspond absolutely to those used in Fig. 146, and if we temporarily introduce the abbreviations

$$Sph = a_1; \quad Cm = (N'_0/l'_0)a_2; \quad Ast = (N'_0/l'_0)^2 a_3; \quad Ptz = (N'_0/l'_0)^2 a_4;$$
$$Dist = (N'_0/l'_0)^3 a_5; \quad Lch = a_6; \quad \text{and} \quad Tch = (N'_0/l'_0)a_7$$

into our equation, written for the coordinates Y_0^* and Z_0 of a marginal ray from the shifted exit pupil, the equation becomes

$$OPD' = \tfrac{1}{4}(Y_0^{*2} + Z_0^2)^2 \cdot Sph + Y_0^*(Y_0^{*2} + Z_0^2)H'_0 \cdot Cm + \tfrac{1}{2}(3Y_0^{*2} + Z_0^2)H'_0{}^2 \cdot Ast$$
$$+ \tfrac{1}{4}(Y_0^{*2} + Z_0^2)H'_0{}^2 \cdot Ptz + Y_0^* H'_0{}^3 \cdot Dist + \tfrac{1}{2}(Y_0^{*2} + Z_0^2) \cdot Lch + Y_0^* H'_0 \cdot Tch$$

and corresponds exactly to equation OP(2)* on page 713. Consequently the multiplying-out of the factors in $Y_0^* = Y_0 + W'_0 \cdot H'_0$ and in Z_0 and the re-arrangements of the terms would necessarily also correspond to the previous work and must give a result analogous to equation (f) on page 714, namely,

$$OPD' = \tfrac{1}{4}(Y_0^2 + Z_0^2)^2 \cdot Sph + Y_0(Y_0^2 + Z_0^2)H'_0[Cm + W'_0 \cdot Sph]$$
$$+ \tfrac{1}{2}(3Y_0^2 + Z_0^2)H'_0{}^2[Ast + 2W'_0 \cdot Cm + W'_0{}^2 \cdot Sph] + \tfrac{1}{4}(Y_0^2 + Z_0^2)H'_0{}^2 \cdot Ptz$$
$$+ Y_0 H'_0{}^3[Dist + \tfrac{1}{2}W'_0 \cdot Ptz + 3W'_0 \cdot Ast + 3W'_0{}^2 \cdot Cm + W'_0{}^3 \cdot Sph]$$
$$+ \tfrac{1}{2}(Y_0^2 + Z_0^2) \cdot Lch + Y_0 H'_0[Tch + W'_0 \cdot Lch]$$

plus a long term in $H'_0{}^4$. This term is easily seen to be the OPD' of the new central ray $A^*B'_h$ and therefore should be omitted for the reasons stated on page 714, namely, so that the OPD' calculated from the new equation may be the true difference of optical path referred to the new central ray.

If we now reintroduce the values of the temporary coefficients Sph, Cm, &c., we have of course $Sph = a_1$ simply. For the coma factor we find

$$Cm + W'_0 \cdot Sph = (N'_0/l'_0)a_2 + W'_0 \cdot a_1 = (N'_0/l'_0)[a_2 + W'_0(l'_0/N'_0)a_1]$$

or on introducing, again in full analogy with the previous work, a new factor corresponding to the former E^*

$$K' = W'_0(l'_0/N'_0) = 1\Big/\Big(\frac{N'_0}{l'_0} - \frac{N'_0}{l'pr_0}\Big),$$

we have $Cm + W'_0 \cdot Sph = (a_2 + K' \cdot a_1)(N'_0/l'_0)$.

[114]   Working out the other coefficients in the same way, and introducing $K' = W'_0(l'_0/N'_0)$ as in the coma factor, we easily obtain

$$OP(6)** \quad OPD' = \tfrac{1}{4}(Y_0{}^2 + Z_0{}^2)^2 a_1 + Y_0(Y_0{}^2 + Z_0{}^2)H'_0(N'_0/l'_0)[a_2 + K'a_1]$$
$$+ \tfrac{1}{2}(3Y_0{}^2 + Z_0{}^2)H'_0{}^2(N'_0/l'_0)^2[a_3 + 2K'a_2 + K'^2 a_1]$$
$$+ \tfrac{1}{4}(Y_0{}^2 + Z_0{}^2)H'_0{}^2(N'_0/l'_0)^2 a_4$$
$$+ Y_0 H'_0{}^3(N'_0/l'_0)^3[a_5 + \tfrac{1}{2}K'a_4 + 3K'a_3 + 3K'^2 a_2 + K'^3 a_1]$$
$$+ \tfrac{1}{2}(Y_0{}^2 + Z_0{}^2)a_6 + Y_0 H'_0[a_7 + K'a_6].$$

This equation is again in the standard form of OP(6)* and the coefficients of the spherical and chromatic terms and the Petzval term are absolutely unchanged. Shifts of the diaphragm in any lens system consequently have no effect of any kind on spherical aberration, chromatic aberration, or Petzval curvature. The coefficients of the other four aberrations are in each case compounded of the original coefficient plus terms involving K' and the original coefficients of all the lower aberrations; they never involve a coefficient of a higher aberration, the terms 'higher' and 'lower' being used in the sense of the numbering of the 'a' coefficients. This highly characteristic and important compound form of the Seidel coefficients which is brought about by a shifting of the exit pupil represents the substance of our 'second fundamental law'. We shall state it thus:

SECOND FUNDAMENTAL LAW

If the coefficients a_1 to a_7 have been determined for an optical system with a given position of the exit pupil, and a longitudinal shift of the diaphragm causes the new exit pupil to lie at a distance $l'pr_0$ from the original one, we calculate

$$K' = 1 \Big/ \left(\frac{N'_0}{l'_0} - \frac{N'_0}{l'pr_0}\right) = W'_0\left(\frac{l'_0}{N'_0}\right).$$

The new coefficients will then be:

a_1, a_4, and a_6 unchanged.
$a_2{}^* = a_2 + K' \cdot a_1$
$a_3{}^* = a_3 + 2K' \cdot a_2 + K'^2 \cdot a_1$
$a_5{}^* = a_5 + \tfrac{1}{2}K' \cdot a_4 + 3K' \cdot a_3 + 3K'^2 \cdot a_2 + K'^3 \cdot a_1$
$a_7{}^* = a_7 + K' \cdot a_6.$

Inversely, we can calculate the value of K' which will bring any one of the starred coefficients to a desired value.

These laws were given in Part I, page 344, in terms of the g-coefficients, but for numerical calculation it will be found simpler to use them in their new form.

THE SEIDEL ABERRATIONS OF A THIN LENS WITH STOP IN CONTACT

The Seidel aberration contributions of a thin combination of thin lenses, forming part of a centred system of k surfaces, and placed in contact with the

diaphragm of that system, were stated explicitly under 'Seidel Aberrations III' [114]
on page 329 of Part I; the aberrations were further discussed in section [60]
under 'The Third Fundamental Law'.

In the following chapters, much use will be made of the aberrations of a thin
lens system with stop in contact, expressed in terms of the seven aberration
coefficients $a_1 \ldots a_7$.   It is therefore of interest to adapt the contribution formulae
given in Part I to this special case.   By comparing the formulae of page 329 of
Part I with those given in section [113] of Part II, we find:

$$a_1 = \sum \text{(spherical } G\text{-sums of the individual lenses)}$$
$$a_2 = \sum \text{(coma } G\text{-sums of the individual lenses)}$$
$$a_3 = 1/2f'$$
$$a_4 = \sum (1/Nf' \text{ of the individual lenses)}$$
$$a_5 = 0$$
$$a_6 = \sum (c.\delta N \text{ of the individual lenses)}$$
$$a_7 = 0$$

CHAPTER XVIII

ANALYTICAL SOLUTIONS FOR SIMPLE SYSTEMS WITH REMOTE STOP

[115] THE utility of the TL solutions for the oblique aberrations becomes greatly increased if we apply the fundamental laws of oblique pencils, more especially the second law which enables us to find the aberrations for a shifted position of the diaphragm. For with the diaphragm in the original position, in contact with the thin system, the fixed astigmatism $a_3 = 1/2f'$ is so enormous that it can only be tolerated in systems covering very small fields of view, like telescope and simple microscope objectives.

The second fundamental law has been proved for the *general case* of any centred lens system; it is therefore directly available for the simple case of a thin system with the original diaphragm in contact. The l'_0 of the general formula then becomes our l'_k, for it represents the distance from the final exit pupil to the paraxial focal plane. To find the physical constants $a_1{}^*$ to $a_5{}^*$ for a shift of the diaphragm by $l'pr_k$, which in the TL case will be the distance of the new diaphragm from the thin system itself, we have to calculate (with $N'_0 = 1$)

$$K' = \cfrac{1}{\cfrac{1}{l'_k} - \cfrac{1}{l'pr_k}}$$

and we then have

a_1 and a_4 unchanged

$a_2{}^* = a_2 + K'a_1$

$a_3{}^* = a_3 + 2K'a_2 + K'^2a_1$

$a_5{}^* = a_5 + \tfrac{1}{2}K'a_4 + 3K'a_3 + 3K'^2a_2 + K'^3a_1.$

In our special TL case we can calculate a_1 and a_2 by the respective G-sums, a_4 as the Petzval sum, $a_3 = 1/2f'$, and $a_5 = 0$ as shown on page 759.

We shall nearly always be aiming at a flat field for the tangential rays, and in accordance with $\dfrac{1}{R'_T} = -N'_0(a_4 + 6a_3)$ in which, for the *shifted* diaphragm, a_3 must be replaced by $a_3{}^*$, we must then aim at making $a_3{}^* = -a_4/6$.

All the solutions hitherto given have aimed at freedom from spherical aberration and coma, which are the worst aberrations in systems with small fields of view. For eyepieces and photographic objectives, the relative importance of the aberrations changes, and when we can correct only two systematically, astigmatism and coma must take first place because they can attain such a magnitude in the outer part of the field as to render any probable remnant of spherical aberration quite insignificant. Moreover, it was shown in the discussion of the fundamental laws that good correction of astigmatism and coma is possible, *in thin systems*, only if under-corrected spherical aberration is admitted.

As a first example of solutions which aim in the first instance at correction of [115] astigmatism and coma, we shall take eyepieces consisting of a reasonably thin achromatic combination, either cemented or with one 'broken contact'.

SIMPLE ACHROMATIC EYEPIECES

The Thin Achromatic Cemented Lens as an Eyepiece

WE must notice first that such a lens leaves us only one liberty when the strength [116] of the two components has been determined so as to give achromatism with the prescribed focal length; we can only bend such a lens as a whole, and therefore can satisfy only *one* condition systematically. To satisfy *two* conditions, we must try changes of glass, or in some cases we may be able to make a sufficient compromise. The analytical method comes in only at the beginning by indicating the approximate shape of the lens and the direction in which the glass must be changed.

As eyepieces of the present type are chiefly useful for telescopes and at high magnifications, we consider this case only. The distant object-glass will supply

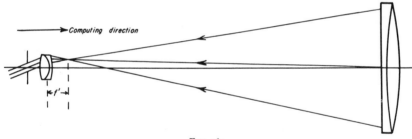

Computing direction

$\kappa f' \rightarrow$

FIG. 161.

cones of rays with nearly parallel principal rays, as shown in Fig. 161, and the eye-piece is expected to turn these into inclined parallel bundles to be focused on the retina of the observer's eye. In the direction of the actual light, the object-glass will be the entrance pupil for the eyepiece; but we will calculate the eyepiece in the reverse direction of the light and then the object-glass will mark the exit pupil and its distance from the eyepiece will be the $l'pr_k$ required for the determination of the constant K' of the second fundamental law. As these thin achromatic eye-pieces are highly insensitive to changes in $l'pr_k$—so long as $l'pr_k$ is at least 20 times the focal length of the eyepiece—we will go to the limit and assume $l'pr_k = \infty$.

Then for our present case $K' = 1 / \left(\dfrac{1}{l'_k} - \dfrac{1}{l'pr_k} \right) = l'_k = f'_{EP}$; and assuming a value of $f' = 1$, we have $K' = 1$.

In the computing direction, the eyepiece receives parallel rays, hence $v_1 = 0$, and as it is crown-in-front, we can take over the formulae in Chapter XI, page 523, for

[116] central passage of the oblique pencil and for unit focal length. These formulae are:

$$a_1 = SGS = 1.198c_2{}^2 + 7.616c_2 + 11.48$$
$$a_2 = CGS = \qquad\quad 0.850c_2 + \ 2.67$$

to which we add $a_3 = 0.500$ by the third fundamental law.

We now apply the second fundamental law:

$$a_2{}^* = a_2 + K' \cdot a_1 \qquad a_3{}^* = a_3 + 2K' \cdot a_2 + K'^2 \cdot a_1$$

and note that it is desirable to have $a_2{}^* = 0$ or nearly so, whilst $a_3{}^*$ should, for a flat tangential field, have the value $-a_4/6$, where a_4 is the Petzval sum. This sum was found by the formula $\Sigma(N-1)c/N$ to be 0.699. Therefore with $K' = 1$, we have to try to satisfy the equations

$$a_2{}^* = a_2 + a_1 = 0 \quad \text{and} \quad a_3{}^* = -0.117 = a_3 + 2a_2 + a_1.$$

As the a-values without star are those collected above, the condition for freedom from coma, $a_2{}^* = 0$, gives by simple addition

$$1.198c_2{}^2 + 8.466c_2 + 14.15 = 0$$

whence $c_2 = -4.354$ or -2.713. The condition for a flat tangential field, namely $a_3{}^* = 0$, gives

$$
\begin{aligned}
a_1 &= 1.198c_2{}^2 + 7.616c_2 + 11.48 \\
2a_2 &= \qquad\qquad\ 1.700c_2 + \ 5.34 \\
a_3 - (-0.117) &= \qquad\qquad\qquad\qquad\quad 0.62
\end{aligned}
$$
$$\rule{6cm}{0.4pt}$$
$$\text{sum} = 1.198c_2{}^2 + 9.316c_2 + 17.44 = 0$$

whence $c_2 = -4.636$ or -3.140. We see that the astigmatic condition gives solutions which do not agree with those given by the sine condition, hence the two conditions cannot be satisfied simultaneously. To decide which of the two solutions is the least objectionable, we can first note that the disagreement for the deeper contact-curvature is $4.636 - 4.354 = 0.282$, whilst for the shallower contact-curvature it is $3.140 - 2.713 = 0.427$ or $1\frac{1}{2}$ times as large. This points in favour of the deeper contact-curvature. A still safer test is to calculate

$$a_2{}^* = a_2 + a_1 = 1.198c_2{}^2 + 8.466c_2 + 14.15$$

for the two solutions for a flat field, $c_2 = -4.636$ and -3.140, for as telescope objectives work at a high ratio of focal length to aperture, curvature of field will be more important than coma and so will have to receive more attention. This gives $a_2{}^* = +0.65$ for $c_2 = -4.636$ and $a_2{}^* = -0.62$ for $c_2 = -3.140$, so that in this respect there is nothing to choose between the solutions.

We can next test the solutions with regard to their spherical aberration by the formula $a_1 = 1.198c_2{}^2 + 7.616c_2 + 11.48$, which of course is not changed by diaphragm shifts and so is directly applicable. It gives $a_1 = +1.92$ for $c_2 = -4.636$ and $a_1 = -0.62$ for $c_2 = -3.140$. In this respect the shallow contact-curvature has a 3:1 advantage. Finally, we can test both solutions for distortion [116] by

$$a_5{}^* = a_5 + \tfrac{1}{2}K' \cdot a_4 + 3K' \cdot a_3 + 3K'^2 \cdot a_2 + K'^3 \cdot a_1.$$

For our special case, $K' = 1$, and as we are working with a thin-lens solution, $a_5 = 0$; hence

$$a_5{}^* = 3(a_3 + a_4/6) + 3a_2 + a_1 = 3(a_3 + a_4/6) + 3a_2{}^* - 2a_1.$$

The value of $(a_3 + a_4/6)$ has already been found to equal $0.500 + 0.117 = 0.617$ and, on using the above calculated values for $a_2{}^*$ and a_1 of the two solutions, we find

for $\qquad c_2 = -4.636: a_5{}^* = 1.851 + 1.95 - 3.84 = -0.04$

for $\qquad c_2 = -3.140: a_5{}^* = 1.851 - 1.86 + 1.24 = +1.23.$

As the ideal is $a_5{}^* = 0$, the solution with the deep contact has a huge advantage in this respect. It would, however, allow of only a small field on account of the short contact radius.

We next test the 'reversed Steinheil objectives' of Chapter XI in the same way. Turning our telescope in the opposite direction (Fig. 162), we get our eyepiece into

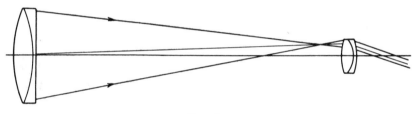

FIG. 162.

the correct position for which the constants in Chapter XI were calculated (page 523) and have

$$a_1 = 1.198c_2{}^2 + 11.014c_2 + 24.85$$
$$a_2 = \qquad\quad 0.850c_2 + \ \ 4.02$$

But in the present case we have the exit pupil at the Ramsden disk of the telescope, and if we again assume the latter to be infinitely long, we have $l'pr_k = f'_{EP} = 1$, and as parallel rays issue from the eyepiece, $l'_k = \infty$. Hence the constant K' is now $1/(1/\infty - 1/1) = -1$, and the previous formulae now give

$$a_2{}^* = a_2 + K'a_1 = a_2 - a_1; \qquad a_3{}^* = a_3 + 2K'a_2 + K'^2 a_1 = a_3 - 2a_2 + a_1.$$

From the above numerical equations, we now obtain

$$-a_2{}^* = 1.198c_2{}^2 + 10.164c_2 + 20.83 = 0$$

for freedom from coma. This gives $c_2 = -5.022$ or -3.462.

For a flat tangential field, we again have $a_3 = 0.500$ and we must ask for

[116] $a_3{}^* = -a_4/6 = -0.117$. Then $(a_3 - a_3{}^*) = 0.617$ and the condition $(a_3 - a_3{}^*) - 2a_2 + a_1 = 0$ gives

$$(a_3 - a_3{}^*) + \quad a_1 = 1.198c_2{}^2 + 11.014c_2 + 25.467$$
$$- 2a_2 = \qquad\quad - 1.700c_2 - 8.040$$

$$\overline{\quad 0 = 1.198c_2{}^2 + 9.314c_2 + 17.427 \quad}$$

This will be seen to be identical with the corresponding equation for the first case, that of the thin cemented achromat, whence for freedom from tangential curvature of field, $c_2 = -4.636$ or -3.140.

The two solutions again disagree by $4.636 - 5.022 = -0.386$ and by $3.140 - 3.462 = -0.322$ respectively, more nearly by equal amounts but by exactly the same average as before. We therefore make the same tests of the solutions for a flat field, namely, $c_2 = -4.636$ and $c_2 = -3.140$. We first calculate $a_2{}^* = a_2 - a_1$ by the equation $a_2{}^* = -1.198c_2{}^2 - 10.164c_2 - 20.83$ and find $a_2{}^* = +0.54$ for $c_2 = -4.636$, and $a_2{}^* = -0.73$ for $c_2 = -3.140$, not very different from the two values of $a_2{}^*$ in the first case.

For the spherical aberration, the formula $a_1 = 1.198c_2{}^2 + 11.014c_2 + 24.85$ gives $a_1 = -0.46$ for $c_2 = -4.636$ and $a_1 = +2.08$ for $c_2 = -3.140$. In this respect the deep contact now has a great advantage, whilst the advantage lay with the shallow contact in the first case.

Finally we test for distortion, and with $K' = 1$ we have the equation

$$a_5{}^* = -3(a_3 + a_4/6) + 3a_2 - a_1 = -3(a_3 + a_4/6) + 3a_2{}^* + 2a_1.$$

With $(a_3 + a_4/6) = 0.617$, we have $a_5{}^* = -1.851 + 3a_2{}^* + 2a_1$, which with

$$a_2{}^* = \begin{cases} +0.54 \\ -0.73 \end{cases} \text{ and } a_1 = \begin{cases} -0.46 \\ +2.08 \end{cases} \text{ gives}$$

for $\quad c_2 = -4.636: a_5{}^* = -1.851 + 1.62 - 0.92 = -1.15$

for $\quad c_2 = -3.140: a_5{}^* = -1.851 - 2.19 + 4.16 = +0.12$.

The advantage lies with the shallow contact curvature, and although this gives a little more coma and very much more spherical aberration, it is the best solution for telescope eyepieces because both the coma and the spherical aberration will be below the tolerances for any f-ratio likely to be encountered. Most astronomical telescopes work at $f/15$ or near that value. But in the case of an eyepiece of 1-inch focal length, which is about the weakest ordinarily employed, and an aperture of $f/10$, the diameter of the Ramsden disk, which is $2SA$, is 0.1 inch and $SA = 1/20$. Now the meaning of a_1 is that the OPD'_p at the paraxial focus is equal to $\frac{1}{4}a_1 \cdot SA^4$, and with $a_1 = 2.08$, this gives $OPD'_p = 0.52 \times (1/20)^4 = 0.52/160\,000$. Since the Rayleigh limit corresponds to one wavelength $= 1/50\,000$ inch of OPD'_p, the a_1-value of our eyepiece amounts to only one sixth of the tolerance. With regard to coma, we have $OSC' = a_2 \cdot SA^2$, and with the coma-constant $a_2{}^* = -0.73$, this gives $OSC' = -0.73/400$. As the eyepiece tolerance for OSC' is $1/400$, the coma amounts to only three-quarters of the tolerance. At the usual $f/15$ aperture, the coma would use up only $3/(4 \times 2\frac{1}{4}) = 1/3$ of the tolerance, and the spherical aberration would amount to $1/(6 \times 2\frac{1}{4} \times 2\frac{1}{4}) = 1/30$ of the tolerance.

The distortion can be expressed as   [116]

$$\delta H'_0 = -H'^3_0 (N'_0/l'_0)^2 \cdot a_5.$$

If we turn this into a percentage by dividing throughout by H'_0 and notice that $N'_0 = 1$ for air and $l'_0 = l'_k$, we see that H'_0/l'_0 is the tangent of the angle between the optical axis and the direction of the distant image, which tangent in eyepieces of this type will be of the order of $1/4 = \tan 14°$, hence

$$\frac{\delta H'_0}{H'_0} = -a_5(\tfrac{1}{4})^2$$

or for the $a_5{}^* = +0.12$ of our final form,

$$\frac{\delta H'_0}{H'_0} = -\tfrac{1}{9} \times \tfrac{1}{16} = -\tfrac{1}{144}.$$

This may be regarded as exceptionally good for an eyepiece, as it means only 0.7 per cent. at the edge of the field in the sense of pincushion distortion. It will be noticed that the forms that were unfavourable as regards distortion had $a_5{}^* = -1.15$ and $+1.23$ or roughly 11 times the above, amounting therefore to the very serious amount of about 8 per cent.!

The discussion has been given in considerable detail because of the importance of this kind of compromise in innumerable cases which occur in actual practice. But the numerical results should not be trusted too much because they apply to the *primary* aberrations of *thin* lenses. When the necessary thicknesses are put in and when the lens is trigonometrically tested at a point near the margin of the field, very considerable changes will be found, and that is the time when this kind of discussion is required. The chromatic tolerance can then also be drawn upon if necessary. The leading principle should always be to smooth out the multitudes of residuals so as to render them as nearly invisible as seems possible.

For the actual design of an eyepiece of this simplest type, we may take it that the lead obtained by our discussion is correct to the extent that the form with the crown facing the telescope object-glass will have about $1/3$ of the total curvature of the crown lens (which was 4.70) on the outside and $2/3$ at the contact face. But to obtain the largest possible field, it will be advisable to increase the V-difference of the glasses, and we will try the crown (Chance 3463) used in Chapter XI, although we shall combine it with the extra dense flint 337 as the densest which may be used with reasonable safety. We can then go straight to trigonometrical calculation, for in these deeply curved lenses the higher oblique aberrations are always so large that a rough preliminary knowledge of the proper form is all that is really useful.

The glasses selected have the following characteristics:

	Nd	$Nf-Nc$	V
Crown:	1.5407	0.00910	59.4
Flint:	1.6469	0.01917	33.7
		Difference =	25.7

[116] For unit focal length of the combination, $c_{crown} = 4.28$. Putting $1/3$ of the crown curvature on the external surface, in accordance with the general discussion, we obtain for our eyepiece in the preferable computing direction $c_2 = +2.85$, $c_3 = -1.43$, or by reciprocals $r_2 = 0.351$ and $r_3 = -0.699$. The value of r_1 is to be determined by the condition for achromatism, and here we must note that, as in all eyepieces, achromatism of *magnification* is the important correction. A little over- or under-correction of colour for the individual image-points is utterly insignificant; in that connexion, it should be recalled that we found in Chapter X of Part I that the ordinary chromatic aberration at the focal point of the usual Huygenian eyepiece is equal to that of a simple flint-glass lens of the same focal length—but nevertheless may be treated as almost negligible in the majority of cases! It is of course utterly impossible that a cemented lens achromatized for principal rays should have anything like this huge chromatic aberration at the individual image-points; therefore we need not bother at all about the ordinary achromatism.

To obtain the desirable *full* field of at least 30°, a very generous thickness of the crown is required; we therefore fix this at 0.300—nearly equal to the contact-radius! For the flint, 0.100 will be ample (Fig. 163).

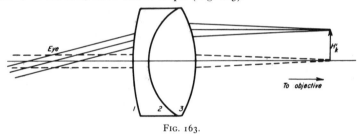

Fig. 163.

As these cemented eyepieces are only intended for high magnifications on telescopes, and are comparatively insensitive to moderate changes in the angle of the principal ray, it is admissible to assume that the latter is parallel to the optical axis on the right, as if it came from an infinitely distant object-glass. This simplifies the calculations considerably. Although a full field of about 35° is usually possible, the outermost part of this will be badly over-corrected and only fit for 'finding' purposes. The calculation should be for a field of 20°–25°, and with unit focal length this will result if we assume $H'_k = 0.200$.

We therefore begin with a right-to-left tracing of a principal ray parallel to the axis and at $Y = 0.200$. It would be theoretically correct to trace this in colours C and F, but it will save time and will introduce no visually discoverable error to use $Ny = Nd + 0.188(Nf - Nc)$ and $Nv = Ny + (Nf - Nc)$; the 'yellow' principal ray is then at once available as ray 'pr' between the outer tangential rays 'a' and 'b' for the determination of the oblique aberrations. To obtain achromatism of magnification, r_1 must be determined so that the 'yellow' and 'violet' principal rays emerge parallel to each other, but it would be waste of time to try to effect this to the last second of arc; it will be amply near enough if the two final Upr_1

agree within about 30 seconds because no observer could detect an error as small [116] as that. Equation Chr.(2) of Part I, Chapter IV, will usually give the best solution in these cases of rays passing the surfaces at very large angles. As we are calculating for about 3/4 of the full field, and as the extreme margin will tend to be heavily over-corrected on account of the deep contact-curvature, it is better to err—for the computed principal ray—on the side of slight *under*-correction; that is, a slightly larger angle for violet than for yellow. In the present case, the first shot with a guessed value of 1.05 for r_1 gave $Upr_y = -11$-40-28 and $Upr_v = -11$-40-4. This represents *over*-correction, and for the reason just stated, a change was considered advisable. Trying $r_1 = 1.000$ gave $Upr_y = -12$-1-2 and $Upr_v = -12$-1-15, and this was accepted as perfectly satisfactory.

The next step should be to add rays 'a' and 'b' in the left-to-right direction. The reason for beginning with the oblique pencil is that if this should—as it easily might—display very heavy coma, the bending selected would be useless and a preceding axial ray-trace would have been a mere waste of time. To obtain the data for rays 'a' and 'b' we must decide upon the SA, in this case the radius of the Ramsden disk, and as large telescopes hardly ever work at more than f/10, the choice was $SA = 0.045$.

Tracing rays 'a' and 'b' from this aperture, the results were $Y'_{ab} = 0.20084$, $X'_{ab} = 0.8704$. Therefore, with the principal ray at $Y = 0.2000$, $Coma'_T = -0.00084$, and as this would be a perfectly admissible amount, an axial ray was added for the same SA-value and gave a final $U' = 2$-42-12 and $L' = 0.8490$. Therefore $X'_T = X'_{ab}$ – final $L' = +0.0241$ or a decidedly hollow field. The astigmatism is therefore markedly over-corrected, and as this can only have arisen at the deep dispersive contact surface, a lengthening of the contact radius is indicated. On account of the huge angles of incidence (nearly 45° at the contact surface for ray 'a') the differential method that is usually advisable would give very inaccurate results and a trial was made with $\delta c = -0.15$. The computations for the radii gave

	original $c_1 =$	1.000	$c_2 =$	2.85	$c_3 = -1.43$
	$+\delta c =$	-0.15		-0.15	-0.15
	corrected $c_1 =$	0.85	$c_2 =$	2.70	$c_3 = -1.58$
or	corrected $r_1 =$	1.176	$r_2 =$	0.370	$r_3 = -0.633$

The instinct developed by long practice suggests that, as the negative higher aberrations would be considerably diminished by the flatter contact surface, a slight lengthening of r_1 as directly determined would probably be beneficial, and the calculation was therefore made with r_1 increased to 1.200. The calculation for the principal ray in 'y' and 'v' then gives $Upr_y = -12$-3-46 and $Upr_v = -12$-4-17, or just about the stated safe amount of chromatic under-correction. Rays 'a' and 'b' then give, with the addition of the axial ray, $Coma'_T = $ nil and $X'_T = -0.0177$, or the reversal of the previous hollow field to one that is round to almost the same extent. A bending midway between the two would therefore be likely to give the desired result.

[116]   It may be added that eyepieces of this type have very decided advantages when the rather small field is not objectionable. Having only two glass-air surfaces, they pass 10 per cent. more light than eyepieces with two separate components and at the same time give only one lot of doubly-reflected false light against 6 lots in a two-component eyepiece. There is also a comfortable distance of the eye point equal to about $8/10$ of the focal length, against only $2/10-3/10$ in Huygenian and Ramsden eyepieces. All forms actually tried give a small negative coma when they are corrected for a flat tangential field. The admissibility of the amount found may be tested by bearing in mind that the eye never sees the full amount of the tangential coma actually calculated; it is usually safe to estimate that the *sagittal coma* $= \frac{1}{3} coma'_T$ is more nearly what is actually appreciated. But, as we see it through the magnifying eyepiece, this magnification must be allowed for.

In our first example, we had $Coma'_T = -0.00084$ and we may estimate that 0.00028 would really be the visible spreading of the image. At the unit focal length this would subtend an angle of the same value in radians, and 0.00028 radians is 58 seconds of arc. That is about the limit of acuity of human sight, and the calculated coma would therefore not be felt to be a serious defect. The estimation of the defect in angle has the advantage of being independent of the scale to which the eyepiece might be executed, as the calculated coma changes in strict proportion to the focal length and therefore always retains the same angular subtense for a given type. But it will vary as the square of the ratio of clear aperture to focal length. Our calculation was for about $f/10.5$. At the usual $f/15$ of astronomical telescopes, the coma would have a little less than half the calculated subtense and would be quite invisible.

A Triple Lens (Cemented or Air-Spaced) as an Eyepiece

[117]   IF instead of a simple cemented lens we use either a triple cemented lens or a thin combination with air-gap, we gain at least one extra liberty and we are then no longer restricted to only one aberrational correction, but may attempt to satisfy two (or *possibly* more) conditions simultaneously. We will examine the possibilities first by the TL primary theory, for although we have already seen that we have to introduce very substantial thicknesses, it may be accepted that eyepieces of the type under consideration never get usefully away from the thin-lens restrictions. Any thin system with the diaphragm in contact has the fixed astigmatic constant $a_3 = 1/2f'$, a fixed Petzval sum a_4, and no distortion. The only two aberrations which can be varied by bending or by a change of type are a_1 and a_2, and we may assume that with the rather more complicated types now to be discussed, we can give to a_1 and a_2 any value within very wide limits.

When the exit pupil is shifted away from the system, the eccentricity correction by K' will come in and we shall have:

$$a_1 \text{ unchanged}; \quad a_2{}^* = a_2 + K'a_1; \quad a_3{}^* = a_3 + 2K'a_2 + K'^2a_1;$$
$$a_4 \text{ unchanged}; \quad a_5{}^* = 3K'(a_3 + a_4/6) + 3K'^2a_2 + K'^3a_1.$$

We have shown that for eyepieces K' is always very nearly unity, but we will maintain generality by retaining an unrestricted K' in the equations. It would, however, be useless to discuss cases which would lead to a field of considerable [117] curvature, and as the extreme range of a_3^* worth considering is between zero and $-a_4/4$, which is small compared with the thin-system a_3, we can without loss of generality solve for the usually best value $a_3^* = -a_4/6$, and that gives us a general condition:

(1) $-a_4/6 = a_3 + 2K'a_2 + K'^2 a_1$ or $(a_3 + a_4/6) + 2K'a_2 + K'^2 a_1 = 0.$

We know moreover that, for our usual unit focal length, $(a_3 + a_4/6)$ will never differ seriously from the value 0.6.

The general condition just adopted immediately excludes the possibility of aplanatism in Abbe's sense for any thin-lens system, for the condition obviously can be fulfilled only if at least one of the constants a_1 and a_2 is finite; hence simultaneous correction of coma and spherical aberration is out of the question when a reasonably flat field is demanded (see page 352 of Part I).

For the rest of the discussion, we must remember from the previous section that, for eyepieces, values of a_5^* (the distortion) exceeding about 0.2 or 0.3 will lead to obviously faulty perspective, that a_2^* may, for $f/10$, almost reach unity, and that —again at $f/10$—a_1 may be allowed to reach two units.

First case. The spherical aberration shall be zero. As $a_1 = 0$, condition (1) gives $2K'a_2 = -(a_3 + a_4/6)$ or $a_2 = a_2^* = -(a_3 + a_4/6)/2K'$. The a_2^* is *in this case* equal to a_2 because a_1 is zero. For eyepieces K' may be taken as unity, hence $a_2^* = -0.6/2 = -0.3$ approximately, which is small enough. To find the resulting distortion, we put $a_1 = 0$ and the value of a_2^* just determined into the equation for a_5^* and find

$$a_5^* = 3K'(a_3 + a_4/6) - 3K'^2(a_3 + a_4/6)/2K' = 1\tfrac{1}{2}K'(a_3 + a_4/6) = \text{roughly } 0.9$$

for $K' = 1$. This is several times the value that was shown to be reasonably tolerable; hence a spherically corrected thin eyepiece with a flat field will have a most objectionable amount of distortion (see page 353, paragraph 2). For landscape lenses K' is always fractional, usually about $1/4$ to $1/10$. Hence for these the required amount of coma would become intolerable since K' is in the denominator of the equation for a_2^*, whilst the distortion might be tolerable. The general conclusion is that spherically corrected thin systems are unfit for use as either eyepieces or landscape lenses.

Second case. The coma should be zero. There would be little sense in discussing $a_2 = 0$ as the diaphragm shift would bring in $a_2^* = K'a_1$; hence we take as the second case that a_2^*, the coma of the finished article, is to be zero.

From $a_2^* = a_2 + K'a_1 = 0$ we obtain $a_1 = -a_2/K'$. Putting this into condition (1) we find

$$(a_3 + a_4/6) + 2K'a_2 - K'a_2 = 0, \quad \text{or} \quad K'a_2 = (a_3 + a_4/6)$$

therefore

$$a_2 = -(a_3 + a_4/6)K'$$

and putting this into $a_1 = -a_2/K'$, we find

$$a_1 = (a_3 + a_4/6)/K'^2.$$

[117] With these necessary values of a_2 and a_1, the equation for $a_5{}^*$ gives

$$a_5{}^* = 3K'(a_3 + a_4/6) - 3K'(a_3 + a_4/6) + K'(a_3 + a_4/6) = K'(a_3 + a_4/6)$$

which in the present case has a value of roughly o.6 for $K' = 1$. The coma and tangential field curvature are now zero for the finished system. The spherical aberration is amply small enough for eyepieces, and the distortion is only 2/3 of that in the first case. For landscape lenses with their fractional values of K', the spherical aberration will exclude *large* apertures, but for wide fields the permissible aperture will be far greater than for the first case on account of the growth of coma with H'_k.

Third case. Zero distortion shall accompany the flat field. If $a_5{}^* = $ o we have the condition (on dividing throughout by K') that

$$3(a_3 + a_4/6) + 3K'a_2 + K'^2a_1 = 0$$

whilst we have by (1)

$$(a_3 + a_4/6) + 2K'a_2 + K'^2a_1 = 0.$$

These two equations give

$$a_2 = -2(a_3 + a_4/6)/K' \quad \text{and} \quad a_1 = 3(a_3 + a_4/6)/K'^2$$

The value of a_1 is approximately 1.8 for $K' = 1$. Substituting the values of a_1 and a_2 into $a_2{}^* = a_2 + K'a_1$, the coma-constant of the resulting system becomes

$$a_2{}^* = (a_3 + a_4/6)/K' = \text{roughly o.6}.$$

For the usual eyepieces, with $K' = 1$, as we are assuming in the present treatment, the coma and spherical aberration will be admissible up to an aperture of about $f/10$, and this state of correction will then be highly desirable. Note that it leads to *positive* coma at the focus whilst the first case, with serious distortion, calls for negative coma. For landscape lenses, with a fractional value of K', this correction would be quite impracticable on account of the huge values of both spherical aberration *and* coma.

This discussion is instructive by displaying the importance of our tolerances from a new point of view. Admitting a comparatively harmless amount of one aberration may render it possible to change the magnitude of another aberration from a highly objectionable value to a tolerable one; and we have shown that even coma, the most undesirable of all aberrations when of visible magnitude, may be deliberately admitted to improve the general quality of an important type of eyepiece.

From the practical point of view, the most important conclusion from the discussion is that these thin eyepieces will be at their best when they display positive coma at the image in the focal plane and that negative coma is a practically certain indication of decidedly serious distortion. We cannot expect the final trigonometrical results to agree *numerically* very closely with the discussion because the higher aberrations and the necessary thicknesses will modify the magnitude of the

aberrations.   But we shall find that the tendencies in the state of correction fully   [117]
bear out our theoretical TL conclusions.

NUMERICAL EXAMPLE

An attempt to turn the triple lenses solved for in general algebraic form in
Chapter XI into a satisfactory achromatic eyepiece did not give a good result.
Using the general solutions for a_1 and a_2 for central passage of the oblique pencils
to obtain the equations for a coma-free eyepiece ($K' = 1$) with a flat tangential
field, namely,

$$a_2{}^* = a_2 + a_1 = 0; \qquad (a_3 + a_4/6) + 2a_2 + a_1 = 0$$

and solving these for $x = c_1 - c_2$ and $y = c_2$, the most promising of the two
resulting solutions was

$$
\begin{aligned}
r_1 &= 1.015 \\
r_2 &= 0.665 \\
r_3 &= -0.313 \\
r_4 &= -0.806
\end{aligned}
$$

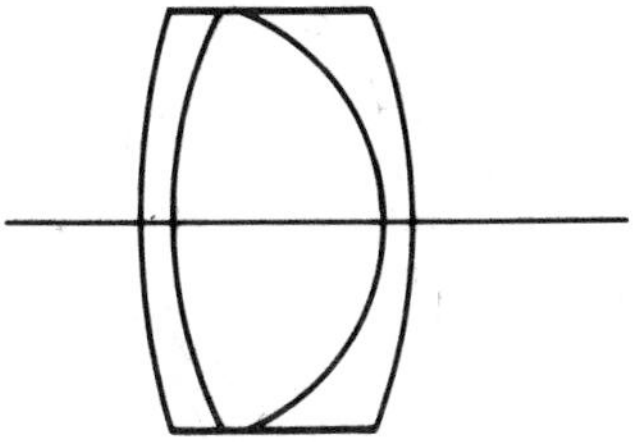

FIG. 164.

but on account of the deep contact surface this gave a very hollow field and considerable negative coma.   Differential correction gave such huge changes as to be
inadmissible (the form resulting was a freak with all four radii negative!).   As a
last resort the starting equations were solved for values of $a_2{}^*$ and $a_3{}^*$ approximately equal to, but of the opposite sign of, the residual aberrations found for the
first solution.   The result was practically a doublet like those treated earlier owing
to the second flint-lens having practically $c_c = 0$, and the formula ($r_1 = 1.695$,
$r_2 = 0.327$, $r_3 = -0.610$) gave no improvement on the original solution owing to
the deep contact surface.   We may therefore take it that the triple *cemented* lens is
not likely to prove superior to the simpler double cemented form.

Conditions become far more favourable when we break up the cemented
combination into two parts, each with positive power, and separated by an air-space,
for we then have the 'optical work' divided between two components and ought to
be able to reach a larger field with reasonable angles of incidence.   The simplest
possible type consists of a simple convex lens combined with an over-corrected
cemented doublet.   But this immediately presents an embarrassing number of
possibilities.   Assuming the eye to be at the left and the object-glass of the
telescope to be at the right, we have the four possible types shown in Fig. 165, and
in each of these the total crown curvature may be divided between the two convex
elements in an infinite variety of proportions.   Preference will naturally be given
to solutions which lead to radii of curvature that are not very short, as otherwise
the field would become small and there would be very little advantage over a simple

[117] cemented lens of the best form and most suitable glass.   Preliminary analytical solutions represent the line of least resistance in arriving at definite conclusions with a moderate amount of work, and these were applied to types (*a*) and (*b*) for a combination of a dense barium crown with extra dense flint.   It was soon found

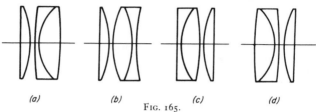

(*a*)   (*b*)   (*c*)   (*d*)
FIG. 165.

that type (*a*) always gave a short contact-radius and was therefore definitely objectionable, whilst type (*b*) gave solutions with an exceptionally long contact-radius, which promised good solutions for a large field.   The other two types have not been tried.   There seem to be possibilities in type (*c*), but optical instinct (which, however, may be wrong!) shies at type (*d*).   All four types do not seem to have ever been tried as closely-spaced achromatic eyepieces, but type (*c*) *with a large air-space* is the favourite form of the 'Kellner' eyepiece or achromatized Ramsden eyepiece as now used almost universally in prismatic binoculars.

The only closely-mounted achromatic eyepieces hitherto made in any considerable numbers are the two containing a triple lens (Fig. 166), the first being used for

(*a*)   (*b*)
FIG. 166.

astronomical eyepieces and the second, Abbe's, for compensating microscope eyepieces.

As a numerical example of a three-element eyepiece, we will take type (*b*) of Fig. 165, using the glasses:

	Nd	$Nf-Nc$	V
Chance 9753:	1.5881	0.00962	61.1
Chance 337:	1.6469	0.01917	33.7

which, for visual achromatism of a thin system of unit focal length, give

$$c_{crown} = 3.79 \qquad c_{flint} = -1.90.$$

Either of these total curvatures may be put into a single component or split up into several components without affecting the achromatism of the combination.   For type (b) the crown curvature is to be divided, and the proportion must be settled before an analytical solution can be made.   Computations were made, first with $c_a = 0.95$, which was selected because it divides about equally the 'optical work' of bending the oblique pencils; but it did not lead to a distribution of curvature among the five radii that seemed favourable.   The indications were such as to suggest that the separate single lens should be strengthened; $c_a = 1.19$ was next tried, and it confirmed the instinctive choice of the right direction, but did not seem to go quite far enough.   The analytical solution was therefore made with $c_a = 1.29$ which, in order to make up the total crown curvature of 3.79, called for $c_b = 2.50$, whilst c_c is of course fixed at the directly calculated $c_c = -1.90$.

The form of solution chosen was to calculate a_1 and a_2 by the G-sums for three bendings of each of the two *separated* components, the second of which was treated as a cemented combination.   The bendings selected for lens (a) were $c_1 = -0.5$, 0, $+0.5$, and for lens (b) were $c_3 = -1$, 0, $+1$, giving for lens (c) on account of the cementing, $c_4 = c_3 - c_b$, and as $c_b = 2.50$, $c_4 = -3.5$, -2.5, -1.5.

The calculation for a_1 and a_2 of each component was then made exactly as in the complete example given previously for the LA' and OSC' contributions of each constituent of the photovisual objective in section [88], excepting that the factor $Y^2 l'_k{}^2$ dropped out of the spherical G-sums and the factor Y^2 out of the coma G-sums.   The results obtained were:

For lens (a) if $c_1 =$	-0.5	0	0.5
a_1 contribution $=$	2.842	1.601	0.790
a_2 contribution $=$	-1.090	-0.780	-0.470

For lens (b) if $c_3 =$	-1.0	0	1.0
a_1 contribution $=$	34.588	21.562	11.866
a_2 contribution $=$	-5.607	-4.406	-3.205

For lens (c) if $c_4 =$	-3.5	-2.5	-1.5
a_1 contribution $=$	-33.604	-21.575	-12.281
a_2 contribution $=$	5.138	4.144	3.152

Combining the contributions of (b) and (c) by addition, we obtain for the cemented component:

a_1 contribution $=$	0.984	-0.013	-0.415
a_2 contribution $=$	-0.469	-0.262	-0.053

The last sets of figures give the spherical and coma coefficients for the lens when passed *centrally* by the oblique pencils.   When the lens is used as an eyepiece on a long (reversed) telescope, we have to apply the second fundamental law with $K' = +1$ to find

$$a_2{}^* = a_2 + a_1 \qquad a_3{}^* = a_3 + 2a_2 + a_1$$

in which a_3 has the fixed thin-lens value $1/2f'$ or in our case 0.500.   For a flat tangential field, $a_3{}^*$ should be $-a_4/6$, and as the Petzval sum of our eyepiece gives $a_4 = 0.657$, we should aim at $a_3{}^* = -0.110$.

[117] As a_2 and a_1 for the complete system are made up of the contributions of the two separated components which are to be bent *independently*, we write

$$a_2{}^* = (a_1{}^a + a_2{}^a) + (a_1{}^{bc} + a_2{}^{bc})$$
$$a_3{}^* = a_3 + (a_1{}^a + 2a_2{}^a) + (a_1{}^{bc} + 2a_2{}^{bc})$$

and plot the (a) and the (bc) contributions for a solution by the matching principle. Now $a_3{}^*$ and a_3 have the fixed values, unaffected by bending, which we have already given. For $a_2{}^*$ we may desire either zero or the moderate positive value which was shown to be desirable in the previous discussion; anyhow, $a_2{}^*$ will also be *fixed*. Hence we obtain for plotting quantities which should be equal and which therefore can be submitted to the matching principle by transposing the last equation into the forms:

$$(a_1{}^a + a_2{}^a) \text{ should be equal to } a_2{}^* - (a_1{}^{bc} + a_2{}^{bc})$$
$$(a_3 + a_1{}^a + 2a_2{}^a) \text{ should be equal to } a_3{}^* - (a_1{}^{bc} + 2a_2{}^{bc}).$$

The placing of a_3 on the left and of $a_2{}^*$ and $a_3{}^*$ on the right is of course arbitrary; we may plot them on whichever side we like provided the *signs* are adjusted accordingly.

In any case, we plot the left-hand sides for the three bendings of the single lens and side by side with them the right-hand sides for the three bendings of the

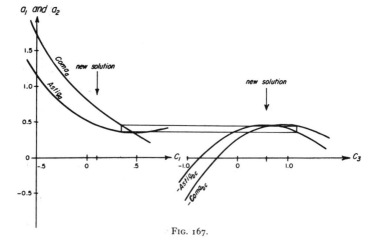

Fig. 167.

doublet. In the graph, shown in Fig. 167, $a_2{}^*$ was taken as zero, so the graph reads directly for zero coma. As we found above, for the single lens,

$$\text{for } c_1 = \quad -0.5 \qquad 0 \qquad 0.5$$
$$a_1{}^a = \quad 2.842 \qquad 1.601 \qquad 0.790$$
$$a_2{}^a = \quad -1.090 \qquad -0.780 \qquad -0.470$$

we have to plot

$$Coma_a = a_1{}^a + a_2{}^a = \qquad 1.752 \qquad 0.821 \qquad 0.320$$
$$Astig_a = 0.500 + a_1{}^a + 2a_2{}^a = \quad 1.162 \qquad 0.541 \qquad 0.350$$

For the doublet, we found

$$\begin{array}{cccc}
\text{for } c_3 = & -1.0 & 0 & 1.0 \\
a_1{}^{bc} = & 0.984 & -0.013 & -0.415 \\
a_2{}^{bc} = & -0.469 & -0.262 & -0.053
\end{array}$$

giving

$$-Coma_{bc} = -a_1{}^{bc} - a_2{}^{bc} = \quad -0.515 \qquad 0.275 \qquad 0.468$$
$$-Astig_{bc} = -0.110 - a_1{}^{bc} - 2a_2{}^{bc} = \quad -0.156 \qquad 0.427 \qquad 0.411$$

The graph shows that, for zero coma and a flat tangential field, $c_1 = +0.333$ and $c_3 = +1.15$. With the selected values of c_a, c_b, and c_c, the TL prescription is therefore

$$\begin{array}{llll}
c_1 = & 0.333 & r_1 = & 3.00 \\
& & & \quad 0.16 \\
c_2 = & -0.96 & r_2 = & -1.04 \\
& & & \quad 0.01 \text{ (air)} \\
c_3 = & 1.15 \text{ or by reciprocals} & r_3 = & 0.870 \\
& & & \quad 0.26 \\
c_4 = & -1.35 & r_4 = & -0.741 \\
& & & \quad 0.08 \\
c_5 = & 0.55 & r_5 = & 1.82
\end{array}$$

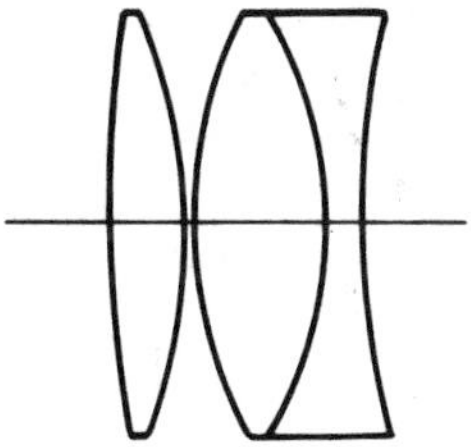

Fig. 168.

Suitable thicknesses have been selected in the usual way.

The trigonometrical correction was begun by tracing a principal ray right-to-left through the system in two colours y and v as usual. Correction for the contraction and expansion of the cone of rays was applied to the principal ray and gave: corrected $r_4 = -0.754$, $r_3 = 0.950$, $r_2 = -1.133$, $r_1 = 3.019$ from the successive L_{pr} of a principal ray entering parallel to the optical axis at distance 0.300 above it. Four-figure calculations gave the final angles of emergence $Upr_y = -17\text{-}38\text{-}30$ and $Upr_v = -17\text{-}40\text{-}40$, and although these differed by 2 minutes of arc, they were accepted as near enough for a first test. When rays a and b and an axial ray with $SA = 0.07$ were traced in the usual left-to-right direction, the final results were

$$Coma'_T = -0.0003 \qquad X'_T = +0.0088 \qquad \text{Distortion} = 6.1 \text{ per cent. at } 17^\circ 40'.$$

Differential correction was then applied to make both coma and field curvature zero. The graph gave the differential coefficients da_2/dc and da_3/dc in the usual way as

$$\frac{da_2{}^a}{dc_1} = -0.9 \qquad \frac{da_3{}^a}{dc_1} = -0.24 \qquad \frac{da_2{}^{bc}}{dc_3} = 0.24 \qquad \frac{da_3{}^{bc}}{dc_3} = 0.44$$

but in order to apply these, the calculated residuals of tangential coma and tangential

[117] astigmatic difference of focus had to be converted into their equivalents in terms of a_2 and a_3.

For the coma, TSA(1) gives $y' = -Coma'_T = -3Y_0{}^2H'_0a_2$, whence $a_2 = Coma'_T/3Y_0{}^2H'_0$. Since $Coma'_T = -0.0003$, $Y_0 = SA = 0.07$, and $H'_0 = 0.300$, the equation gives $a_2 = -0.067$.

For the tangential astigmatism, we have by page 725, $\delta l'_{0_T} = -3H'_0{}^2a_3$, and as $\delta l'_{0_T}$, like X'_T, is subject to the sign-convention of analytical geometry, we have $a_3 = -X'_T/3H'_0{}^2$. Since $X'_T = +0.0088$ and $H'_0 = 0.300$, the equation gives $a_3 = -0.033$. Hence the equations for the differential correction are

$$-0.067 - 0.98\delta c_1 + 0.24\delta c_3 = 0$$
$$-0.033 - 0.24\delta c_1 + 0.44\delta c_3 = 0,$$

giving $$\delta c_1 = -0.064 \quad \text{and} \quad \delta c_3 = +0.039.$$

On applying these to the previous TL formula, we find the corrected TL prescription to be

$$c_1 = 0.269, \quad c_2 = -1.021, \quad c_3 = 1.189, \quad c_4 = -1.311, \quad c_5 = 0.589.$$

The last value gives $r_5 = 1.698$ as reciprocal for the opening of the new trigonometrical test. Since the remaining radii will be changed by the correction for contraction and expansion of the cone of principal rays, it is preferable to use the above *curvatures* directly in obtaining the corrected radii. The latter were thus found as

$$r_1 = 3.744, \quad r_2 = -1.072, \quad r_3 = 0.923, \quad r_4 = -0.777, \quad r_5 = 1.698$$

and gave with the thicknesses previously fixed,

$$Coma'_T = -0.0001 \quad X'_T = +0.0006 \quad \text{Distortion 6 per cent. at } 17° 40'.$$

The correction aimed at has been attained with remarkable precision, but there is heavy distortion in accordance with the theory in the previous sections.

This solution was worked out before it had been discovered that comparatively harmless amounts of *positive* coma would greatly diminish the distortion. The old graph in Fig. 167 was therefore later examined from this point of view and a new solution was picked off which gave about the maximum of positive coma obtainable for the selected type, namely,

$$c_1 = 0.07, \quad c_2 = -1.22, \quad c_3 = 0.60, \quad c_4 = -1.90, \quad c_5 = 0.$$

As, owing to the flat last surface, the principal ray would go through the greater part of the system nearly parallel to the optical axis, correction for contraction of the cone could be dispensed with. Tracing y and v principal rays through right-to-left, it was found that $c_1 = 0$, that is, a plane next to the eye as well as next the objective of the telescope, gave achromatism of magnification with $Upr_y = Upr_v = -16\text{-}48\text{-}20$ for $H'_k = 0.300$. The tracing left-to-right of rays (*a*) and (*b*) and of an axial pencil at $SA = 0.05$ then gave

$$Coma'_T = +0.00043 \quad X'_T = +0.0048 \quad \text{Distortion 2.6 per cent at } 16.8°.$$

At $SA = 0.07$, which was previously used, the coma would be twice as large or $[117]$ $+0.0009$, but even that would be quite inappreciable in an eyepiece of unit focal length.   Yet the change of $Coma'_T$ from -0.0003 to $+0.0009$ has lowered the distortion from 6 per cent. to 2.6 per cent. or by more than half, so that it is rather less than that of a Huygenian eyepiece, for the latter has about 3 per cent. at $16°$.

A complete new solution using borosilicate crown and the same flint gave

$$Coma'_T = +0.00084 \quad X'_T = +0.0034 \quad \text{Distortion 1.5 per cent. at } 15.9°$$

or another very considerable reduction of the most obstinate aberration of thin achromatic eyepieces.

PHOTOGRAPHIC LANDSCAPE LENSES

A simple cemented lens combination has very decided advantages in landscape $[118]$ photography on account of the presence of only two glass-air surfaces.† It is a well-known fact that every glass-air surface reflects at least $4\frac{1}{2}$ per cent. of the incident light and transmits the remainder.   At the first surface of any system, this reflected light is merely lost and means nothing more than a slight increase in exposure.   On the other hand, the light reflected at a second glass-air surface is sent back towards the first, and the latter again reflects some 5 per cent. of it, now towards the sensitive plate.   This doubly-reflected light is decidedly objectionable for it is not focused, and thus it produces a general illumination and fogs the image.

A cemented landscape lens produces only this one lot of doubly-reflected light, and as its average intensity is 5 per cent. of 5 per cent. or 0.25 per cent. of the useful light reaching the plate, the fogging effect is almost negligible. The gradation of tones in the image consequently is practically true to nature, even in the deepest shadows for which the exposure is sufficient to affect the plate, and shadow-*detail* is well and truly brought out.

In a separated doublet lens there are six separate lots of the objectionable doubly-reflected light, which therefore amounts to about $1\frac{1}{2}$ per cent. of the average intensity of the useful, focused, light but naturally to a far higher fraction of the light in the shadows of the landscape.   Consequently a strong general illumination is added to the true light values in the shadows, and the contrast in the latter is greatly diminished and may be almost totally lost.   This was noticed from the earliest days of photography and long kept the simple landscape lens in high favour with all serious and discriminating workers.   The modern craze for fast snapshot work with easily portable hand-cameras and the wholesale production of pictorially worthless under-exposures which goes with it has pushed the comparatively slow landscape lens temporarily into the background, but it seems certain that before long the pendulum must swing back towards a higher standard of quality in landscape photography.   A study of the possibilities of the landscape lens is therefore well worth-while, in addition to its being a natural example of the application of the fundamental laws.

The vast majority of the old landscape lenses were very much worse than they

† This was written before the introduction of anti-reflection coatings.

[118] needed to be owing to blind insistence on smallness of the spherical aberration. As we saw in the discussion of the second and third laws, a considerable residue of under-corrected spherical aberration is absolutely necessary in a reasonably thin landscape lens if, in accordance with its intended use on a field covering 30° or more—usually about 45°—reasonably *uniform* definition over the whole of that field is made the primary requisite; for then coma, astigmatism, and curvature of field become by far the most serious aberrations and must be given precedence over the spherical aberration. Even a simple non-achromatic meniscus lens can be adapted for the production of excellent negatives by due regard to the fundamental laws of oblique pencils.

As freedom from coma is the first condition to be fulfilled by a landscape lens, we found that the modified eccentricity-constant K' must satisfy the condition:

$$K' = -a_2/a_1.$$

If we introduce this value into the equation for the astigmatism $a_3{}^*$ modified by diaphragm shift, we obtain

$$a_3{}^* = a_3 - 2\frac{a_2{}^2}{a_1} + \frac{a_2{}^2}{a_1} = a_3 - \frac{a_2{}^2}{a_1}$$

or

$$\frac{a_2{}^2}{a_1} = a_3 - a_3{}^*.$$

This equation enables us to calculate the astigmatic state of correction of any lens whose a_1, a_2, and a_3 are known for some particular position of diaphragm, as, for example, when the diaphragm is shifted so as to produce freedom from coma. When a_1 and a_2 of a lens or 'thin' system have been determined for central passage of the oblique pencils by the analytical method of Chapter XVII, then we know that for such a system $a_3 = 1/2f'$ and our equation enables us to find the value of $a_3{}^*$ which will result. For freedom from astigmatism, $a_3{}^*$ should be zero and we would have to select that form of the lens or system which makes $a_2{}^2/a_1 = a_3 = 1/2f'$. But we showed in the discussion of the first fundamental law that in general a flat tangential field represents the best compromise and that it calls for $a_3{}^*$ (our subsequent symbol for the astigmatic constant with the shifted diaphragm position) equal to $-a_4/6$, where a_4 represents the Petzval sum of the system. Hence for a flat tangential field we should select that lens form which satisfies the equation

$$a_2{}^2/a_1 = a_3 + a_4/6.$$

This simple principle gives good approximations in the case of simple lenses and of ordinary (or 'old') achromats of moderate thickness and not excessive curvature values; it fails more or less seriously in the case of 'new' achromats on account of their pronounced meniscus form, which renders them extremely sensitive to change of thickness and so causes the results with the necessary finite thickness to disagree greatly with those deduced analytically for zero thickness.

Even for simple lenses and old achromats, the inaccuracy introduced by neglecting thickness renders it impossible to make a purely analytical solution for the thin-lens values of the a_1 &c. Such a solution is tempting for, by the methods of

Chapters VI and XVII, we obtain a_1 as a quadratic equation in terms of c_1 or c_2, [118] and a_2 as a linear equation in c_1 or c_2, and transposing the equation of condition

$$a_2^2/a_1 = a_3 - a_3^*,$$

in which $(a_3 - a_3^*)$ has a constant value, into the form

$$a_2^2 - a_1(a_3 - a_3^*) = 0,$$

a new quadratic equation in c_1 or c_2 will result which could easily be solved by the usual simple methods. But unfortunately it very frequently gives imaginary solutions when the desirable flat tangential field is asked for. It is therefore necessary to use a semi-empirical method by calculating a_2^2/a_1 for a few bendings and selecting that bending which yields a value close to that of the desired $(a_3 - a_3^*)$.

(A) A Simple Convex Lens

If we calculate the spherical and coma G-sums of a simple lens of $N = 1.517$, $f' = 1$, assuming parallel incident light, we find $c = 1.9342$ and

$$[SGS] = a_1 = 4.304 - 3.901c_1 + 1.1593c_1^2$$
$$[CGS] = a_2 = -1.467 + 0.8295c_1.$$

As $f' = 1$, we have $a_3 = 0.5$, and as the Petzval sum for a thin simple lens is $1/Nf'$ or in our case $1/(1 \times 1.517)$, we have $a_4 = 1/1.517 = 0.659$. Therefore the value of a_2^2/a_1 which is desirable for a flat tangential field is

$$a_2^2/a_1 = a_3 + a_4/6 = 0.5 + 0.11 = 0.610.$$

By taking a series of bendings and calculating the corresponding values of a_1 and a_2, we easily find

c_1	a_1	a_2	a_2^2/a_1
-7	88.417	-7.274	0.597
-4	38.457	-4.785	0.595
-1	9.364	-2.296	0.563
$+2$	1.139	0.192	0.032
$+5$	13.781	2.681	0.522
$+8$	47.291	5.169	0.565

We see that the important ratio in the last column approximates to the desired value of $+0.610$ for deep bendings in either direction. However, carrying the calculated values still further, we should find that the desired value is never reached, but that for very extreme bendings the values diminish again; therefore a solution of the quadratic equation referred to gives two imaginary solutions. Evidently the nearest approach to the desired value is found for negative values of c_1, and as it is undesirable to have excessive spherical aberration, we shall try a value near $c_1 = -4$; this would give $c_2 = -5.9342$, or radii of the meniscus roughly in the proportion of $2:3$. Lenses of this type are actually used for the least expensive

[118] small hand-cameras, and to obtain a focal length of suitable order, we will test our solution trigonometrically on the following meniscus:

$$r_1 = -1.5 \text{ inch}$$
$$\text{thickness } 0.12 \text{ inch}$$
$$r_2 = -1.0 \text{ inch.}$$

The most expeditious way of carrying out the calculation consists in tracing a paraxial and a marginal ray from the infinitely distant axial object-point and finding by Sine Theorem II (Part I, page 371) the position of the exit pupil which will secure freedom from coma. Carrying this out with the rather excessive semi-aperture $Y = 0.250$ inch and working with four-figure logs, we find:

$$\text{final } l' = 5.512, \qquad \text{final } \log u' = 8.6684$$
$$\text{final } L' = 5.112, \qquad \text{final } \log \sin U' = 8.6965.$$

These give $l' - L' = 0.400$, and $1 - (u'/\sin U') = 0.0627$. Hence

$$l'_{pr} = l' - (l' - L')/(1 - u'/\sin U') = -0.868.$$

Tracing a paraxial principal ray right-to-left from this virtual exit pupil, we find the real entrance pupil and position for the actual diaphragm at $l_{pr} = -0.6346$ inch to the left of the first (concave) face of our meniscus.

We then calculate the three rays a, pr, and b of an oblique pencil at $U_{pr} = -15°$ through this iris with the adopted semi-aperture $Y = SA = 0.250$. The results are $X'_{ab} = 5.1281$ and the tangential curvature of field $X'_T = X'_{ab} - \text{final}$ $L' = +0.016$ inch. The tangential field therefore is hollow or over-corrected to the extent of about $1/60$ inch, not round as the analytical solution suggested by the insufficient value of a_2^2/a_1. This difference is due to the thickness of the meniscus lens, and it demonstrates the importance of trigonometrical verification in designing all these landscape lenses.

The coma comes out to have the insignificant value of -0.0006 inch and once more demonstrates the precision with which the optical sine theorem determines the position of the diaphragm which will yield freedom from coma.

The correction for freedom from coma is thus practically perfect, and the astigmatic correction is so close to the best possible compromise that it may be accepted as final. The lens, however, could not be used at the large aperture for which it was calculated, for the longitudinal spherical aberration $LA' = 0.400$ inch. On the other hand, the tolerance for longitudinal spherical aberration

$$LA'_p = 4 \text{ wave-lengths}/\sin^2 U'$$

works out, with $\log \sin U' = 8.6965$ as stated above and 4 wave-lengths $= 0.00008$ inch, at

$$LA'_p = 0.0324 \text{ inch.}$$

The actual aberration is thus 12 times the tolerance. As the tolerance refers to a difference of optical path and since spherical aberration expressed as OPD' grows with the fourth power of the aperture, the latter would have to be reduced to $\sqrt[4]{1/12} = 0.54$ of that used in the calculations, which was $2 \times 0.25 = 0.5$ inch.

The diaphragm would therefore have to be reduced to an aperture of 0.27 inch to [118] bring the spherical aberration to the Rayleigh limit. Photographic objectives, however, are never worked to anything approaching their theoretical resolving power and a considerably increased aperture would therefore be permissible, probably something in the neighbourhood of 0.4 inch.

The want of chromatic correction would be the principal outstanding defect of our simple meniscus. But if the best photographic focus is found by trial, the resulting chromatic diffusion of the image would only amount to about 1/100 of the clear aperture, or less than 0.005 inch, which would admit of the securing of images which would be reasonably sharp when viewed at 10 inches from the eye. The defect would, however, increase in the outer part of the field, as owing to the eccentric passage of the oblique pencils, there is also a chromatic difference of magnification.

(B) An Old-Achromat Lens

Evidently the next step in improving the landscape lens is to render it achromatic. For photographic purposes the achromatism should be such as to unite the D and G' rays, but we shall illustrate the application of our designing principles first by the visually achromatized cemented combinations of Chapter XI. To pick suitable bendings we must again use the criterion employed in the case of the simple meniscus: a_2^2/a_1 should be equal to the desired value of $a_3 - a_3*$. Assuming that a flat tangential field is to be aimed at, this means again

$$a_2^2/a_1 = a_3 + a_4/6.$$

Our analytical a_1 and a_2 were calculated for focal length $f' = 1$, so $a_3 = 1/2f' = 0.500$. But we must determine a_4 of our achromatic lens. The latter had $N_d = 1.5407$ for the crown and 1.6225 for the flint, and the respective total curvatures were 4.70 and -2.47; hence by the first form of the Petzval sum

$$a_4 = \frac{0.5407}{1.5407} \times 0.470 + \frac{0.6225}{1.6225} \times (-2.47) = 1.650 - 0.948 = 0.702.$$

We therefore have to look for bendings which give $a_2^2/a_1 = a_3 + a_4/6 = 0.500 + 0.702/6 = 0.500 + 0.117 = 0.617$. On page 523 we found for crown-in-front cemented objectives

$$a_1 = 1.198c_2^2 + 7.616c_2 + 11.48$$
$$a_2 = 0.849c_2 + 2.668$$

and by extending the series of values that were originally calculated, we easily find

$c_2 =$	-9	-7	-5	-1	0	$+1$
$a_1 =$	39.97	16.87	3.35	5.06	11.48	20.29
$a_2 =$	-4.973	-3.275	-1.577	1.819	2.668	3.517
$a_2^2/a_1 =$	0.619	0.636	0.742	0.654	0.620	0.610

We see that the terms with negative coma (a_2) give excessive values of the critical ratio and would therefore lead to a hollow field of the tangential pencils, but that

[118] $c_2 = 0$ gives almost exactly the desired value. This form is therefore worth trying, and the calculation may be made as an additional exercise. It will, on account of the positive value of a_2, lead to the unusual situation that the diaphragm is behind the lens, resulting in pincushion distortion.

We next try the critical ratio on the reversed Steinheil objectives (page 523), for which we found the values of a_1 and a_2 to be

$$a_1 = 1.198c_2{}^2 + 11.014c_2 + 24.85$$
$$a_2 = 0.849c_2 + 4.018.$$

These equations lead to the following results:

$c_2 =$	-9	-7	-6	-5	-3	-1	0	$+1$
$a_1 =$	22.76	6.454	1.894	-0.270	2.590	15.03	24.85	37.06
$a_2 =$	-3.623	-1.925	-1.076	-0.227	1.471	3.169	4.018	4.867
$a_2{}^2/a_1 =$	0.577	0.574	0.611	-0.191	0.835	0.668	0.650	0.639

(Note that $a_2{}^2/a_1$ becomes infinite at the two values of c_2 (-3.974 and -5.219) for which $a_1 = 0$ and it is negative between those values. It is zero when a_2 is zero; see Fig. 169.)

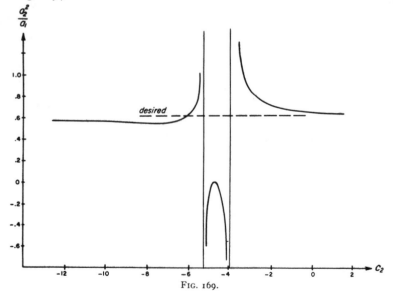

Fig. 169.

As a first example, $c_2 = 0$ was chosen, in spite of the rather large value of the ratio. This is the same lens as the one advised above for private study, but in the reversed position with the flint on the side of the distant objects. It was decided

to calculate for $f' = 10$ inches; the original c-values had therefore to be reduced   [118]
to one-tenth.

The contact-surface being flat, and the total curvature of the first lens, the flint, being -0.247, we have $r_1 = -4.05$ and $r_2 = \infty$, r_3 being left to be determined by the achromatic condition. The three-ray method was adopted, with initial $Y = 0.3$ inch.

$$N_y \text{ of crown} = [0.18820] \quad \text{of flint} = [0.21105]$$
$$N_v \text{ of crown} = [0.19075] \quad \text{of flint} = [0.21565].$$

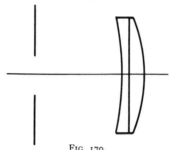

FIG. 170.

The thickness of the flint lens was chosen as 0.1 inch and that of the crown lens as 0.25 inch. The tracing of the three rays led to $r_3 = -2.258$ (Fig. 170) and gave the final

$$L'_y = 10.7701 \quad l'_y = 11.0034 \quad L'_v = 10.7677 \quad l'_y - L'_y = 0.2333 \text{ inch}$$
$$\log \sin U'_y = 8.45820 \quad \log u'_y = 8.45034 \quad u'/\sin U' = 0.98206.$$

These gave by the sine theorem the requisite position of the virtual exit pupil for freedom from coma at $l'_{pr} = -2.0013$. Tracing a paraxial ray from this position in the right-to-left direction gave the distance of the real diaphragm at $l_{pr} = -1.3660$ from the first lens surface. With the adopted semi-aperture of 0.3 inch, an oblique pencil at $-15°$ gave $L_a = -2.4856$, $L_{pr} = -1.3660$, and $L_b = -0.2464$. Tracing these three rays through the three surfaces gave the final data:

$$L'_a = -3.90719 \quad L'_{pr} = -2.01219 \quad L'_b = -0.51836$$
$$U'_a = -10\text{-}43\text{-}59 \quad U'_{pr} = -12\text{-}15\text{-}29 \quad U'_b = -13\text{-}47\text{-}48.$$

The closing formulae then gave

$$Y'_{ab} = 2.8162 \quad Coma'_T = +0.0001$$

or again an extraordinarily close correction of coma. The closing calculation further gave

$X'_{ab} = 10.9501$, and by difference with the above L'_y of the axial pencil,

$X'_T = +0.1800$ as the hollowness of the tangential field. This sense was to be expected from the excessive value of $a_2^2/a_1 = 0.65$ compared with the theoretical

[118] value of 0.617. Nevertheless the magnitude is greater than it should be, in accordance with the general statement already made that the trigonometrical test of a meniscus of finite thickness always gives a reduced roundness or increased hollowness of field as compared with the analytical prediction for an infinitely thin lens.

We can form a fairly definite judgement as to the permissibility of $X'_T = 0.18$ by determining the Petzval curvature of the lens at the calculated distance Y'_{ab} from the optical axis. Our actual lens, with $r_3 = -2.258$, has the total curvatures:

$$\text{of the flint lens} = -0.247$$
$$\text{of the crown lens} = 0 - (1/-2.258) = +0.443.$$

Note that the latter figure is much smaller than the 0.470 which would be right for an infinitely thin lens. With the true values we find

$$a_4 = \frac{0.5407}{1.5407} \times 0.443 + \frac{0.6225}{1.6225} \times (-0.247) = 0.1555 - 0.0948 = 0.0607.$$

Note the difference against one tenth of the thin-lens value of 0.702 (for unit focal length) used earlier!

We can now calculate closely enough the $\delta l'_{0_{Ptz}}$-value by the equation on page 722, namely,

$$\delta l'_{0_{Ptz}} = -\tfrac{1}{2} N'_0 H'_0{}^2 a_4$$

by putting $N'_0 = 1$ for air and $H'_0 =$ our Y'_{ab} as a near enough approximation. This gives

$$\delta l'_{0_{Ptz}} = -\tfrac{1}{2}(2.8162)^2 \times 0.0607 = -0.241$$

and means that if our lens were corrected for freedom from astigmatism the field would be rounded, at 2.816 from the axis, to the extent of -0.241 inch. In the discussion of the first fundamental law we concluded that $X'_T = -\tfrac{1}{2}$ of the Petzval curvature represented the maximum of over-corrected astigmatism which could be justified. Our value 0.180 is decidedly in excess of the maximum of 0.1205 now determined, and our lens therefore has more astigmatic over-correction than it should have. This could only be got rid of by shifting the diaphragm, but that would lead to the appearance of coma and would take us out of the frying pan into the fire! The only conclusion therefore is that the glass is not suitable.

(C) A Front Landscape Lens

There is a second flint-in-front solution in Fig. 169 at $c_2 = -6$, but the computation will be worked out for a slightly increased bending at $c_2 = -7$. This gives a combination of two deep meniscus lenses for, with flint in front, c_2 becomes $+7$ and therefore $c_1 = c_2 + c_a = 7 - 2.47 = 4.53$. Again adopting $f' = 10$ for the trigonometrical test, we obtain $r_1 = 1/0.453 = 2.208$ and $r_2 = 1/0.7 = 1.429$. In this case thicknesses of 0.12 and 0.30 were adopted in order to secure sufficient diameter for the deeply curved lenses. The third radius was determined by the $(d'-D')$ method for a semi-aperture of 0.5 inch. It gave $r_3 = 3.642$ and with it

Final $L' = 10.5222$ final $l' = 10.7430$ log sin $U' = 8.64950$ log $u' = 8.63869$

and with these, by the optical sine theorem, $l'_{pr} = +1.7600$. The real diaphragm [118] therefore must stand behind the lens.   By a paraxial right-to-left calculation, the virtual entrance pupil was located at $l_{pr} = 2.6953$ and gave for an oblique set of rays a, pr, and b, at initial $U = -15°$, Fig 171:

$$L_a = 0.8293 \qquad L_{pr} = 2.6953 \qquad L_b = 4.5613$$

and on emergence from the third surface

$$L'_a = 0.44965 \qquad L'_{pr} = 1.7477 \qquad L'_b = 2.78701$$
$$U'_a = -16\text{-}49\text{-}45 \qquad U'_{pr} = -19\text{-}9\text{-}26 \qquad U'_b = -21\text{-}30\text{-}6$$
$$Y'_{ab} = 3.0449 \qquad Coma'_T = +0.0011 \qquad X'_T = -0.0060.$$

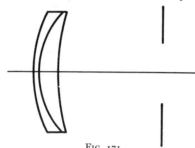

FIG. 171.

There is just a minute residual roundness of the tangential field, which need not worry us, and the coma also is far below the limit at which it could be recognized in a photograph.   This lens, however, is of the heterodox type with the diaphragm behind it, and it exhibits the corresponding pincushion distortion.   There are real objections to this type, and our solution is therefore chiefly of theoretical interest.

(D) NEW ACHROMATS AS LANDSCAPE LENSES

It was stated in the introduction of this chapter that the older achromatic land-scape lenses were less satisfactory than they might have been because a fetish was made of spherical correction.   It was not known that there exists in these lenses a close connection between the various aberrations such that good correction of coma and astigmatism calls for considerable spherical under-correction unless the lens is made very thick.   The consequence was that the comparatively good defini-tion in the centre of the field deteriorated rapidly towards the margin and produced an ill-balanced general effect which could only be avoided by stopping down to $f/32$ or even less.

Now an important requirement of any photographic lens is that the distance from the diaphragm shall be small, for otherwise a very large diameter will be required in order to secure a field as large as is usually expected, which is at least $60°$.   The modified eccentricity constant K' must therefore be small.   We found on page 778 that in a thin coma-free lens we have

$$a_3 - a_3{}^* = K'^2 a_1,$$

[118] and as $a_3 - a_3^*$ is of a fairly constant value, it follows that a small K' can only be secured by a very large a_1, that is, by heavy spherical under-correction. Moreover, we found that the most desirable value of $a_3 - a_3^*$ is equal to $a_3 + a_4/6$, and this shows that a small value of a_4 is the only means of keeping down the spherical under-correction. As 'new achromats' give this small, or even zero, value of a_4, we can draw the important conclusion that for landscape lenses the new achromat must be more favourable than the old achromat even with regard to spherical correction; the fact that the spherical aberration cannot be corrected in simple new achromats thus becomes a positive virtue as far as doublet landscape lenses are concerned.

The simple analytical preparation applied in the case of the simple meniscus and of old achromats is of very little use when new achromats are substituted, for on account of the low V-difference, the combinations become thick and they also always assume a very pronounced external meniscus form. Both circumstances lead to profound disagreements as regards focal length, achromatism, and oblique aberrations between the thin-lens theory and the thick-lens practice. A series of trigonometrical trials is therefore nearly always to be preferred.

For an example, the glasses

	Nd	$Nf - Nc$	$Nf - Nd$	$Ng' - Nf$	V
Chance 4873	1.6118	0.01037	0.00733	0.00590	59.0
Chance 458	1.5472	0.01196	0.00848	0.00707	45.8

were selected, giving, for ordinary photographic correction,

Nf for crown $= 1.61913 = [0.20928]$, and for flint $= 1.55568 = [0.19192]$

$Ng' - Nd$ for crown $= 0.01323$, and for flint $= 0.01555$.

Hence $V' = (Nf - 1)/(Ng' - Nd) = 46.80$ for the crown and 35.74 for the flint. For flint-in-front, the thin-lens paraxial formulae for achromatism give for a focal length of 10 inches.

$$c_a = -0.5814 \quad \text{and} \quad c_b = 0.6833.$$

Experience suggests that a bending which produces a contact-surface with a moderate positive curvature is usually arrived at. Hence a trial was made with the formula

$c_1 = -0.450 \qquad r_1 = -2.222$

$\qquad\qquad\qquad\qquad\qquad 0.10$

$c_2 = +0.1314 \qquad r_2 = 7.610$

$\qquad\qquad\qquad\qquad\qquad 0.25$

$(c_3 = -0.5519) \qquad r_3$ by solution

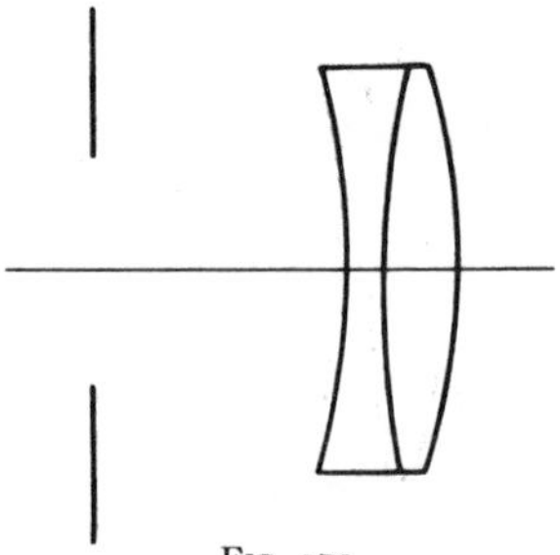

FIG. 172.

The first step again consists in tracing a paraxial and a marginal ray from the

distant object-point and finding the position of the exit pupil for zero coma by [118]
the Sine Theorem II.   The semi-aperture was selected as 0.35 inch.   Using the
$(d' - D')$ method, r_3 was found to be -1.984 (note that the thin-lens solution
indicated $-1/0.5519 = -1.812$), and the final data were

$$l' = 12.0930 \quad L' = 11.6549 \quad \log u' = 8.48413 \quad \log \sin U' = 8.49862$$

which give $l'_{pr} = -1.2594$ and $f' = 11.480$.

Tracing a paraxial ray from the virtual exit pupil so determined towards the
object space, the diaphragm was located at $l_{pr} = -0.8176$ by four-figure logs,
which are amply sufficient for this part of the work.   It was decided that this
solution for the diaphragm position should be taken on trust and that instead of
tracing the rays a, pr, and b, only a principal ray should be traced and the astigmatic
foci should be determined for a small aperture by the s and t formulae in section
[67].   The distant extra-axial object-point was in this case taken at $U_{pr} = -20°$
and led to a principal emerging ray under angle $U'_{pr} = -17$-5-48.   The subsequent calculations led to the final result:

$$s'_3 = \quad 12.402 \qquad t'_3 = \quad 12.389$$

and on calculating (22) and (24),

$$X'_s = \quad 11.818 \qquad X'_t = \quad 11.805 \quad (l'_k = 12.093)$$
$$H'_s = \quad 4.0263 \qquad H'_t = \quad 4.0223$$
$$R'_s = \quad -29.47 \qquad R'_t = \quad -28.09$$

As X'_s and X'_t have come out very nearly equal, we see that our lens is almost
free from astigmatism, but with a residual in the sense of under-correction because
the t-focus is the shorter one.   The field is moderately round for both astigmatic
fans, and a modification of the design is therefore called for.   By experience, lenses
of the new achromat type require bending towards a more pronounced meniscus
form and a corresponding shallower contact curvature in order to flatten the field
by over-corrected astigmatism.   A new bending should be tried with

$$c_1 = -0.53 \qquad r_1 = -1.887$$
$$c_2 = +0.0514 \qquad r_2 = +19.5$$

and will probably give a greatly improved result.

We can determine the Petzval radius of curvature of our worked example by the
usual formula $\Sigma(N' - N)/NN'r = 0.0353$, whence $R'_{Ptz} = -2.46f'$.   For a simple
crown lens, the factor would be $-Nf'$ or, say, $-1.517f'$, and for an old achromat,
about $-1.4f'$.   This shows that the substitution of a new achromat has greatly
reduced the curvature of the stigmatic field and that this type is vastly superior
as regards definition in the outer part of the field of view.

SPHERICALLY-CORRECTED ANASTIGMAT LANDSCAPE LENSES

CEMENTED combinations of at least three components can be simultaneously [119]
corrected for spherical aberration, for anastigmatic flatness of field, and

|119| approximately for coma. The first lens of the type was the Zeiss Convertible Anastigmat, with the following formula taken from von Rohr's *Theorie und Geschichte des photographischen Objektivs*, page 374:

$r_1 = -14.2$

       1.7 $Nd = 1.52246$

$r_2 = -5.5$

       0.5   1.56724

$r_3 = +20.7$

       1.4   1.61120

$r_4 = -12.6$

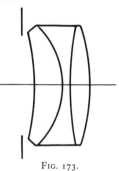

Fig. 173.

Semi-aperture $= 3.5$;   Distance of diaphragm $= 1.0$.

The dimensions are in millimetres and should give 100 mm. equivalent focal length. This lens had only *reduced* astigmatism but no anastigmatic zone. It could be improved by lowering the index of the meniscus next to the diaphragm and also that of the central lens, so as to keep the N-difference at the deep dispersive contact about the same whilst considerably increasing that at the shallow collective contact, for the latter is the chief source of the anastigmatic correction.

The Holostigmat came next, characterized by inverse sequence of the collective and dispersive contacts (Fig. 174). Suitable indices are—from the diaphragm

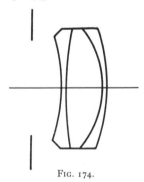

Fig. 174.

outwards—1.47 to 1.50 for the biconcave, 1.57 to 1.58 for the biconvex, and 1.623 (either barium or ordinary flint) for the meniscus. Suitable radii are approximately -1, $+2$, -0.52, and -1.04.

Considerably later Zeiss produced a quadruple lens with the following formula: [119]

$r_1 = -14.7$

       1.3 $Nd = 1.51743$

$r_2 = +23.4$

       1.7   1.61002

$r_3 = -11.2$

       1.7   1.51156

$r_4 = -7.4$

       1.3   1.58254

$r_5 = -15.4$

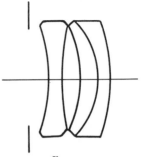

FIG. 175.

Semi-aperture = about 4.7;   Distance from diaphragm = 1.9.

In designing a triplet we must assume the first two radii and thicknesses in some reasonable way and then change the shape of the third lens until the last radius, which gives chromatic correction, also produces spherical correction, at any rate within, say, twice the Rayleigh limit, under-correction being usually preferable to over-correction.   We then calculate $OSC' = 1 - (u'_k l'_k/\sin U'_k \cdot L'_k)$; if this is of the order of the usual limit (0.0025), it is worth-while to test for flatness of field; but if OSC' greatly exceeds the limit, the coma would be too large to promise success.   It would be of no use to employ the solution for l'_{pr}, because, on account of the close approach to spherical correction, it would give not only a very uncertain value but also a wildly impossible one.   A certain residue of coma, especially in the more central parts of the field, is in fact almost inevitable in these triple and quadruple landscape lenses.

With regard to astigmatism and curvature of field, valuable preliminary information will be given by the relation between r_1 and the last radius.   If the latter is decidedly *shorter* than r_1, good anastigmatic correction is out of the question and it will be better to modify the combination further until the last radius is sufficiently near r_1 and if possible longer than r_1.   Then we assume a reasonable diaphragm position, about $0.1r_1$ for the Zeiss form and about $0.2r_1$ for those with a biconcave element next the diaphragm, trace a principal ray at $U_{pr} = -25°$ to $-35°$, and then trace s and t along this ray.   If a decidedly round or hollow mean field results, it will be best to change the thickness of the lens before proceeding further, increasing the total thickness if the field proves round, diminishing it if the field is hollow.   When a reasonable approach to flatness of mean field has been reached, we may add a second principal ray at a different angle in order to ascertain the location of the anastigmatic zone (if any!), and we should add rays a and b to one of the principal rays to determine the true coma.   When a reasonably good approximation has been secured, progress towards the final solution will be quick and systematic by the usual interpolation tricks.

Lenses of this type will give their best results only if fitted with what is usually called a 'reflection ring'; the name is meant to suggest (to the user of the lens)

[119] that the ring merely cuts off reflections from the metal-work in front of the diaphragm (Fig. 176), but in reality it performs more responsible duties demanded by the residual imperfections of the lenses: If the mount did not extend beyond the diaphragm, then at all large diaphragm openings, pencils not symmetrical with

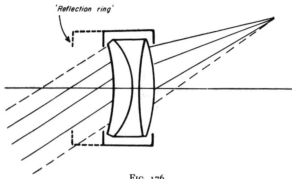

FIG. 176.

regard to the centre of the diaphragm would be admitted owing to the one-sided vignetting effect of the lens (Fig. 176). As the dotted reflection ring in this diagram clearly shows, it supplements the vignetting effect of the lens, but at the cost of a greater reduction in illumination; that is the part of the story which is mercifully withheld from the buyer of the lens!

CHAPTER XIX

SYMMETRICAL PHOTOGRAPHIC OBJECTIVES

A large and important class of photographic objectives derives great advantage both [120] to the computer and to the user from the 'symmetrical principle', which consists in placing two similar lenses or lens combinations symmetrically with reference to the centre of the limiting diaphragm of the system.

(*A*) HOLOSYMMETRICAL SYSTEMS

Strictly speaking, the full benefit of the symmetrical principle is obtained only if the symmetry is complete, i.e., if the two components are exactly similar and are made to the same scale and if object and image are also of equal size and at equal

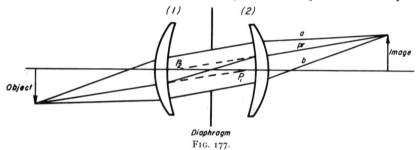

Diaphragm
FIG. 177.

distances from the diaphragm plane. Such a 'holosymmetrical system' will therefore be considered first (Fig. 177).

(1) *Distortion.* If we follow a principal ray, crossing the optical axis midway between the two similar combinations, outward in both directions, it is obvious that its two branches will meet the various corresponding surfaces at the same angles and distances from the axis and that the two branches will therefore emerge parallel to each other and will intersect the axis, when produced rearwards, at points P_1 and P_2 equidistant from the centre of the system. Consequently the object and image planes, being also equidistant from the centre of the system, will be cut at equal distances from the optical axis whatever the inclination of the principal ray may be. There will therefore be perfect equality of object and image for principal rays at any obliquity, and therefore there will be no distortion.

(2) *Coma.* If in the above diagram we consider the two rays parallel to each other, which pass through the upper and lower limits of the diaphragm and therefore at equal distances from the centre of symmetry, it follows from the symmetry that the lower ray in its progress towards the right must meet the successive surfaces at the same distance from the optical axis and under the same angles as does the upper ray in progressing to the left; the same reasoning applies to the remaining parts of the two rays. Therefore the figures formed by these two rays at the left and

[120] the right of the diaphragm are congruent, and the two rays must therefore intersect at equal distances from the central diaphragm and from the optical axis. From either of the intersecting points so defined, we can assume any number of other rays starting out towards the lens system. Among these there must be one which will pass through the centre of the diaphragm. To this particular ray the reasoning of the preceding paragraph (on distortion) applies, and it proves that this ray must pass through the other, conjugate, intersecting point of rays a and b. We therefore, on account of the symmetry, obtain three approximately equidistant rays which after refraction cut one another in the same point and are therefore free from coma. Our perfectly symmetrical system is therefore free from coma for unit magnification, regardless of the individual correction of the two similar components. These may have heavy coma individually, in which case the middle ray will not be parallel to rays a and b in the space between the components.

(3) *Chromatic Difference of Magnification*, or different-sized images in different colours. The absence of this defect in a holosymmetrical lens is easily proved, for the preceding arguments apply equally to any colour of the light; hence we obtain the same magnification (-1) for all colours. In most types of lenses, a white principal ray travelling from the centre of the system through one component is spectrally decomposed in such a way that the different-coloured principal rays cut the image plane at different distances from the optical axis; hence there are residues of the above aberration in simple landscape lenses and in the vast majority of unsymmetrical lenses. The symmetrical arrangement completely removes it.

We must now consider how far the above invaluable properties of a strictly symmetrical arrangement remain intact when some of the conditions are not fulfilled.

(B) A Symmetrical System Working at Unequal Conjugate Distances

Let the system still be holosymmetrical, but consider the case when object and image are at different distances from the centre of the system.

(1) *Distortion*. The two branches of any principal ray in the object and image spaces will still be parallel to each other, and object and image would be similar (and therefore the image would be free from distortion) if all the principal rays at varying obliquities in the object space cut the optical axis in the same point P_1 and all those in the image space cut the axis at the same point P_2, for then all the triangles formed by optical axis, principal ray, and object-plane would be in one and the same proportion to their corresponding triangles in the image space. Evidently this is equivalent to demanding that the single components must be free from spherical aberration for the principal rays crossing the centre of the system, and that is therefore the condition for complete freedom from distortion in our case B.

(2) *Coma*. When the object- and image-planes are at different distances from the centre of the system, the image-forming rays a and b will no longer be parallel in the space between the two components; our argument as to freedom from coma therefore breaks down, but it is easy to see that a very considerable advantage will still result from the symmetrical arrangement, and in fact it is found in practice

that there is very little coma in holosymmetrical objectives (excepting those of high [120] aperture ratio and large separation) for almost any relation between object and image.

(3) *Chromatic difference of magnification.* The reasoning applied in the case of distortion still applies; strict freedom from this defect demands that the single components shall be achromatic with reference to the centre of the system. Although this condition cannot usually be strictly fulfilled, the resulting defect is nearly always totally insensible.

(C) HEMISYMMETRICAL SYSTEMS

These are systems in which the two components are similar in every way, but are made to a different scale. By similar arguments to those already employed (except that the figures are proved to be similar instead of congruent) it is easily shown that:

(1) *Distortion* will be strictly and unconditionally absent for a relation of object to image in the same proportion as that between the adjacent single components. In all other cases, it is bound up with the condition of freedom from spherical aberration of the single component with regard to the centre of the system, just as in the case of holosymmetrical systems.

(2) *Coma* is only diminished, not corrected. But there is usually one particular pair of conjugate object- and image-planes for which there is complete correction of coma. In the case of symmetrical systems of high aperture, it is sometimes important to find by trigonometrical trial by the three rays a, pr, and b that proportion of scale between the front and back components which will thus completely correct for coma in the most important use of the lens—usually for distant objects.

(3) *Chromatic difference of magnification* in hemisymmetrical systems requires achromatism of the single component with reference to the centre of the system for its strict correction. The residual of this aberration is almost invariably insensible.

Symmetrical systems also offer great advantages to the computer as to spherical and chromatic correction of the axial pencils. If this has been established for the single components for a bundle of parallel rays coming from the direction of the diaphragm, it is usually practically perfect for the combined system under all conditions under which it is likely to be employed.

Naturally the highest degree of benefit is obtained when the single components are themselves reasonably free from coma, for then we obtain practically the full advantage of the symmetrical type even with strongly hemisymmetrical combinations. Single lenses of this type lend themselves to the construction of convertible sets which can be used either singly or in any combination of two, with great advantage to the owner of such a set. The two Zeiss anastigmats specified in the previous chapter are of this favourable type.

The properties of the symmetrical type demonstrated above have been completely worked out only in comparatively recent years. For a long time, the conditions for freedom from distortion were not fully realized; on the contrary, it was taken as a matter of course that any symmetrical lens must be completely free from distortion for any conjugate distances of object and image.

[120]   The freedom from spherical aberration of the single components with reference to the diaphragm and its images (the latter being the pupils of the system), which we recognized as an additional condition for perfect freedom from distortion, can become of real importance in certain circumstances, such as the reproduction of maps to varying scales, and it will then profoundly affect the type of the lens.

In the case of the simple meniscus types with which we began our study of photographic lenses, this condition can be easily fulfilled by selecting the distances from the diaphragm and correspondingly modifying the thickness.

The calculation of holosymmetrical objectives is greatly simplified by the automatic removal of distortion, coma and chromatic difference of magnification which results from the principle of symmetry. We can carry the calculation close to completion by dealing with the back component only. We select that bending of the component which corrects its spherical aberration for parallel rays coming from the side of the diaphragm, and we then seek that distance of the diaphragm which secures the desired astigmatic correction, usually a flat tangential field. Two components so corrected will make a perfect combination at a magnification of -1. For other magnifications, the variations of the corrections are usually so small that the tracing of a pencil from a distant object right through the complete system either reveals no objectionable aberrations or calls only for a slight final adjustment of the bending and the position of the diaphragm.

RECTILINEAR LENSES

[121] THE components of rectilinear lenses, which are of the type shown in Fig. 178, are built up of two meniscus lenses cemented together and are usually of very moderate thickness. As they are to be spherically corrected, we have $a_1 = 0$; the coma is therefore independent of the diaphragm position and the astigmatism a_3 with diaphragm in contact is changed by diaphragm-shift according to the law

$$a_3{}^* = a_3 + 2K'a_2.$$

As a compact shape of the complete objective is desirable, the diaphragm must not be very far from the lens and that means a small value of K'. Obviously correction of the big central astigmatism of a thin lens will then be possible only if a_2 is large, and it is this requirement which decides the choice of glass. With the glass combination used in Chapter XVIII, the spherically-corrected forms had only a small fractional value of a_2 whilst $a_3 = 0.5$; this combination of glass is therefore out of the question. We must choose one which leads to an aberration-parabola dipping deeply below the horizontal axis of zero aberration. With the old glasses, this object was attained by using light flint of about 1.57 index and dense flint of about 1.61 index. Later on, ordinary light crown in combination with light barium flints was frequently substituted. The real requirement is a low V-difference accompanied by a considerable N-difference. Suitable glasses for trial would be

	Nd	V
Chance 407	1.5787	40.8
Chance 572	1.6182	36.4

and it would probably be best to find the spherically-corrected form of the rear [121]
component by the trigonometrical method of Part I, but at full aperture and with
determination of r_3 by the $(d' - D')$ method.   The full aperture should be selected
so that the angle of incidence of the marginal ray at the contact surface is about 35°,
and the first trial may always be made with $c_1 = -\frac{1}{2}c_a$.   If this comes out over-
corrected, c_1 must be suitably increased, say by $-0.1c_a$, and vice versa.   By
linear or parabolic interpolation the spherically-corrected form will be quickly
found.

The diaphragm position is probably most easily found by tracing three parallel
rays at about $U_{pr} = -25°$, one cutting the axis at the first lens surface, and the other
two cutting the axis at half the lens diameter and at the whole lens diameter,
respectively, to the left of the first surface.   Calculate the intersection-distance
L'_{ab} of rays (1) and (2), and also that of rays (2) and (3); these will give two values of
the tangential curvature of field for a diaphragm at one quarter and three-quarters,
respectively, of the lens diameter from the first surface.   Linear interpolation
will then give the approximate diaphragm position for a flat tangential field.   The
rest is merely a matter of finishing touches.   As an example, we take from von
Rohr's *Theorie und Geschichte des photographischen Objektivs*, page 298, the back
component of one of Steinheil's holosymmetrical 'Aplanats';

$r_1 = -38.3$

$\qquad$ 2.00 $Nd = 1.58027$

$r_2 = -11.0$

$\qquad$ 1.05   1.61912

$r_3 = -26.3$

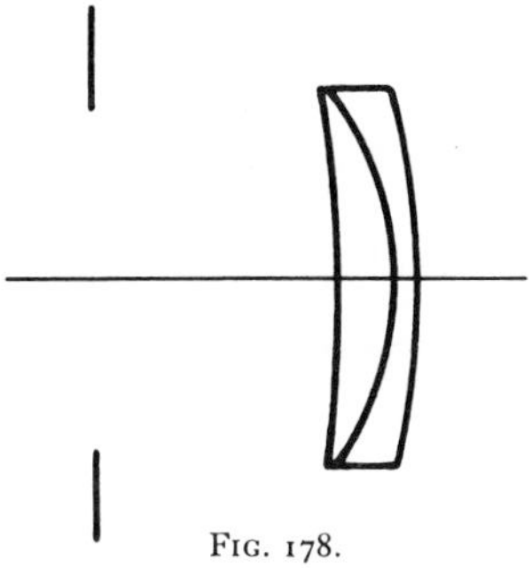

FIG. 178.

Semi-aperture, about 6.0 (make a drawing first!).   Diaphragm distance, 9.0.

CONVERTIBLE ANASTIGMATS

THE principal use, at the present time, of well-corrected landscape lenses consists [122a]
in selling them in sets of two, three, or even more of graduated focal lengths, all
fitting the same mount, so that any one can be used alone or any two together as a
hemisymmetrical doublet.   As was shown in the section on symmetrical objectives,
hemisymmetrical objectives will be free from coma only if the single components
are themselves coma-free, and as a good landscape lens has the latter property to a
fair approximation, it is highly suitable for this purpose.   Two such lenses of focal
lengths 2.0 and 1.4 will together (owing to the separation!) give a focal length about
1.0, thus giving a choice of three focal lengths in almost uniform geometrical
progression, so that almost any view, whether of wide or narrow angular extent,

[122a] may be made to fill a given plate. Three lenses in somewhat closer succession give three different doublets or a choice of six focal lengths.

There is one feature in these sets which prevents them from giving the full perfection which could be attained with fixed mounts, each corrected for the best result on the given plate size, and that arises from the fact that an anastigmat really is entitled to its name for only one zone in the field. It is easily shown that if this zone is laid into its best position for the doublet of a two-lens set, it lies far outside the field of the single components that is actually used. Specifically, the doublet will usually work with a total field of 50° to 60° and the anastigmatic zone should therefore lie near half of this value, or $U'_{pr} = 25°$, say. The single components will cover roughly 40° and 27° respectively, but *their* anastigmatic zones will, like their fields, be measured on the diaphragm side of the lens, where the principal rays have a bigger slope than on the other side. Therefore, if the ray entering the doublet at $U_{pr} = -45°$ emerges with $U'_{pr} = -35° \, 45'$, a total field of $2 \times 35\frac{3}{4} = 71\frac{1}{2}°$ of the doublet would correspond to the same principal ray as a field of 90° of the single rear component, whereas the single component would work with the *smaller* field when used alone. It follows that if—as is usual—the single components are anastigmatically corrected with a view to the best result in the doublet, then they will be anastigmatically under-corrected when used alone on the same plate.

It will be seen to be an important point to bear in mind that when single components are designed with a view to their use in symmetrical doublets, then the anastigmatic zone must be measured by the angle of the *emerging* and not by that of the entering principal ray.

NON-CONVERTIBLE SYMMETRICAL ANASTIGMATS

[122b] WHEN it is *not* required that the single components of a symmetrical doublet should be well corrected for coma so as to admit of use as landscape lenses by themselves, we are rid of the restriction of reasonable freedom from coma in the single component and we can use this added liberty to obtain more favourable results with regard to spherical aberration, astigmatism, and flatness of field. The procedure then is to seek first the form of rear component which gives spherical correction and achromatism. We then trace three rays at equal intervals from a suitable extra-axial object-point, usually at $U_{pr} = -30°$ to $-35°$, locate the intersections of ab and of bc, and by linear interpolation find the diaphragm position giving a flat tangential field if possible. If both pairs of rays give a round field, the lens must be made thicker, in the converse case, thinner. Closer study by the s and t method for two zones of the field must then follow, with further modifications of the construction so as to attain favourable location of the anastigmatic zone and smallness of the departures from the ideal flat field at $\frac{2}{3}$ of that zone. Addition of the rays a and b to the more oblique principal ray will disclose any serious magnitude of the H'^2 spherical aberration. Finally, an axial and oblique pencil from infinity must be traced through the complete, at first holosymmetrical, doublet. If this discloses an objectionable residue of coma, the easiest way of removing this consists in cautious reduction of scale of the second component (by about 10 per cent. for a first shot). If very high correction of distortion is of im-

portance, then the single component must be tested for freedom from spherical [122b] aberration of the *principal rays* (Fig. 179) before proceeding to the study of the doublet. This will be found a very awkward restriction, and something may have to be sacrificed in other respects in order to satisfy this rather unusual condition. Nearly always it means a diaphragm *very close* to the first surface. Dr. Rudolph's

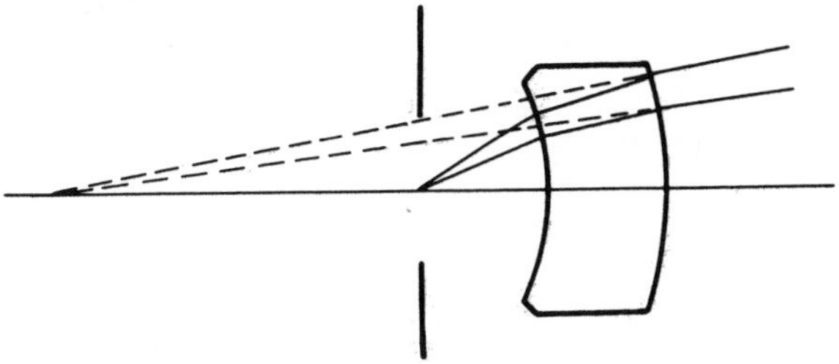

FIG. 179.

'Ortho Protar' is an example (from Gleichen, *The Theory of Modern Optical Instruments*, English ed. of 1921, p. 273) with the following prescription:

$r_1 = -19.04$		*Nd*
	1.06	1.49833
$r_2 = +30.31$		
	3.64	1.58950
$r_3 = -\ 9.15$		
	1.10	1.62210
$r_4 = -20.83$		

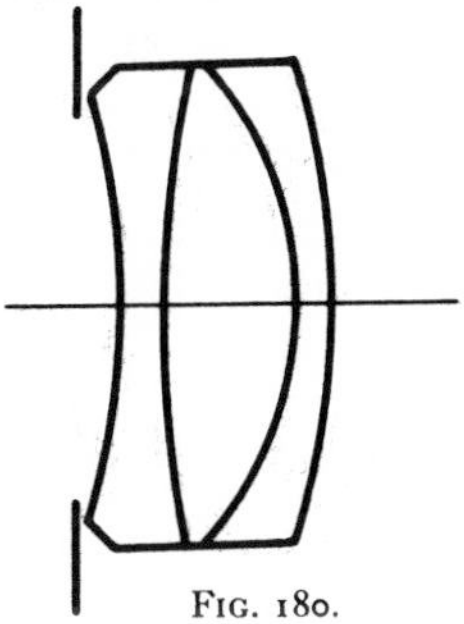

FIG. 180.

Semi-aperture, about 5; Diaphragm distance, 1.41.

The following two examples are of lenses which do not satisfy the condition for complete correction of distortion:

The Goerz Dagor, Fig. 181 (a)

$r_1 = -20.5$		*Nd*
	1.9	1.51497
$r_2 = -\ 8.3$		
	0.8	1.56804
$r_3 = +22.7$		
	3.1	1.61310
$r_4 = -19.1$		

Semi-aperture, about 5.8.
Diaphragm distance, 2.3.

A Gauss-type lens, Fig. 181 (b)

$r_1 = -3.04$		*Nf*
	0.46	1.5733
$r_2 = -\ 6.40$		
	0.09 (air)	
$r_3 = -14.90$		
	0.54	1.5184
$r_4 = -\ 3.85$		

Semi-aperture, 1.5.
Diaphragm distance, 0.76.

[122b]

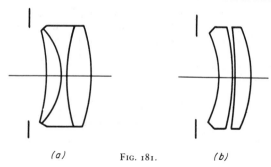

(a)   FIG. 181.   *(b)*

SYMMETRICAL ANASTIGMATS OF EXCEPTIONALLY WIDE ANGLE

[123] IN order to appreciate the curious departure from the usual types in the few exist-ing types of extreme wide-angle objectives, we must first study the question of *tolerance for astigmatism.* In an astigmatic pencil (Fig. 182), the sagittal rays are out of focus with reference to the tangential rays by the whole $s' - t'$ difference, and this evidently holds for any point between S and T; but for points at the left of T,

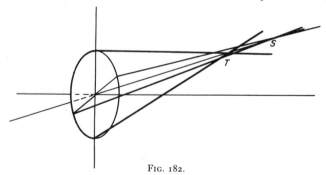

FIG. 182.

the S-rays will be out of focus by a *greater* amount than $s' - t'$, and for points at the right of S, the T-rays will be out of focus by more than $s' - t'$. Hence positions of the focusing screen outside the S-to-T range are definitely worse than positions within that range. Our focal-range tolerance was fixed at *twice* the permissible out-of-focus distance; therefore the Rayleigh tolerance for the $s' - t'$ distance is half of the focal range or $\frac{1}{2}$ wave-length$/\sin^2 U'_m$. The tolerance for ordinary spherical aberration is four times the focal range. From that fact follows the important deduction that a given amount of astigmatic difference of focus is *eight* times as serious in its effect on definition as an equal amount of ordinary spherical aberration. If that fact were better known, there would be more good photographic objectives about!

The few extreme-angle objectives are illustrative of this relation; they give [123] exceptionally low astigmatism with good flatness of field, but have an amount of spherical under-correction which is simply staggering to people who still make a fetish of spherical correction.

The eight-times ratio follows from the Rayleigh limit and is undoubtedly right for amounts of aberration up to perhaps twice the Rayleigh limit. For the much larger amounts of both aberrations admissible in the majority of photographic objectives, the relations are more complicated and are not yet fully ascertained. But it may be taken as quite certain that the ratio is always very high, and with respect to *resolving power* for bold detail probably even higher than eight. On the other hand, for rendering delicate contrasts, this ratio may possibly prove to be lower than eight because astigmatism does not tend to scatter light into a general and extended halo, as spherical aberration does.

We may at any rate take it that if we succeed in diminishing the maximum of astigmatism in a given lens at the expense of not more than four times as much addition to the longitudinal spherical aberration, then we have *improved* the objective with reference to definition for its field as a whole.

Now the astigmatic zones of ordinary so-called anastigmats are due to the presence of a considerable amount of secondary astigmatism growing as H'^4 which can only be reduced to the minimum effect represented by the zones by playing out against it an appropriate amount of primary H'^2-astigmatism and a considerable residue of Petzval curvature. The higher astigmatism arises from the very oblique passage of the extra-axial pencils through one or more of the refracting surfaces, and these great angles of incidence are inevitable if spherical correction is insisted upon because spherical correction is practically impossible with deep meniscus lenses. If we drop the demand for spherical correction, then we can take advantage of the obvious fact that the oblique aberrations arising at any one surface can be reduced to any required extent by so placing the diaphragm and so choosing the radii of curvature that pencils of almost any obliquity will pass nearly radially. It

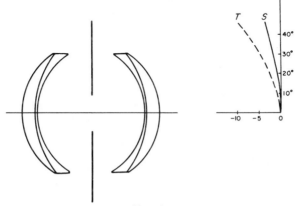

FIG. 183.

[123] is this simple conclusion which is acted upon in the extreme wide-angle objectives, and it apparently points out the only way in which satisfactory correction can be secured *beyond* the usual $\pm 45°$ limit of available field.   As under this condition of nearly radial passage of the oblique pencils there will be very little secondary H'^4-astigmatism, the sagittal and tangential fields will have very nearly simple parabolic curvature, and this by close fulfilment of the Petzval condition and by small departures from strictly radial passage can be almost perfectly flattened into the ideal plane.

The first objective on this principle was the 'Pantoskop', designed in 1865 (25 years before the first modern anastigmat!) by Emil Busch, who was primarily a manufacturer of spectacle lenses and such like in the German centre of that industry—Rathenow.   The formula given by von Rohr (page 281) is, for $f' = 100$ of the complete objective,

Back component (holosymmetrical), Fig. 183

Diaphragm distance, 6.2

$$r_4 = -7.63 \qquad Nd$$
$$0.19 \quad 1.6079 \text{ (flint)}$$
$$r_5 = -10.39$$
$$1.74 \quad 1.5331 \text{ (extra light flint)}$$
$$r_6 = -7.45$$

Semi-aperture of stop, about 2 or a little less.   Lens diameter, about 13.

The first remarkable property of this objective is that the Petzval sum is so nearly zero (0.00006) that it seems certain that the lens data were deliberately fixed with full knowledge and observance of the theorem, a most remarkable fact!   The data evidently are not the best possible, for von Rohr's calculations of a number of s and t pencils at angles up to $45°$ through the complete objective show a round field for t' and s', everywhere almost exactly at a $3:1$ distance from the ideal image plane. This means that a small alteration would yield a close approximation to a flat anastigmatic field.   It was found that a reduction of the diaphragm distance from 6.2 to 6.0 reversed the field curvature, making it hollow by about half the amount that it was originally round; hence a diaphragm distance of 6.08 should be about right.   The lens is certainly remarkable, and it would be excellent practice to try to raise it to its best performance by moderate changes.   The reasons for its comparatively small success were probably (1) that it was too slow for most purposes at its normal aperture of $f/32$ with the very insensitive wet collodion plates, and (2) that in 1865 a spectacle factory could not produce, centre, and mount the tremendously deep and frail meniscus lenses with a sufficiently close approach to the high precision demanded; the great majority of the specimens would therefore fall far short of the calculated performance.   There does not seem any room for doubt that the Busch Pantoskop was the first, and a highly creditable first, true anastigmat.

The only other extreme-angle objective was issued by Goerz about the beginning

of the present century (computer E. von Höegh) under the name of the 'Hypergon'. [123]
The prescription given by Gleichen (p. 278) is:

Back component (holosymmetrical),

Diaphragm distance, 6.9

$r_3 = -8.63$

2.2   Borosilicate crown, $Nd = 1.5105$

$r_4 = -8.57$

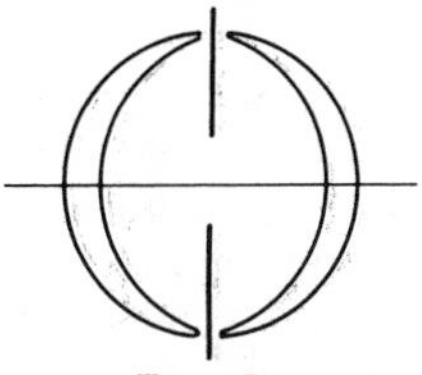

Fig. 184.

This lens also has a very low Petzval sum. It shares with the Pantoskop the
characteristic of very heavy positive spherical aberration, but unlike its predecessor
it is not even achromatic! At $f/32$ or thereabouts it covered a total field of about
130° with very uniform and quite good definition. The illumination in the image
dropped so greatly in the outer part of the field, however, that part of the exposure
had to be given with a rotating star-shaped diaphragm in front of the lens.

CHAPTER XX

UNSYMMETRICAL PHOTOGRAPHIC OBJECTIVES

[124] THE design of unsymmetrical systems is difficult and laborious because systematic attention must be paid throughout to the three aberrations (coma, transverse chromatic, and distortion) which the symmetrical principle removes automatically and usually with ample completeness.

Formally, the various solutions given in previous chapters represent methods for finding the proper form of *any* lens system, but the work implied would be so extensive as to render these methods practically useless without further specialization. This will be in the direction of a breaking-up of the formidable direct solution for seven unknowns, *viz.*, the seven primary aberrations (with a strong prospect of failure owing to imaginary roots coming in), into several simpler steps involving not more than two or three unknown quantities at a time. Several examples will be given of this method, and if properly understood, they will serve as guides in seeking a proper subdivision in other cases.

SYSTEMS WITH TWO SEPARATED COMPONENTS

IF an optical system is to be free from all primary aberrations, it must satisfy seven conditions, for the five Seidel aberrations for brightest light must all be zero and there must also be achromatism of the focal point and achromatism of magnification. One additional datum is used up in adjusting the system for focal length. Hence we must have a total of *eight* variable data in a lens system if all the conditions are to be satisfied by a purely mathematical solution.

The curvatures of the surfaces are practically the only data which can be executed in an almost unlimited range of positive and negative values and which thus satisfy the mathematical conception of a freely disposable variable. Thicknesses and separations are limited (in our usual left-to-right computing direction) to positive and, moreover, to reasonably modest values; for that reason, it is always risky to introduce these data into an analytical solution for there is a considerable probability that the values required for *formally* satisfying the equations to be solved are negative. The same objection applies in a still higher degree to refractive indices, which are limited to a range of only about $+1.47$ and $+1.65$, a ridiculously small part of the mathematical range from $-\infty$ to $+\infty$; hence any attempt to entrust the selection of suitable refractive indices to a formal equation or set of equations is almost certain to lead to impossible results. For these reasons, it is always desirable to frame the scheme of a solution so that only surface curvatures are put at the disposal of the mathematical mill. Even then we may be disappointed by being offered imaginary solutions or radii so excessively short as to reduce the possible aperture to a uselessly small size.

It will be seen that such considerations prevent the purely mechanical solution of the more complicated optical problems by solving a set of hard-and-fast equations. Success largely depends on preliminary study leading to a promising

rough scheme which can then be tested by an analytical solution with considerable [124]
prospects of success.

(*A*) Two Thin Positive Components at Finite Separation

IF we begin with the simplest case of two thin cemented lenses (Fig. 185), the [125]
disposable data are six radii and one separation, or seven in all; therefore we are
short by one liberty and cannot hope to produce complete correction by a straightforward solution.   In general we shall have to sacrifice one condition; this in the
old days was always the Petzval condition because no practical way was known of
meeting it anyhow.

In this case of two thin components, the designing problem is greatly simplified
by considering first the problem of achromatism.   Ordinary longitudinal chromatic
aberration could be corrected by combining an over-corrected first component with
a suitably under-corrected second, or vice versa, but that would lead to different
magnifications for different colours (Fig. 186).   The different-coloured components

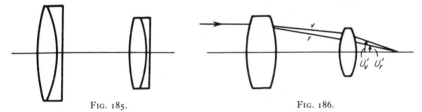

Fig. 185.   Fig. 186.

of an entering white ray would emerge from the second component at different
incidence heights; they would therefore arrive at the achromatic focus with $U'_r \neq$
U'_v and that, by the sine condition, would mean a different magnification in different
colours.   Hence *the two components of such a system must be separately achromatized*
if a large field is to be covered.   As was shown in the microscope chapter, this
condition can be violated to a considerable extent when the field is small; but we are
now aiming at the large field of photographic objectives and of eyepieces, and the
condition is then highly important.

Limiting ourselves to strict achromatism of the first component, we can now
adopt a suitable focal length for it; $f'_A = 1$ is usually the most convenient choice.
On selecting glass, we can then study the bendings of this component, generally
most rapidly by the TL-method, but the primary aberration method or the
strict trigonometrical one may sometimes be preferable or even necessary if a deep
meniscus form is likely to result.   The problem is then reduced to finding:

(1) The right bending of the selected first component,
(2) The right separation,
(3) The proper strength $1/f'_B$ of the second component,
(4) The correct bending of the second component, which also must be strictly
     achromatic.

By these four liberties we can correct the outstanding aberrations: Spherical, Coma,

[125] a chosen compromise as to Curvature of Field, and Distortion. The number of conditions is thus reduced from seven to four.

To complete the solution, we now take advantage of the fundamental laws of oblique pencils: As our final system is to be free from spherical aberration *and* coma, shifts of the diaphragm will have no effect on these two aberrations nor on the astigmatism, and we can place the diaphragm wherever we like. We choose the position in contact with the second component; then the latter will have the fixed astigmatic constant $a_{3B} = 1/2f'_B$ and *no distortion*. Hence—and this is the crucial point of the reasoning—the system as a whole will be free from distortion if the first component delivers to the centre of the second component pencils which are free from distortion. Therefore correction of the distortion depends on the selection of such a bending of the first component and such an associated separation d'_A that for $l'pr_A = d'_A$ the first component gives $a^*_{5A} = 0$, and this condition supplies the key to the final solution. If $a_{1A}, a_{2A}, a_{3A}, a_{4A}$ are the aberration constants for any bending of the first component when the diaphragm is in contact with it, we shall have $a_{5A} = 0$ and hence

$$a^*_{5A} = 3(a_{3A} + \tfrac{1}{6}a_{4A})K' + 3a_{2A}K'^2 + a_{1A}K'^3,$$

and equating this to zero, we obtain a quadratic equation in K' from which two values of K' are obtained for each bending to which this solution is applied. From these values, the corresponding $l'pr_A = d'_A$ is immediately obtainable by the definition of K'. Any selected bending of the first component can therefore be used only with the $l'pr_A$ found for it. It will be found that only bendings with a moderate concavity of the third surface give real solutions. For the first trial, a bending should be selected which gives a reasonably short separation d'_A. For this bending, we then calculate

$$a^*_{3A} = a_{3A} + 2K'a_{2A} + K'^2a_{1A}.$$

As this combines directly by algebraic addition with the $a_{3B} = 1/2f'_B$ of the second component and must give a flat field, we have now a means of fixing the requisite focal length of the second component. We must have

$$a^*_{3A} + 1/2f'_B = -a_4/6$$

for the whole system. The a_{4A} is known and has already been used. The a_{4B} will depend on f'_B and may be put equal to $1/kf'_B$, where k can be deduced from the relation of a_{4A} to f'_A; alternatively, it may usually be taken as about 1.4 for the kinds of glass (crown and light flint) usually employed in systems of this kind. We therefore have an equation

$$a^*_{3A} + 1/2f'_B = -a_{4A}/6 - 1/6kf'_B,$$

which gives a solution for f'_B. A second component of this focal length placed at the separation obtained from the condition $a^*_{5A} = 0$ will therefore give a flat field free from distortion, *no matter into what form the second component is bent* to correct spherical aberration and coma. It therefore only remains to trace an axial pencil trigonometrically through the first component and then to solve for that bending of the second component which will give spherical correction and fulfilment of the

sine condition.   If we break the contact of the second component, that solution [125]
will always be possible and will lead to the Petzval or Dallmeyer type of portrait
lens (Fig. 188) or the two possible crosses or hybrids according to the crown-flint
sequence in the two components.   For a cemented rear component, we cannot
expect an exact solution at the first shot for *both* spherical aberration *and* coma; we
must then try another bending of the first component in the same manner and
eventually interpolate.

The work must of course be completed by direct trigonometrical test of the
solution for an oblique pencil; but this should be traced through the actual position
of the diaphragm (about midway between the components is the usual position).
The solution will provide all the data for differential correction.

An actual numerical example will best illustrate the steps that must be taken to
design a lens of this type.   We shall design a lens similar to the original Petzval
portrait lens, using two glasses which are close to those actually used by Petzval,
namely:

	Nd	V	$Nf-Nc$
Crown:	1.517	60.0	0.00861
Flint:	1.575	41.5	0.01385

We arbitrarily assign unit focal length to the front component, whence its c_a and c_b
for paraxial achromatism may be determined in the usual way to be $c_a = 6.278$,
$c_b = -3.902$.

As we have seen, the equation to be used in solving for K' is

$$3(a_{3A} + a_{4A}/6) + 3a_{2A}K' + a_{1A}K'^2 = 0,$$

in which a_1 is the sum of the spherical G-sums, a_2 is the sum of the coma G-sums,
$a_3 = 1/2f'$, and $a_4 = \Sigma(1/Nf')$ as usual.   Using the stated glass indices, and with
an object at infinity, the G-sum formulae give for the cemented front component
with stop in contact:

$$a_1 = 25.989 + 11.779c_2 + 1.2163c_2{}^2$$
$$a_2 = 4.1447 + 0.8586c_2$$
$$a_3 = 0.500$$
$$a_4 = 0.715$$

At this point it is necessary to choose a likely bending for the front component,
and as this choice is arbitrary within limits, we shall start by dividing c_a equally
between c_1 and c_2, giving:

$$c_1 = 3.139 \qquad r_1 = 0.3185$$
$$c_2 = -3.139 \qquad r_2 = -0.3185$$
$$\text{TL } c_3 = +0.763 \qquad (r_3 = 1.3106)$$

With this value of c_2, the five aberration coefficients of the chosen front component
are found to be $a_1 = 0.9993$, $a_2 = 1.4496$, $a_3 = 0.500$, $a_4 = 0.715$, and $a_5 = 0$.
On inserting these values into the quadratic equation for K', we obtain the two
solutions:

$$K' = -0.480 \quad \text{or} \quad -3.871$$

[125]   Since $1/K' = 1/l' - 1/l'_{pr}$, with $l' = 1$, we find

$$d' = l'_{pr}r_A = 0.324 \quad \text{or} \quad 0.794$$

respectively. This is the value of the air-space which will give freedom from distortion when the diaphragm is in contact with the rear system.

We must next determine the power of that rear component which, when situated at this 'distortionless' position, will give a flat tangential field. We begin by calculating

$$a^*_{3A} = a_{3A} + 2K'a_{2A} + K'^2a_{1A} = 0.500 + 2.8992K' + 0.9993K'^2,$$

and for the two values of K' this becomes

$$a^*_{3A} = -0.6614 \quad \text{or} \quad +4.2513$$

respectively. Again following the argument outlined above, and assuming that $k = 1.4$, we can now solve for the focal length f'_B of the second component to give a flat tangential field, by

$$a^*_{3A} + \frac{1}{2f'_B} = -\tfrac{1}{6}a_{4A} - \frac{1}{6kf'_B},$$

giving the two values of f'_B corresponding to the two values of K' as

$$f'_B = 1.1414 \quad \text{or} \quad -0.1416$$

respectively. The first solution is the one we are seeking, as the second would represent an extreme telephoto lens. With $f'_A = 1$ and $f'_B = 1.1414$ and using the calculated air-space $d' = 0.324$, the focal length of the whole system is found by

$$1/F' = 1/f'_A + 1/f'_B - d'/f'_A \cdot f'_B$$

to be $F' = 0.628$. The rear image-distance or 'back focus' is 0.4245.

The construction of the rear component is now arbitrary, provided only that it has the focal length and position just determined, and provided that it is achromatic and is capable of correcting the spherical aberration and OSC' of the whole system. We must therefore begin our solution by tracing a marginal and paraxial ray through the front component at a suitable aperture, which in the original Petzval design was about $f/3.5$. Thus the clear aperture of the front component must be $0.628/3.5 = 0.180$, leading to the choice of suitable thicknesses of 0.035 and 0.013 for the crown and flint elements respectively. The marginal ray provides the necessary data for exact achromatism of the front component by solving for r_3 by the $(d' - D')dN'$ method, which in this example came out to be 1.2008 instead of the thin-lens value of 1.3106.

For the rear component, the original design of Petzval used a closely air-spaced doublet with flint leading, as shown in Fig. 188 (a). We can conveniently solve for the bendings of these two elements by an application of the matching principle in the final image space. The ordinates of our matching graphs will represent respectively the contributions of the front component plus the rear negative element, and of the rear positive element alone, the abscissae being c_5 for bendings of the rear negative element and c_6 for bendings of the rear positive element.

To determine the G-sums of the two elements of the rear component, we must first ascertain the values of $c_c = -3.4186$ and $c_d = +5.5002$ for achromatism, and we note that the object-distance for lens c is equal to the focal length of the front component minus the air-space, namely, $1.0 - 0.324 = 0.676$. Hence the 'v'_2' for lens c and the 'v_1' for lens d will be given by

$$v'_c = v_d = v_c + (N-1)_c\, c_c = -0.4867.$$

The spherical G-sums are then found to be

$$(SGS)_c = -2.2306c_5{}^2 + 10.8153c_5 - 20.1265$$
$$(SGS)_d = 3.2963c_6{}^2 - 26.9512c_6 + 79.2978$$

and the coma G-sums are likewise

$$(CGS)_c = -1.6068c_5 + 4.0314$$
$$(CGS)_d = 2.3590c_6 - 10.0217$$

On page 329 of Part I, we found that the G-sums can be converted into contributions in the final image-space by the relationships

$$SC' = l'_k{}^2 \frac{SA^4}{SA_k{}^2} \cdot \Sigma(SGS) \qquad CC' = H'_k \cdot SA^2 \cdot \Sigma(CGS).$$

For a semi-field of $10°$, $H'_k = 0.628 \tan 10° = 0.11073$. We have seen that the SA_A of the front component is 0.09, and a simple triangle indicates that the SA_B of the rear component will be

$$SA_B = SA_A \cdot \left(\frac{f'_A - d'}{f'_A} \right) = 0.06084.$$

The contributions of the *front* component are therefore

$$SC'_A = (0.4245)^2 \frac{(0.09)^4}{(0.06084)^2} \Sigma(SGS)_A = 0.003192$$
$$CC'_A = (0.11073)(0.09)^2 \Sigma(CGS)_A = 0.001300.$$

These are to be added to the contributions of lens c, namely,

$$SC'_c = -0.001488c_5{}^2 + 0.007215c_5 - 0.01343$$
$$CC'_c = -0.0006586c_5 + 0.001652$$

giving

$$SC'_{(A+c)} = -0.001488c_5{}^2 + 0.007215c_5 - 0.01024$$
$$CC'_{(A+c)} = -0.0006586c_5 + 0.002952.$$

These equations are now used to plot the bending curves on the left-hand side

[125] of the graph (Fig. 187) with c_5 as abscissae, the right-hand side being plotted for lens d by

$$SC'_d = l'^2_k \cdot SA^2_d(SGS)_d = 0.002199c_6^2 - 0.01798c_6 + 0.05290$$

$$CC'_d = H'_k \cdot SA^2_d(CGS)_d = 0.000967c_6 - 0.004108.$$

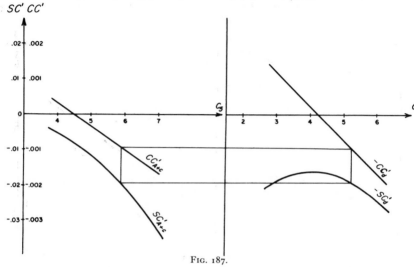

Fig. 187.

The plotting data are found for several bendings as follows:

$c_5 =$	4	5	6	7
$SC'_{(A+c)} =$	-0.00519	-0.01136	-0.02052	-0.03265
$CC'_{(A+c)} =$	0.000318			-0.001658

$c_6 =$	3	4	5	6
$SC'_d =$	0.01875	0.01616	0.01797	0.02418
$CC'_d =$	-0.001207			$+0.001694$

The data for the left-hand curve are plotted with the correct signs, but those for the right-hand curve are plotted with reversed signs so that the various contributions will be equal and opposite at the four corners of the rectangle. These points were found from the graph to fall at

$c_5 = 5.93$ $r_5 = 0.1686$ | $c_6 = 5.25$ $r_6 = 0.1904$
hence $c_4 = 2.5114$ $r_4 = 0.3981$ | TL,$c_7 = -0.2502$ $(r_7 = -3.997)$

Assuming that the diameter of the rear component is the same as that of the front, the same thicknesses will be suitable, and the complete lens specification [125] becomes:

$r_1 =$ 0.3185

    0.035

$r_2 = -0.3185$

    0.013

r_3 (by $d-D$) =   1.20078

air-space   0.265 $\left\{\begin{array}{l}\text{to give a distance of 0.324 between the} \\ \text{adjacent principal points of the front and rear} \\ \text{components}\end{array}\right.$

$r_4 =$ 0.3981

    0.013

$r_5 =$ 0.1686

    0.0034 to allow edge contact of the lenses

$r_6 =$ 0.1904

    0.035

r_7 (by $d-D$) $= -3.579$

On continuing the marginal and paraxial rays from the front component through this rear component, the spherical aberration is found to be $-0.001\ 106$, and the OSC', $-0.003\ 94$, which is equivalent to $Coma'_s = -0.000\ 424$ at a $10°$ field.

If we assume that the unit length is 10 inches, which will make the focal length of the system 6.28 inches, the Rayleigh limit of spherical aberration will be 0.0038. The amount actually found is -0.011 for this focal length, so we should try to improve the LA' (and also the OSC') by a differential correction. This may be easily performed by differentiating the four equations for $SC'_{(A+c)}$, SC'_d, $CC'_{(A+c)}$, and CC'_d, by which we find the slopes of the curves at the four corners of our matching rectangle as

$$\frac{\partial SC'}{\partial c_5} = -0.0104 ; \quad \frac{\partial SC'}{\partial c_6} = 0.00511 ; \quad \frac{\partial CC'}{\partial c_5} = -0.000659 ; \quad \frac{\partial CC'}{\partial c_6} = 0.000967$$

(once more, for unit focal length). Using these slopes and making a solution for $\Delta LA' = +0.001$, $\Delta Coma'_s = +0.00042$, we find the required differential bendings of the two rear components as

$$\Delta c_5 = 0.175 \qquad \Delta c_6 = 0.554.$$

These changes give the corrected solution as $r_4 = 0.3721$, $r_5 = 0.1638$, and $r_6 = 0.1722$, the last radius r_7 being found by the $(d'-D')dN$ solution to be $+2.8125$. The air-space for edge contact is now 0.00155. Using these values we find:

$$LA' = +0.000\ 69; \qquad OSC' = +0.001\ 17\ (Coma'_s = +0.000\ 126).$$

These corrections have obviously gone too far, as we actually required only 55 per cent. as large a change of LA', and only 75 per cent. as large a change of

[125] coma. Reducing the required $\Delta LA'$ and $\Delta Coma'$ by these two proportions and making a new solution, we find:

$$r_4 = 0.3728 \qquad \text{Results:}$$
$$\qquad 0.013 \qquad f' = \quad 0.6237$$
$$r_5 = 0.1641$$
$$\qquad 0.00206 \qquad l'_7 = \quad 0.3945$$
$$r_6 = 0.1756$$
$$\qquad 0.035 \qquad LA' = \; -0.000\,06$$
$$r_7 \, (\text{by } d-D) = 4.4559$$
$$OSC' = \; -0.000\,30$$

The final test of this solution is to trace a $10°$ principal ray through the system, chosen so as to cross the axis at a point approximately midway between the two components, and to calculate the sagittal and tangential fields along it. The results of this final calculation bear out the analytical predictions with surprising exactness and reveal that the lens has a flat tangential field and is remarkably free from distortion:

$$X'_s = \; -0.005\,78$$
$$X'_t = \; -0.000\,63$$
$$Ptz = \quad 1.306\,0$$
$$X'_{Ptz} = \; -0.007\,91$$
$$Dist' = \; +0.000\,07 \; (0.06 \text{ per cent.})$$

The following three examples of photographic objectives of unsymmetrical type with comparatively thin components are taken from von Rohr's *Theorie und Geschichte des photographischen Objektivs*, pages 250, 215, and 302. They are all stated for a focal length of 100, and have an aperture of about $f/3$ (Fig. 188):

	(a) *Petzval Portrait,* 1841		(b) *Dallmeyer Portrait,* 1866		(c) *Steinheil Portrait Aplanat,* 1875	
$r_1 =$	52.9		50.0		51.9	
		5.8		6.2		2.1
$r_2 =$	-41.4		-43.5		25.1	
		1.5		1.6		5.2
$r_3 =$	436.2		250		150	
		46.6		33.3		56.4
$r_4 =$	104.8		-533.3		275.2	
		2.2		5.0		6.45
$r_5 =$	36.8		-33.3		-25.1	
		0.7		1.7		2.1
$r_6 =$	45.5		-29.8		-72.9	
		3.6		1.5		
$r_7 =$	-149.5		-59.7			
	Crown 1.517		Glass probably similar		Crown 1.5147	
	Flint 1.575		to Petzval's		Flint 1.5750	

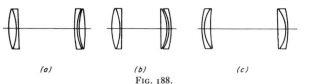

(a)   *(b)*   *(c)*

Fig. 188.

(B) A SIMPLE TELEPHOTO LENS

IN this type of lens, the positions and powers of the two components are usually [126] determined at the outset by the given conditions, namely (a) the telephoto effect, that is the length of the system from front vertex to film plane expressed as a fraction of the focal length, and (b) the separation between the two components.

In most telephoto lenses, the telephoto effect lies between about 0.75 and 0.85, the lower value having the great advantage of compactness, especially in long focal lengths, but the higher value leading to lenses with a flatter field with less over-correction of the Petzval sum.   As regards separation, it can be readily shown that the power of the negative component reaches a mathematical minimum when it is placed midway between the positive component and the final image-plane; how-ever, this optimum condition leads to rather a long lens, which is undesirable in the longer focal lengths although quite acceptable in small lenses for miniature or motion-picture cameras.

For an example, we shall adopt the classical arrangement of a positive and negative component of equal (unit) focal length, the adjacent principal points of which are to be separated by a distance $d'_A = 0.5$.   In this case, the focal length of the system will be 2.0, the thin-lens back focus will be 1.0, and the total length will be 0.75 of the focal length.   A suitable aperture for such a lens is $f/5.6$, to cover a semi-field of about $14°$.

In making our preliminary analytical solution, we may place the diaphragm at any convenient position, and we shall imagine it to be situated at the front (positive) thin component.   The principal ray will then be refracted only at the rear com-ponent, giving a value of $l'_{pr} = -0.333$ in our example (Fig. 189).   Thus for lens A, the stop will be in contact, whilst for lens B, the value of K' corresponding to the assumed stop position will be

$$K' = \frac{1}{\dfrac{1}{l'} - \dfrac{1}{l'_{pr}}} = \frac{1}{1 - \left(\dfrac{1}{-0.333}\right)} = \frac{1}{1+3} = 0.25.$$

Now the sag X'_T of the tangential field is made up of two contributions, namely, $3H'_0{}^2[a_3 + a_4/6]_A$ from lens A and $3H'_0{}^2[a_3 + 2K'a_2 + K'^2a_1 + a_4/6]_B$ from lens B. We shall take $a_3 = 1/2f'$ and $a_4 = \sum(1/Nf') = 1/1.4f'$ as before, and since the two focal lengths are now equal and opposite, we have $a_{3A} = -a_{3B}$ and $a_{4A} = -a_{4B}$. Hence the sag of the tangential field becomes simply

$$X'_T = 3H'_0{}^2[2K'a_{2B} + K'^2a_{1B}]$$

and for this to be zero with $K' = 0.25$ we have the final condition:

$$(a_1/a_2)_B = -8.$$

[126] It is thus possible to secure a flat tangential field with any form of the front component and any choice of glasses in the rear component by merely selecting that bending of the rear component for which $a_1/a_2 = -8$. This includes the special

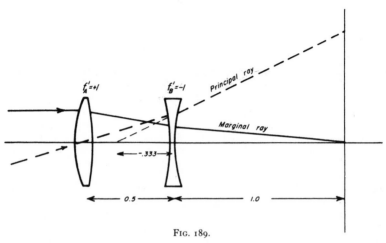

FIG. 189.

case in which the rear component is aplanatic, when of course the front must be aplanatic also.

As regards distortion, the front component contributes nothing, but lens B gives $Dist' = -H'_0{}^3 \cdot a^*_5/l'_0{}^2$, where

$$a^*_{5B} = 3K'[a_3 + a_4/6]_B + 3K'^2a_{2B} + K'^3a_{1B}.$$

Now in our example the data for the rear component are $a_3 = -0.5$, $a_4 = -0.7143$, $a_3 + a_4/6 = -0.6190$, $K' = 0.25$, hence

$$a^*_5 = 0.015625a_{1B} + 0.1875a_{2B} - 0.4638.$$

In the case when both components are aplanatic, $a_1 = a_2 = 0$ and $a^*_5 = -0.4638$. At an obliquity of $10°$, $H'_0 = 2\tan 10° = 0.3526$; $l'_0 = 1.333$, whence $Dist' = (0.3526)^3 \times (0.4638)/(1.333)^2 = 0.01144$, which is 3.2 per cent. of the image height in the pincushion sense. As many of the early telephoto lenses were made with well-corrected components, the positive member often being a Petzval portrait lens, normally used on the camera, distortion of the pincushion type and of serious magnitude became an accepted and inevitable property of telephoto lenses. However, by departing from aplanatism of the separate components and deliberately introducing a large amount of negative spherical aberration into the rear, it is possible to reduce the distortion radically, and many recent systems have been designed which show no greater distortion than an ordinary photographic objective.

We can readily see that the thin-lens condition for absence of distortion can be [126] combined with that for a flat tangential field as follows:

$$a_2 + 0.125a_1 = 0 \qquad \text{for a flat field}$$

and $\qquad 0.1875a_2 + 0.015625a_1 = 0.4638 \qquad$ for zero distortion.

Solving these equations simultaneously gives $a_1 = -59$, $a_2 = +7.4$ as the most desirable values for the spherical and coma G-sums of the rear component.

For an $f/5.6$ lens of focal length 2.0, the SA_A of the front component is $2/11.2 = 0.1785$, and the SA_B of the rear is half of this or 0.08925. By the formula $a_1 = \Sigma SC' \cdot u'_0{}^2 / SA^4$, in which $u'_0 = 0.08925$ and SA^4 for the rear component is 0.00006345, we calculate $\Sigma SC'$ for the rear as being -0.4699. Similarly, for coma we note that $a^*_2 = \Sigma CC'/H'_0 \cdot SA^2 = a_2 + K'a_1 = 7.4 - 59 \times 0.25 = -7.35$, whence $\Sigma CC' = -0.02064$.

To work out the design in detail, it is best to ignore achromatism at first because we shall have to use the choice of glass as a degree of freedom. We shall merely refer to the two types of glass by their approximate refractive indices in F light, namely, 1.522 for the crown and 1.612 for the flint in both components, these values being chosen because there are several types listed with approximately these index values and a variety of dispersions. For each component we must now use as our degrees of freedom the c_a and c_2, solving for c_b by

$$\frac{(1/f') - (N_a - 1)c_a}{(N_b - 1)}$$

for use in working out the G-sums. The coma G-sum, being linear, can be equated to the desired value of a_2 to give a solution for the bending parameter, after which the spherical G-sum, a_1, can be found and compared with its desired value. A few trials will give the desired powers and bendings of each component, ignoring thickness effects for the present.

This procedure has been followed, but the details will be omitted here. First, the following values of c_c were selected and the corresponding values of c_d for a focal length of minus unity were found. In each case the value of c_5 was solved to make the coma-sum a_2 equal to $+7.4$, and the corresponding value of the spherical sum a_1 was found:

c_c	c_d for focal length	c_5 for coma	a_1 calculated
−7.0	+4.3366	0.5785	−74.42
−7.5	4.7630	1.2516	−71.37
−8.5	5.6160	2.6322	−63.82

The last trial was adopted for further work. Actually we are not interested primarily in the a_1 and a_2 of the rear component, but only in its contributions to the tangential field curvature and distortion which should have the values respectively equal and opposite to the corresponding contributions from the front component with stop in contact, namely,

$$(X'_T C')_B = -3H'_0{}^2[a_3 + a_4/6]_A \quad \text{and} \quad DC'_B = 0.$$

[126] Since $H'_0 = 0.3526$, $a_{3A} = 0.5$, and $a_{4A} = 0.71$, we see that

$$(X'_T C')_B = (PC' + 3 \cdot AC')_B = -0.2306.$$

The third trial above led to the following rear component, after including suitable thicknesses:

$$r_4 = -0.1704$$
$$0.03$$
$$r_5 = +0.3799$$
$$0.11$$
$$r_6 = -0.4229 \quad \text{(by solution for } f' = -1\text{)}.$$

A paraxial ray and a paraxial principal ray were traced through this rear system with the desired $l = +0.5$ and $l_{pr} = -0.5$ relative to the front principal point, and the Seidel aberration contributions were determined for the thick system as

$$AC'_B = -0.1431; \quad PC'_B = -0.0633; \quad DC'_B = -0.0115,$$

giving $\qquad\qquad X'_T C' = +0.4926.$

As this rear component is of a strongly meniscus form, we ought not to be surprised to find that the tangential field is much more hollow than we had intended and that the distortion is not zero. The next step is to make small trial changes of the r_4 and r_5, using r_6 always to maintain the focal length of minus unity, and to determine coefficients which will enable us to secure the desired values of $X'_T C'$ and DC'. A few trials led to the corrected rear component:

$$r_4 = -0.2165 \qquad\qquad SC' = -0.1969$$
$$0.03 \quad N = 1.522 \qquad CC' = -0.0112$$
$$r_5 = +0.4690 \qquad\qquad AC' = -0.0603$$
$$0.11 \quad N = 1.612 \qquad PC' = -0.0547 \Big\} X'_T C' = +0.2356$$
$$r_6 = -0.6061 \qquad\qquad DC' = -0.0037$$

The field curvature and distortion now having about the desired values, it was decided to accept this TL solution for the rear component and to proceed to build up a front component having $SC' = +0.1969$ and $CC' = +0.0112$. The corresponding G-sums desired for the front were therefore

$$a_1 = SC' \cdot u'^2_0 / SA^4 = 0.1969 \times 0.007966/(0.1785)^4 = 1.5450$$
$$a_2 = CC'/H'_0 \cdot SA^2 = 0.0112/0.3526/(0.1785)^2 = 0.9969.$$

Following the former procedure and solving each trial bending for the desired value of a_2, the following TL system was derived:

$$c_a = 4.5; \qquad c_b = -2.2042; \qquad c_2 \text{ for coma} = -1.7232; \qquad a_1 = 1.4224.$$

On applying thicknesses and making small corrections, the following front system was finally found:

$$r_1 = 0.3601 \qquad\qquad SC' = +0.2220$$
$$0.14 \quad N = 1.522 \qquad CC' = +0.0112$$
$$r_2 = -0.5803 \qquad\qquad AC' = +0.0810$$
$$0.03 \quad N = 1.612 \qquad PC' = +0.0408 \Big\} X'_T C' = -0.2838$$
$$r_3 = 1.6353 \text{ (by solution for} \qquad DC' = +0.0008$$
$$f' = +1\text{)}.$$

The actual air-space between the two components was found to be $0.5 + l'pp_3 - lpp_4$ [126]
$= 0.5 - 0.15782 + 0.09233 = 0.43452$.

Taking a likely final diaphragm position at $l'pr_3 = 0.2$ and tracing true $10°$ and $14°$ principal rays through the whole system, the tangential and sagittal fields were found to be as follows:

Angle	Dist'	X'_s	X'_t
$10°$	-1.4%	-0.0217	-0.0607
$14°$	-2.4%	-0.0231	-0.1000

The tracing of an $f/5.6$ marginal ray gave $LA'_6 = +0.00513$ and $OSC'_6 = -0.00128$.

The trigonometrical correction of this type of analytical solution is best made in steps because not all the coefficients are sufficiently linear for triple differential solutions to be useful. It was decided to use a bending δc_A of the whole front and a change δc_2 in r_2 alone to correct the LA' and OSC' of the whole system, the l'_{pr} in the OSC' expression being found by a paraxial principal ray traced from the true stop position at $l'pr_3 = 0.2$. By a few trials the coefficients were found to have the approximate values:

$$\frac{\partial LA'}{\partial c_A} = +0.53 \quad \frac{\partial LA'}{\partial c_2} = +0.17 \quad \frac{\partial OSC'}{\partial c_A} = -0.020 \quad \frac{\partial OSC'}{\partial c_2} = -0.020.$$

Application of these ratios led to an improved front component resulting in a small residual of over-corrected spherical aberration and zero coma (actually by chance only c_2 needed to be changed) but which had almost the same distortion and field curvature as before:

$$r_1 = \quad 0.3601 \qquad\qquad LA' = -0.0066$$
$$0.14$$
$$r_2 = -0.5565 \qquad\qquad OSC' = +0.00014$$
$$0.03$$
$$r_3 = \quad 1.6655 \text{ (for } f' = 1)$$

A $10°$ principal ray gave $Dist' = -1.41$ per cent., $X'_T = -0.0549$.

To improve the field, the rear component was now given a small bending by $\delta c_B = -0.3$, which naturally re-introduced some spherical aberration and coma; these were again removed by application of the almost linear coefficients above, giving as our next approximation:

$$r_1 = \quad 0.3457 \qquad\qquad LA' = -0.0043$$
$$0.14$$
$$r_2 = -0.5018 \qquad\qquad OSC' = -0.00005$$
$$0.03$$
$$r_3 = \quad 1.4393$$
$$0.44081 \text{ (air)}$$
$$r_4 = -0.2033$$
$$0.03$$
$$r_5 = \quad 0.5458$$
$$0.11$$
$$r_6 = -0.5328$$

The $10°$ principal ray gave: $Dist' = -1.28$ per cent., $X'_T = -0.0048$.

[126]   This was considered close enough to justify the insertion of real glasses in place of the nominal glasses used so far.   Reference to the 1926 Chance catalogue, which has been used for all the examples in this book, shows that the following glasses have refractive indices close to the adopted values.   We may assume that the V-number of each glass will remain constant as the index varies; hence the dN of each glass was first adjusted to give its equivalent value for an index of 1.522 or 1.612 respectively.

	Crowns				Flints		
Glass	Nf	dNf	Adjusted dN for $N = 1.522$	Glass	Nf	dNf	Adjusted dN for $N = 1.612$
646	1.51426	0.00778	0.00790	8894	1.61812	0.01148	0.01137
6493	1.52167	0.00794	0.00794	4153	1.61537	0.01412	0.01404
7742	1.52518	0.00812	0.00807	4626	1.63489	0.01635	0.01576
1961	1.51833	0.00804	0.00810	5953	1.61730	0.01650	0.01636
4990	1.51575	0.00807	0.00817	1034	1.61552	0.01655	0.01646
1203	1.52148	0.00839	0.00840	1068	1.62057	0.01689	0.01666
605	1.52352	0.00844	0.00842	3743	1.62434	0.01713	0.01679
9322	1.52466	0.00851	0.00847	370	1.62365	0.01716	0.01684
9663	1.52652	0.00858	0.00851	572	1.63033	0.01759	0.01708
9071	1.52546	0.00866	0.00860	361	1.63371	0.01786	0.01725
3071	1.52767	0.00867	0.00858	360	1.63487	0.01794	0.01729
1066	1.52115	0.00881	0.00882				
569	1.52161	0.00899	0.00900				

The $(d' - D')$ values of the two components were next determined and, using the adjusted dN values, possible pairs of glasses were selected to make each sum $\Sigma(d' - D')dN'$ separately equal to zero.   When several choices were available, those having Nf as close as possible to the original values were taken.   The radii of curvature were then adjusted in such a way that the slope-angles of the paraxial ray in each lens or air-space were maintained at their previous values.   In this way the Gaussian optics of the lens remained unchanged.   The marginal ray was traced to investigate the LA' and OSC', small changes in the c_A and c_2 being made if necessary to restore those aberrations to their desired values.   The $(d' - D')$-sum was also calculated to verify the correctness of the choice of glasses.   Principal rays traced at $10°$ in D and G' light enabled the transverse chromatic aberration to be found, and the astigmatism was also calculated along principal rays in F light at $10°$ and $14°$ respectively.   The final lens and all its aberrations may be summarized as follows, Fig. 190:

	Glass	Nd	V

$r_1 = $ 0.3471

   0.14   7742 1.5194 63.0

$r_2 = $ −0.5030

   0.06   1034 1.6041 37.8

$r_3 = $ 1.4356

   0.4399 (air)

$r_4 = $ −0.2030

   0.03   1066 1.5149 57.9

$r_5 = $ 0.5714

   0.11   1034 1.6041 37.8

$r_6 = $ −0.5360

$LA' = $ −0.0101 at $f'/5.6$

$OSC' = $ +0.00018

$LZA' = $ +0.0066

$\Sigma(d' - D')dN' = $ −0.24 wave-lengths
$f' = $ 2.0

Field	Dist'	X'_S	X'_T	Tch'
10°	−1.29%	+0.0077	−0.0032	+0.00116
14°	−2.17%	+0.0219	+0.0211	+0.00305

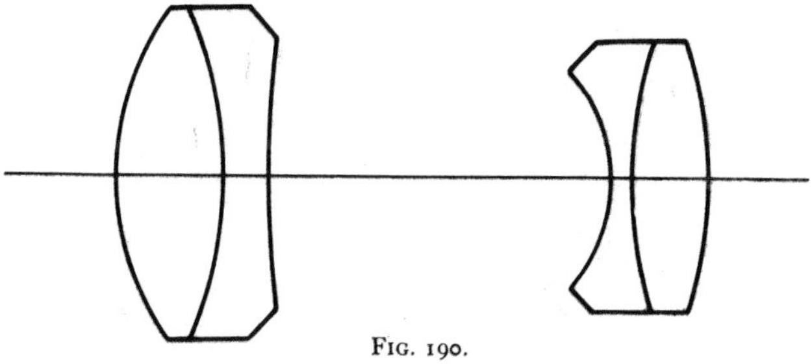

FIG. 190.

The outstanding fault of this objective is the large residue of transverse chromatic aberration in the under-corrected sense, and the next move to perfect the design would therefore be a redistribution of the chromatic correction between the two components. A reasonable trial would be to select the dispersions of the glasses so as to make the $(d' - D')$-sum of the front component about $+4$ wave-lengths and that of the rear component -4 wave-lengths (assuming the decimetre as the unit). Moreover, as the field has a decidedly hollow tendency, a small positive bending (say $\delta c_B = +0.03$) of the rear component might also be tried. By following these suggestions and touching up the other aberrations in accordance with the plan already outlined, a very satisfactory design for a telephoto lens could be achieved.

SYSTEMS WITH THREE SEPARATED COMPONENTS

A COOKE TRIPLET OBJECTIVE

THIS interesting design was first developed by H. Dennis Taylor in 1893 under [127] the name of the 'Cooke' lens, and it represents the simplest possible anastigmat (Fig. 191). It has eight degrees of freedom, namely, three powers, three bendings,

[127] and two air-spaces, which are just sufficient to secure the desired focal length and to 'correct' the seven primary or Seidel aberrations.   The types of glass may be chosen arbitrarily within reasonable limits.   This lens, with its various modifications, has become the most popular type of anastigmat for moderate-priced cameras, both on account of its great simplicity and economy in manufacture and because of its many satisfactory qualities.

By 'correcting' an aberration is meant, of course, reducing its magnitude to any desired value, positive or negative, which may be needed to compensate the inevitable higher-order aberrations, so as to yield the best compromise over the aperture and field for which the lens is to be used.   The complete elimination of any *primary* aberration is generally useless as the higher orders of that aberration would then be present in their full magnitude; invariably a residual of each

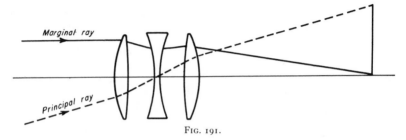

Marginal ray

Principal ray

FIG. 191.

primary is deliberately designed into the lens so that the complete aberration found by actual ray-tracing will show the best possible balance of over- and under-correction.

The design problem, therefore, involves first a decision as to the probable required magnitudes of all the primary aberration residuals; then by use of the thin-lens equations given in section [61], it is possible to solve for the three lens powers, the three bendings, and the two air-spaces to give these assumed residuals; and lastly, after inserting finite thicknesses into the elements, the actual tracing of a sufficient number of rays will enable us to determine the true values of each of the seven aberrations.   If any of these is too large or too small, an error of judgment is indicated in the selection of the primary residuals, and judicious changes in these must be made until ray-tracing reveals that the desired state of correction has been reached.

With a triplet of this type, any one change in construction will in general affect all aberrations; nevertheless, by proceeding in a systematic way, some changes may be used to affect only single aberrations, and the very difficult general problem may be reduced to a series of logical and simple steps.   If these are carefully followed, even a beginner in lens design will be able to reach a satisfactory construction with relatively little computational labour.

To begin the design, therefore, the three powers and two separations of the lens elements are selected to control (*a*) the focal length, (*b*) the Petzval sum, (*c*) the longitudinal chromatic aberration, and (*d*) the transverse chromatic aberration, since these four properties depend only on the powers and separations. The three [127] bendings will then be left for adjusting the spherical aberration, coma, and astigmatism. The seventh aberration, namely distortion, presents a difficulty as it depends on both the powers and the bendings, but as it depends mainly on the powers, we shall assign it to the preliminary TL section of the work, remembering that its final value will vary somewhat with the bendings actually adopted.

THE PRELIMINARY THIN-LENS STUDY

In the thin-lens stage of the design, we first decide on the three types of glass, select two of our five degrees of freedom arbitrarily, and then solve for the remaining three by the conditions for (a) the focal length of the system, generally $f' = 10$, (b) the desired primary longitudinal chromatic aberration Lch', and (c) the desired primary transverse chromatic aberration Tch', which is usually but not always zero. We assume that the diaphragm is situated at the middle lens of the system.

The formulae used for this solution can be derived in various ways, but the algebra becomes simplified if we assign arbitrary values for u'_a and u'_b, the slopes of the paraxial ray in the two air-spaces of the system, assuming a very distant object and assuming that $SA_a = 1$. If we add to these data the power of the system, $\phi = u'_0$, we can solve algebraically for the heights of incidence of the paraxial ray at the second and third lenses, SA_b and SA_c respectively, to satisfy the conditions for longitudinal and transverse chromatic aberration. Having ascertained these values, the powers and air-spaces of the system are found by the following relationships:

$$(1) \quad \begin{aligned} \phi_a &= u'_a \\ \phi_b &= -(u'_a - u'_b)/SA_b \\ \phi_c &= (\phi - u'_b)/SA_c \end{aligned} \qquad \begin{aligned} d'_a &= (1 - SA_b)/u'_a \\ d'_b &= (SA_b - SA_c)/u'_b \end{aligned}$$

a. *Transverse chromatic aberration.* As we have seen in Seidel Aberrations V, on page 359 of Part I, the contribution of each element to the transverse chromatic aberration is given by

$$TchC'^* = LchC' \cdot Q \cdot u'_0$$

and by equation (e) on page 327, we found that for each element

$$LchC' = SA^2 \cdot c \cdot \delta N / u'_0{}^2 = SA^2 \cdot \phi / V u'_0{}^2,$$

the value of Q being Ypr/SA at each element (page 358). Since the diaphragm is assumed to be in contact with the middle element, the total transverse chromatic aberration will be given by:

$$Tch' = SA_a{}^2 \phi_a Q_a / V_a u'_0{}^2 + SA_c{}^2 \phi_c Q_c / V_c u'_0{}^2$$

Hence
$$\begin{aligned} Tch' \cdot V_a \phi^2 &= \phi_a Q_a + SA_c{}^2 \phi_c Q_c (V_a / V_c) \\ &= \phi_a Ypr_a + SA_c \phi_c Ypr_c (V_a / V_c). \end{aligned}$$

[127] But since the diaphragm is at the middle lens, $Ypr_c = -Ypr_a(d'_b/d'_a)$. Hence

$$Tch' \cdot V_a\phi^2/Ypr_a = u'_a - (SA_b - SA_c)(u'_a/u'_b)(\phi - u'_b)(V_a/V_c)/(1 - SA_b),$$

giving
$$SA_c = SA_b - \frac{u'_a - Tch' \cdot \phi^2(V_a/Ypr_a)}{\left(\dfrac{u'_a}{u'_b}\right)\left(\dfrac{\phi - u'_b}{1 - SA_b}\right)\left(\dfrac{V_a}{V_c}\right)}.$$

For $Tch' = 0$ this becomes simply

(2) $$SA_c = SA_b - u'_b \cdot \frac{1 - SA_b}{\phi - u'_b} \cdot \frac{V_c}{V_a}.$$

It is often more convenient to specify the ratio of the air-spaces, $J = d'_a/d'_b$, than the slope-angle u'_b. These two quantities are related as follows:

$$J = d'_a/d'_b = (1 - SA_b) \cdot u'_b/(SA_b - SA_c) \cdot u'_a.$$

But our condition for zero transverse chromatic aberration, equation (2), shows that

$$(SA_b - SA_c) = (V_c/V_a) \cdot (SA_b - 1) \cdot u'_b/(\phi - u'_b).$$

Hence for this case,

(3) $$u'_b = \phi - J \cdot u'_a(V_c/V_a).$$

b. Longitudinal chromatic aberration. As we have seen, the longitudinal chromatic aberration of the whole system is given by

$$Lch' = \Sigma[SA^2\phi/Vu'_0{}^2].$$

We now insert into this the value of ϕ_b and ϕ_c given in equation (1); also $SA_a = 1$ and $u'_0 = \phi$. This gives the solution

(4) $$SA_b = \frac{(u'_a - u'_b) - Lch'(\phi^2 V_a)}{(u'_a - u'_b)(V_a/V_b) - (\phi - u'_b)(V_a/V_c) - u'_b}.$$

We first solve this equation for SA_b, then equation (2) for SA_c, and lastly equation (1) for the powers and spaces of the lens itself.

c. The Petzval Sum. Since the ray-slopes u'_a and u'_b are chosen arbitrarily, likely values of these variables must be selected with care. However, for each thin-lens solution we generally calculate the Petzval sum $\Sigma(\phi/N)$, and a simple graphical solution will enable us to choose a particular value of u'_a to yield any desired value of the Petzval sum with any specified value of the air-space ratio J. A direct algebraic solution for u'_a is possible, but as it leads to an awkward cubic equation, a few trials are generally preferable.

d. An example. As an example of the use of these formulae, we shall design a triplet lens using the following glasses:

		Nd	Ng'	dN	V
(a) and (c) Chance DBC	1565	1.6105	1.62384	0.01334	46.76
(b) DF	360	1.6225	1.64539	0.02289	28.19

The modified V-number, $\bar{V} = (Ng'-1)/(Ng'-Nd)$, is used since we are aiming at [127]
$D-G'$ achromatism.   For a focal length $f' = 10$, the lens power $\phi = 0.1$, and the
desired Lch' is set at 0.020 as a likely value.   We start by assuming the value
$u'_a = 0.2$, and we shall study the case in which the air-spaces are equal, that is,
$J = 1$, and we assume that a zero value of the transverse chromatic aberration is
desired.   Equation (3) shows that for $J = 1$, $u'_b = -0.1$, and equation (4) then
yields the solution $SA_b = 0.7305$.   Reference to equation (2) tells us that $SA_c =$
0.8652, whence the air-spaces become 1.348 and the lens powers become
$\phi_a = 0.2000$, $\phi_b = -0.4107$, and $\phi_c = 0.2314$.   Using these values, the Petzval
sum is found to be 0.0161.   A much better value for the Petzval sum of a lens of
this type would be about 0.035; hence we shall try the effect of reducing the slope-
angle u'_a to 0.17.   Repeating our solution now gives the Petzval sum as 0.0369, and
a simple interpolation leads us to the final value of $u'_a = 0.1719$.   Equations (3),
(4), and (2) now give us $u'_b = -0.0719$, $SA_b = 0.7695$, and $SA_c = 0.8659$.
From these values, by equation (1), we get

$$\phi_a = \quad 0.1719 \quad f'_a = \quad 5.8173$$
$$d'_a = 1.341$$
$$\phi_b = -0.3169 \quad f'_b = -3.1595$$
$$d'_b = 1.341$$
$$\phi_c = \quad 0.1985 \quad f'_c = \quad 5.0398$$

A parallel paraxial ray entering at $SA_a = 1.4286$ ($f/3.5$), and a paraxial principal
ray entering at $Upr = -0.3640$ ($-20°$) and passing through the middle of the
negative element, are then traced, giving the data needed in the G-sum analysis as:

Lens	SA	Ypr	Q	$v_1 = 1/l$
(a)	1.4286	−0.6345	−0.4441	0
(b)	1.0992	0	0	0.2235
(c)	1.2370	0.6345	0.5129	−0.0831

$$H'_0 = 3.640 \qquad u'_0 = 0.1428 \qquad u'^2_0 = 0.02041$$

This completes the preliminary solution for powers and spacings, and we now
proceed with the determination of the three separate bendings to correct primary
spherical aberration, coma, and astigmatism.

   e. The G-sum analysis.   To follow the procedure outlined in section [61], the
next step must be the computation of the SC', CC', and AC' of the three thin
elements, first assuming the stop to be in contact with each, and then correcting the
contributions by use of the appropriate value of Q in each case.   The formulae
are given on page 329 of Part I as:

$$SC' = (SA^4/u'^2_0)\cdot[SGS]; \quad CC' = H'_0 SA^2\cdot[CGS]; \quad AC' = H'^2_0/2f'.$$

The stop-shifts are included by the further expressions:

$$SC'^* = SC'; \quad CC'^* = CC'+SC'\cdot Qu'_0; \quad AC'^* = AC'+CC'\cdot(2Q/u'_0)+SC'\cdot Q^2.$$

[127] Using the refractive indices for the G' line, the G-sums for our numerical example give, assuming the stop to be in contact with each lens:

$$SC'_a = \quad 39.14c_1{}^2 - 20.53c_1 + 3.512; \qquad CC'_a = \quad 1.032c_1 - 0.2858;$$
$$AC'_a = \quad 1.1388$$

$$SC'_b = -25.11c_3{}^2 - \; 7.596c_3 - 2.498; \qquad CC'_b = -1.121c_3 - 0.1570;$$
$$AC'_b = -2.0994$$

$$SC'_c = \quad 25.42c_5{}^2 - \; 9.281c_5 + 1.608; \qquad CC'_c = \quad 0.893c_5 - 0.1656;$$
$$AC'_c = \quad 1.3152$$

Hence the true contributions become:

$$SC'_a{}^* = SC'_a; \quad CC'_a{}^* = -2.4830c_1{}^2 + 2.3344c_1 - 0.5086;$$
$$AC'_a{}^* = \quad 7.7184c_1{}^2 - 10.464c_1 + 3.6082$$

$$SC'_b{}^* = SC'_b; \quad CC'_b{}^* = -1.121c_3 - 0.1570;$$
$$AC'_b{}^* = -2.0994$$

$$SC'_c{}^* = SC'_c; \quad CC'_c{}^* = \quad 1.8625c_5{}^2 + 0.2130c_5 - 0.0478;$$
$$AC'_c{}^* = \quad 6.688c_5{}^2 + 3.9708c_5 + 0.5491$$

The total aberrations are then found by adding together the starred contributions:

(1) $\quad LA' = \quad 39.14c_1{}^2 \quad -20.53c_1 \quad -25.11c_3{}^2 -7.596c_3 \quad +25.42c_5{}^2$
$$-9.281c_5 + 2.622$$

(2) $\quad Coma'_S = -2.483c_1{}^2 + \; 2.3344c_1 - \; 1.121c_3 +1.8625c_5{}^2 + 0.2130c_5 - 0.7134$

(3) $\quad Ast'_S = \quad 7.7184c_1{}^2 - 10.4645c_1 + \; 6.688c_5{}^2 + 3.9708c_5 \; +2.0579.$

Since equation (3) contains no term in c_3, it can be easily solved for c_5 in terms of c_1 by completing the squares; we then solve (2) for c_3 in terms of c_1 and c_5:

(3a) $\qquad 2.5861c_5 + 0.7677 = \sqrt{2.0783 + Ast'_S - (1.8833 - 2.7782c_1)^2}$

(2a) $\quad c_3 = 2.0824c_1 - 2.2150c_1{}^2 + 0.1900c_5 + 1.6615c_5{}^2 - 0.6364 - (Coma'_S/1.121).$

By means of equations (3a), (2a), and (1), we can find the values of c_3, c_5 and LA' for any given value of c_1 and for any given residual of Ast'_S and $Coma'_S$. The calculation of a few solutions will enable us to plot a curve, Fig. 192, connecting LA' with c_1. From this curve, the value of c_1 can be read off for any desired LA'-residual, and hence the c_3- and c_5-values can also be found.

In our numerical example, the residuals $Ast'_S = -0.08$, $Coma'_S = 0$, and $LA' = +0.14$ give the two possible graphical solutions:

	Left solution			Right solution	
$c_1 =$	0.2640	$r_1 =$	3.787	$c_1 =$ 0.3742	$r_1 =$ 2.672
$c_2 =$	−0.0116	$r_2 =$	−86.20	$c_2 =$ 0.0986	$r_2 =$ 10.141
$c_3 =$	−0.2363	$r_3 =$	−4.231	$c_3 =$ −0.1070	$r_3 =$ −9.345
$c_4 =$	0.2548	$r_4 =$	3.924	$c_4 =$ 0.3841	$r_4 =$ 2.603
$c_5 =$	0.0211	$r_5 =$	47.39	$c_5 =$ 0.1417	$r_5 =$ 7.057
$c_6 =$	−0.2972	$r_6 =$	−3.364	$c_6 =$ −0.1766	$r_6 =$ −5.662

The two systems are closely similar, and have the appearance shown above the [127] graph in Fig. 192. We should of course follow up both solutions and compare their relative corrections, but for the sake of the example we shall concentrate on the first.

f. The introduction of finite thicknesses. The next step is to assign suitable thicknesses by a scale drawing of the lens to its final focal length, and it is convenient to make the front and rear elements of the same diameter, the negative being slightly smaller. On the assumption that the unit of length is the centimetre, the diameters of the positive lenses would be about 3.0 (for an $f/3.5$ objec-

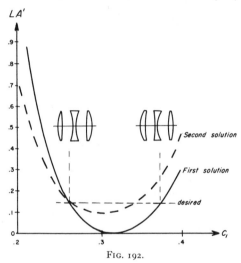

FIG. 192.

tive), and suitable thicknesses are 0.55, 0.2, and 0.6 respectively for the three elements. The second radius of each element is then adjusted to hold the focal lengths of the individual elements at their original TL-values of 5.8163, -3.1551, and 5.0362 respectively. The air-spaces should be chosen so as to maintain the principal-point separations at their TL-values of 1.3412 by the relation air-space $= d + l'pp_1 - lpp_2$ in each case.

A paraxial ray is first traced as a check, and then the first three aberrations (Petzval and transverse and longitudinal chromatic) are determined for the thick system by actual ray-tracing for an aperture of $f/3.5$ and a field angle of 20°. The Petzval sum rises from its TL-value of 0.0355 to 0.0382 as a result of the finite lens thicknesses, but this is still considered acceptable. The transverse chromatic aberration at 20°, determined by a pair of coloured principal rays chosen so as to cross the lens axis at the front principal point of the negative element, is given by $H'_d - H'_{g'} = +0.00072$, which is negligibly small. The longitudinal chromatic aberration, found by $\Sigma(d' - D')(N'_g' - N'_d)$ amounts to $-0.000\ 236\ 2$ or -4.72 wave-lengths in centimetre measure. The marginal ray in G' used for the

[127] chromatic computation yields two other data, namely, the $LA' = -0.0328$ and the $OSC' = +0.0030$ corresponding to $Coma'_S = +0.011$ at 20°. The astigmatic foci computed along the 20° principal ray gives $X'_S = +0.142$ and $X'_T = +0.638$, or a markedly hollow field.

g. The second approximation. To improve the longitudinal chromatic aberration, a little study of the $(d' - D')$ calculation reveals that *if* the dN of the middle lens had been 0.02219 instead of its true value of 0.02289, $\Sigma(d' - D')dN'$ would have been zero. This would have made the TL Lch'-sum equal to 0.0408 instead of 0.0205, and it was therefore decided to solve for this value in the next solution.

With $Lch' = +0.0408$, $Ptz = 0.035$, and a focal length 10 as before, the element focal lengths and separations come out to be:

$$f'_a = 6.0671$$
$$d'_a = 1.51$$
$$f'_b = -3.2707$$
$$d'_b = 1.51$$
$$f'_c = 5.1509$$

The G-sums now give the following equations of condition:

$$2.8915c_5 + 0.690 = \sqrt{1.9255 + Ast'_S - (1.919 - 3.138c_1)^2}$$
$$c_3 = 2.3008c_1 - 2.6661c_1{}^2 + 0.1458c_5 + 1.9219c_5{}^2 - 0.6375 - (Coma'_S/1.03)$$
$$LA' = -18.87c_1 + 37.53c_1{}^2 - 8.524c_5 + 22.97c_5{}^2 - 6.17c_3 - 22.0c_3{}^2 + 2.621.$$

Because of the large aberration residuals observed in the first system, it is well to alter the desired TL residuals to:

$$LA' = +0.16; \quad Coma'_S = -0.01; \quad Ast'_S = +0.08.$$

The curve connecting LA' with c_1, plotted on the same graph as before, gives for the left-hand solution:

$$c_1 = 0.2580 \quad r_1 = 3.876$$
$$c_3 = -0.1937 \quad r_3 = -5.162$$
$$c_5 = 0.0658 \quad r_5 = 15.20$$

After introducing thicknesses and solving for the second radius of each element, the tracing of a true $f/3.5$ marginal ray and an $f/5$ zonal ray, and a pair of 20° principal rays in D and G' as before, gives the aberrations of the whole lens as:

$$Ptz = 0.0385$$
$$\Sigma(d' - D')dN' = +1.55\lambda$$
$$H'_d - H'_{g'} = +0.00264 \qquad Dist' = +0.54\%$$
$$LA' = -0.0214 \qquad X'_S = -0.0443$$
$$OSC' = -0.00044 \qquad X'_T = -0.0523$$
$$LZA' = +0.0300$$

h. The third approximation. In making the third approximation, it is clearly necessary to improve the transverse chromatic aberration, as 0.001 represents a reasonable tolerance for this defect whilst the actual value is over twice as great. [127] Let us therefore set the desired TL residuals at:

$$Ptz = \quad 0.035 \qquad LA' = +0.16$$
$$Lch' = \quad 0.035 \qquad Coma'_S = -0.01$$
$$Tch' = -0.002 \qquad Ast'_S = +0.07$$

The solution for powers and spaces gives, for a focal length of 10:

$$f'_a = \quad 5.7689$$
$$\qquad\qquad\qquad 1.433$$
$$f'_b = -3.2228$$
$$\qquad\qquad\qquad 1.433$$
$$f'_c = \quad 5.2380$$

The G-sum calculation gives for the three bending parameters:

$$r_1 = 3.846; \qquad r_3 = -4.644; \qquad r_5 = 23.21,$$

and after solving for the second radius in each case to maintain powers, and using the proper air-spaces, the following true aberrations are found:

$$Ptz = \quad 0.0389$$
$$\Sigma(d'-D')dN' = -0.03\lambda$$
$$H'_d - H'_{g'} = -0.00112 \qquad Dist' = -0.59\%$$
$$LA' = -0.0643 \qquad X'_S = -0.0287$$
$$OSC' = +0.00206 \qquad X'_T = +0.0230$$
$$LZA' = +0.0208$$

The correction of the transverse chromatic aberration is evidently overdone, but it is acceptable. The principal remaining faults are over-corrected spherical aberration, positive coma, and a somewhat hollow tangential field. By repeating the last TL solution for $LA' = +0.17$, $Coma'_S = -0.015$, and $Ast'_S = +0.08$, only the second and third elements need be bent, and we obtain the following final solution:

$$r_1 = \quad 3.846 \qquad\qquad\qquad\qquad Ptz = \quad 0.0389$$
$$\qquad\qquad 0.55$$
$$r_2 = -52.931 \qquad\qquad\qquad \Sigma(d'-D')dN' = +0.01\lambda$$
$$\qquad\qquad 1.0487 \text{ (air)}$$
$$r_3 = \quad -4.764 \qquad\qquad\qquad H'_d - H'_{g'} = -0.00099$$
$$\qquad\qquad 0.20$$
$$r_4 = \quad 3.753 \qquad\qquad\qquad\qquad LA' = -0.0475$$
$$\qquad\qquad 1.0619 \text{ (air)}$$
$$r_5 = \quad 22.056 \qquad\qquad\qquad\quad OSC' = -0.00035$$
$$\qquad\qquad 0.60$$
$$r_6 = \quad -3.796 \qquad\qquad\qquad\quad LZA' = +0.0246$$

Field	Dist'	X'_S	X'_T
14°	−0.21%	−0.0410	+0.0438
20°	−0.54%	−0.0387	−0.0101

The spherical aberration and astigmatism are shown graphically in Fig. 193.

[127]   These corrections are all excellent for a lens of this aperture, field, and focal length.   By chance, the distortion has come out to be quite small; however, if it were greater than about one half per cent., it would be worth making a small change in the ratio of the separations d'_a to d'_b to correct it.   Increasing d'_b would

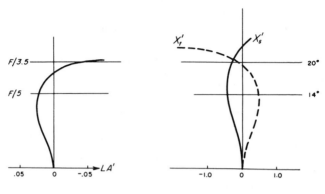

(a) Spherical aberration   FIG. 193.   (b) Astigmatism

have the effect of strengthening the rear component to maintain the transverse chromatic correction, and hence it would make the distortion more positive (barrel). An increase in d'_b by a few per cent. would be worth trying, but it would be necessary of course to repeat the entire design since a change in the rear space will affect every aberration.

APPENDIX I

BIBLIOGRAPHY OF A. E. CONRADY

'Uber das Rotationsvermögen einiger Terpenderivate' (with O Wallach). *Liebigs Annalen der Chemie*, **252**, 141–157 (1889).
'Berechnung der Atomrefractionen für Natriumlicht.' *Zeit. für Phys. Chemie*, **3**, 210–227 (1889).
'Theories of microsopical vision: a vindication of the Abbe theory.' *Jour. Roy. Micro. Soc.*, 610–633 (1904).
'On the chromatic correction of object glasses.' *M. N. Roy. Astron. Soc.*, **64**, 182–188 (1904).
'On the chromatic correction of object glasses' (second paper). *M. N. Roy. Astron. Soc.*, **64**, 458–460 (1904).
'Note on a suggested method of determining the declination of stars.' *M. N. Roy. Astron. Soc.*, **64**, 673–675 (1904).
'An experimental proof of phase-reversal in diffraction-spectra.' *Jour. Roy. Micro. Soc.*, 150–152 (1905).
'On the application of the undulatory theory to optical problems.' *Jour. Roy. Micro. Soc.*, 401–411 (1905).
'The optical sine condition.' *M. N. Roy. Astron. Soc.*, **65**, 501–509 (1905).
'On the spherical correction of object glasses.' *M. N. Roy. Astron. Soc.*, **65**, 594–608 (1905).
'Theories of microscopical vision' (second paper). *Jour. Roy. Micro. Soc.*, 541–553 (1905).
'Mr. Reynolds' dark slides.' *M. N. Roy. Astron. Soc.*, **66**, 329 (1906).
'Note on an early criticism of the Abbe theory.' *Jour. Roy. Micro. Soc.*, 645–647 (1906).
'Determination of the properties of objectives.' *Jour. Roy. Micro. Soc.*, 620–626 (1907).
'The refracting telescope from the practical point of view.' *Jour. Brit. Astron. Assoc.*, **20**, 299–308 (1910).
'Resolution with dark-ground illumination.' *Jour. Quekett Micro. Club II*, **11**, 475–480 (1912).
'The unpublished papers of J. J. Lister.' *Jour. Roy. Micro. Soc.*, 27–55 (1913).
'Anastigmatic eyepieces.' *M. N. Roy. Astron. Soc.*, **78**, 445–447 (1918).
'The five aberrations of lens systems.' *M. N. Roy. Astron. Soc.*, **79**, 60–66 (1918).
'Decentred lens systems.' *M. N. Roy. Astron. Soc.*, **79**, 384–390 (1919).
'Star discs.' *M. N. Roy. Astron. Soc.*, **79**, 575–593 (1919).
'The five aberrations of lens systems' (second paper). *M. N. Roy. Astron. Soc.*, **80**, 78–91 (1919).
'Optical systems with non-spherical surfaces.' *M. N. Roy. Astron. Soc.*, **80**, 320–328 (1920).
'Notes on microscopical optics.' (Joint Symposium R.M.S., Opt. Soc., &c.) *Jour. Roy. Micro. Soc.*, 60–66 (Dec. 1920).

'The present state of photographic optics.' (Traill-Taylor lecture). *Phot. Jour.*, **60**, 296–305 (1920).

'A study of the balance.' *Proc. Roy. Soc.*, **101**, 211–224 (1922).

'Microscope optics.' Chapter in Glazebrook's *Dictionary of Applied Physics* IV, pp. 202–237. (London, Macmillan, 1923.)

'Photographic optics.' Chapter III in *Photography as a Scientific Implement*, pp. 48–102. (New York, Van Nostrand, 1923.)

'On three Huygens lenses in the possession of the Royal Society of London' (with R. A. Sampson). *Proc. Roy. Soc. Edin.*, **49**, 289–299 (1929).

Applied Optics and Optical Design (Part I). (London, Oxford University Press, 1929.)

1925 April 27 · Evening · Boro 4 EHF

a L to R. b

Contribution of jth surface to the final LA'_d =

$$SA_K^2 \cdot a_j \, \frac{N'_j \ell'^2_j u'^4_j}{N_K \ell'^2_K \cdot d'^4_K} + 2\gamma_K H'_K a_j q'_j \frac{\ell'^2_j u'^2_j}{\ell'_K u'_K} + H'^2_K a_j q'^2_j \frac{N'_K}{N'_j}$$

Contribution of jth surface to the final TA'_d =

$$SA_K^2 H'_j \, a_j q'_j \frac{\ell'^2_j u'^2_j}{\ell'^2_K \cdot u'^2_K} + 2\gamma_K H'^2_K a_j q'^2_j \frac{N'_K}{N'_j} \cdot \frac{1}{\ell'_K}$$

Above: A portion of a logarithm computation for Part Two, *Applied Optics and Optical Design.*
Below: Facsimile of a portion of Professor Conrady's tidy manuscript.

BIOGRAPHICAL MEMORANDUM

'CLOSE to the frontier separating Germany from Holland there lies a little town called Üdem. Certainly from sometime in the seventeenth century and possibly still longer the school in that town was under the headmastership of my male ancestors, all without exception living to a good age and handing over to the next Conrady, or rather, in the earlier times, Conradi. The terminal letter was apparently changed during the eighteenth century, in connection with some hoped-for inheritance from a Dutch branch of the Conradi family, first to Conradij according to the Dutch custom of treating ij as interchangeable with y, but eventually to the present spelling. My father always stuck to the ij ending, and I myself also up to the early days of the World War.'

'Less is known of my ancestry on the mother's side. There was and probably still is, in Dinslaken, a distillery run in the first half of the nineteenth century by my maternal grandfather rejoicing in the most unusual and probably Dutch name of Scriverius. . . . My mother, whose only Christian name was Mathilde, received her early education in my paternal grandfather's school, but being well-to-do was sent for a year to a finishing school in the upper Rhine in accordance with custom. . . . In the late '50's my father was appointed private music teacher to Mathilde Scriverius; they were married on January 9, 1859, when my father secured the headmastership of the principal elementary school in my birthplace, Burscheid, in the beautiful mountainous country which lies on the right bank of the Rhine opposite Cologne.'

Thus begins a long autobiographical sketch written at my request not long after my father's retirement in 1931. The first paragraph of the two quoted is especially interesting because it will explain to his old acquaintances who were familiar with his earlier very distinctive signature the origin of the modified 'y' and also the occasion and implied reason for the change to the anglicized form.

Alexander Eugen Conrady, born January 27, 1866, was the fifth in a family of seven children. He was a shy, rather delicate, and studious child who later carried a very critical and highly sensitive nature under a mask of great reserve. He attended schools in Burscheid and Barmen, and in 1884 went to Bonn University where he studied mathematics under Lipschitz, physics under Clausius, chemistry under Kekulé, and astronomy under Schönfeld.

By this time his father had deserted school teaching for more remunerative fields, and was agent for a number of foreign concerns in an effort better to support a large and growing family. In 1886 his scientifically-minded son at Bonn was persuaded to interrupt his courses to spend a period in England at Stanfield, Brown and Co., of which his father was German agent. It was necessary for someone mechanically minded to learn how to set up some button machines, and later to supervise their installation in various German towns. This seemingly unfortunate interruption gave my father his first glimpse of life in England, and more important still, it began his lifelong friendship with Mr. Geo. W. Brown to whom Part I of his *Applied Optics and Optical Design* was dedicated.

In 1887 Conrady returned to Bonn for another year and a half as research assistant to the organic chemist Wallach, his work consisting chiefly of physical measurements. One paper he wrote during that period, on atomic refractions, provided a series of standard data not superseded until 1912.

A growing distaste for the existing German regime, doubtless emphasized in his mind as a result of his appreciation of more desirable conditions found in England, and also a most discouraging series of breakdowns in health, led him to give up a promising academic career for a period of travel as a mining chemist. He went first to Nova Scotia, and later to North Carolina, to supervise some of Mr. Brown's many enterprises, after which he took up residence in England, continuing his business associations with Mr. Brown in one way or another for many years. For a period he conducted a small factory in Keighley, Yorkshire, to make models of new machines, and later to make dynamos, arc lamps and other new electrical devices.

In 1896 he took another long trip for Mr. Brown, this time to the Vaal River in South Africa where he was required to correct some obstinate mechanical difficulty with diamond dredging machinery. He was away from England four months, after which he returned to his Keighley works, later moving to London. Throughout this period the financial problems of running a business were a great trial to his scientific nature. The businesses attracted notice but were monetary failures. Amongst others, Messrs. Watson and Sons of optical fame at times sought the mechanical skills of the small Conrady model factory. Meanwhile, more or less as a hobby and side-line, he had been studying and experimenting with the design and manufacture of telescope and microscope objectives, an interest which was to lead to most unforeseeable results, but doubtless at the time detracted from the financial success of his company.

In 1901 he married Annie Bunney of Harefield, Middlesex, and in 1902 after the birth of the first of four daughters, Hilda, he became naturalized and remained to the end a loyal and almost fanatically devoted citizen of his adopted country. Also at this period he took another decisive step which finally relieved him of the burdens of business management and began the long and fruitful period in the world of optics: Messrs. Watson and Son engaged him on a full-time basis as scientific adviser and lens designer.   His work with them was concerned largely with microscope objectives, but it included photographic and telescope objectives, and during World War I the optical parts of various service instruments including submarine periscopes.

In 1917 when the Technical Optics Department was established at the Imperial College of Science and Technology, Conrady was appointed Professor of Optical Design.   He had an excellent academic background together with much practical knowledge and long experience to draw upon, but it remained for him to learn how to order and present a great deal of information on a hitherto largely untaught subject.   He fortunately proved to be an instinctive and careful teacher and for fifteen years his lectures were the inspiration of a new school of young lens designers, now scattered over the world and themselves teaching his methods and procedures to the rising generation.   In 1929 he published Part I of his book *Applied Optics and Optical Design*, which was the first and apparently remains the only detailed text-book on the subject written with the needs of the complete beginner and the self-taught constantly in mind.   During the Second World War it was eagerly studied by physicists, engineers and astronomers in various parts of the world who were unexpectedly called upon to design new devices employing optical systems.   Part II of the book most unfortunately has been long delayed.   My father's poor health and the emergencies of World War II prevented its completion before his death in 1944.   However he left a well-advanced manuscript in his remarkably clear handwriting from which this posthumously published Part II has been compiled.

His optical interests led him into many and varied associations, and to membership in scientific societies appealing to the optically minded, namely the Royal Astronomical Society, the Royal Microscopical Society, the Quekett Club, and the Royal Photographic Society.   Although generally speaking of an unsociable nature, he took great pleasure in attending most regularly the meetings of the Royal Astronomical Society's Dining Club.   He was awarded the Traill-Taylor medal of the R.P.S. in 1920.

HILDA G. CONRADY KINGSLAKE

INDEX

This comprehensive index covers the two volumes of the book. Volume II begins at page 519, and references to it are given in bold face type.

A CATALOG OF SELECTED
DOVER BOOKS
IN SCIENCE AND MATHEMATICS

A CATALOG OF SELECTED

DOVER BOOKS

IN SCIENCE AND MATHEMATICS

QUALITATIVE THEORY OF DIFFERENTIAL EQUATIONS, V.V. Nemytskii and V.V. Stepanov. Classic graduate-level text by two prominent Soviet mathematicians covers classical differential equations as well as topological dynamics and erqodic theory. Bibliographies. 523pp. 5⅜ × 8½. 65954-2 Pa. $10.95

MATRICES AND LINEAR ALGEBRA, Hans Schneider and George Phillip Barker. Basic textbook covers theory of matrices and its applications to systems of linear equations and related topics such as determinants, eigenvalues and differential equations. Numerous exercises. 432pp. 5⅜ × 8½. 66014-1 Pa. $8.95

QUANTUM THEORY, David Bohm. This advanced undergraduate-level text presents the quantum theory in terms of qualitative and imaginative concepts, followed by specific applications worked out in mathematical detail. Preface. Index. 655pp. 5⅜ × 8½. 65969-0 Pa. $10.95

ATOMIC PHYSICS (8th edition), Max Born. Nobel laureate's lucid treatment of kinetic theory of gases, elementary particles, nuclear atom, wave-corpuscles, atomic structure and spectral lines, much more. Over 40 appendices, bibliography. 495pp. 5⅜ × 8½. 65984-4 Pa. $11.95

ELECTRONIC STRUCTURE AND THE PROPERTIES OF SOLIDS: The Physics of the Chemical Bond, Walter A. Harrison. Innovative text offers basic understanding of the electronic structure of covalent and ionic solids, simple metals, transition metals and their compounds. Problems. 1980 edition. 582pp. 6⅛ × 9¼. 66021-4 Pa. $14.95

BOUNDARY VALUE PROBLEMS OF HEAT CONDUCTION, M. Necati Özisik. Systematic, comprehensive treatment of modern mathematical methods of solving problems in heat conduction and diffusion. Numerous examples and problems. Selected references. Appendices. 505pp. 5⅜ × 8½. 65990-9 Pa. $11.95

A SHORT HISTORY OF CHEMISTRY (3rd edition), J.R. Partington. Classic exposition explores origins of chemistry, alchemy, early medical chemistry, nature of atmosphere, theory of valency, laws and structure of atomic theory, much more. 428pp. 5⅜ × 8½. (Available in U.S. only) 65977-1 Pa. $10.95

A HISTORY OF ASTRONOMY, A. Pannekoek. Well-balanced, carefully reasoned study covers such topics as Ptolemaic theory, work of Copernicus, Kepler, Newton, Eddington's work on stars, much more. Illustrated. References. 521pp. 5⅜ × 8½. 65994-1 Pa. $11.95

PRINCIPLES OF METEOROLOGICAL ANALYSIS, Walter J. Saucier. Highly respected, abundantly illustrated classic reviews atmospheric variables, hydrostatics, static stability, various analyses (scalar, cross-section, isobaric, isentropic, more). For intermediate meteorology students. 454pp. 6⅛ × 9¼. 65979-8 Pa. $12.95

RELATIVITY, THERMODYNAMICS AND COSMOLOGY, Richard C. Tolman. Landmark study extends thermodynamics to special, general relativity; also applications of relativistic mechanics, thermodynamics to cosmological models. 501pp. 5⅜ × 8½. 65383-8 Pa. $11.95

APPLIED ANALYSIS, Cornelius Lanczos. Classic work on analysis and design of finite processes for approximating solution of analytical problems. Algebraic equations, matrices, harmonic analysis, quadrature methods, much more. 559pp. 5⅜ × 8½. 65656-X Pa. $11.95

SPECIAL RELATIVITY FOR PHYSICISTS, G. Stephenson and C.W. Kilmister. Concise elegant account for nonspecialists. Lorentz transformation, optical and dynamical applications, more. Bibliography. 108pp. 5⅜ × 8½. 65519-9 Pa. $3.95

INTRODUCTION TO ANALYSIS, Maxwell Rosenlicht. Unusually clear, accessible coverage of set theory, real number system, metric spaces, continuous functions, Riemann integration, multiple integrals, more. Wide range of problems. Undergraduate level. Bibliography. 254pp. 5⅜ × 8½. 65038-3 Pa. $7.00

INTRODUCTION TO QUANTUM MECHANICS With Applications to Chemistry, Linus Pauling & E. Bright Wilson, Jr. Classic undergraduate text by Nobel Prize winner applies quantum mechanics to chemical and physical problems. Numerous tables and figures enhance the text. Chapter bibliographies. Appendices. Index. 468pp. 5⅜ × 8½. 64871-0 Pa. $9.95

ASYMPTOTIC EXPANSIONS OF INTEGRALS, Norman Bleistein & Richard A. Handelsman. Best introduction to important field with applications in a variety of scientific disciplines. New preface. Problems. Diagrams. Tables. Bibliography. Index. 448pp. 5⅜ × 8½. 65082-0 Pa. $10.95

MATHEMATICS APPLIED TO CONTINUUM MECHANICS, Lee A. Segel. Analyzes models of fluid flow and solid deformation. For upper-level math, science and engineering students. 608pp. 5⅜ × 8½. 65369-2 Pa. $12.95

ELEMENTS OF REAL ANALYSIS, David A. Sprecher. Classic text covers fundamental concepts, real number system, point sets, functions of a real variable, Fourier series, much more. Over 500 exercises. 352pp. 5⅜ × 8½. 65385-4 Pa. $8.95

PHYSICAL PRINCIPLES OF THE QUANTUM THEORY, Werner Heisenberg. Nobel Laureate discusses quantum theory, uncertainty, wave mechanics, work of Dirac, Schroedinger, Compton, Wilson, Einstein, etc. 184pp. 5⅜ × 8½. 60113-7 Pa. $4.95

INTRODUCTORY REAL ANALYSIS, A.N. Kolmogorov, S.V. Fomin. Translated by Richard A. Silverman. Self-contained, evenly paced introduction to real and functional analysis. Some 350 problems. 403pp. 5⅜ × 8½. 61226-0 Pa. $7.95

PROBLEMS AND SOLUTIONS IN QUANTUM CHEMISTRY AND PHYSICS, Charles S. Johnson, Jr. and Lee G. Pedersen. Unusually varied problems, detailed solutions in coverage of quantum mechanics, wave mechanics, angular momentum, molecular spectroscopy, scattering theory, more. 280 problems plus 139 supplementary exercises. 430pp. 6½ × 9¼. 65236-X Pa. $10.95

ASYMPTOTIC METHODS IN ANALYSIS, N.G. de Bruijn. An inexpensive, comprehensive guide to asymptotic methods—the pioneering work that teaches by explaining worked examples in detail. Index. 224pp. 5⅜ × 8½.   64221-6 Pa. $5.95

OPTICAL RESONANCE AND TWO-LEVEL ATOMS, L. Allen and J.H. Eberly. Clear, comprehensive introduction to basic principles behind all quantum optical resonance phenomena. 53 illustrations. Preface. Index. 256pp. 5⅜ × 8½.
65533-4 Pa. $6.95

COMPLEX VARIABLES, Francis J. Flanigan. Unusual approach, delaying complex algebra till harmonic functions have been analyzed from real variable viewpoint. Includes problems with answers. 364pp. 5⅜ × 8½.   61388-7 Pa. $7.95

ATOMIC SPECTRA AND ATOMIC STRUCTURE, Gerhard Herzberg. One of best introductions; especially for specialist in other fields. Treatment is physical rather than mathematical. 80 illustrations. 257pp. 5⅜ × 8½.   60115-3 Pa. $4.95

APPLIED COMPLEX VARIABLES, John W. Dettman. Step-by-step coverage of fundamentals of analytic function theory—plus lucid exposition of 5 important applications: Potential Theory; Ordinary Differential Equations; Fourier Transforms; Laplace Transforms; Asymptotic Expansions. 66 figures. Exercises at chapter ends. 512pp. 5⅜ × 8½.   64670-X Pa. $10.95

ULTRASONIC ABSORPTION: An Introduction to the Theory of Sound Absorption and Dispersion in Gases, Liquids and Solids, A.B. Bhatia. Standard reference in the field provides a clear, systematically organized introductory review of fundamental concepts for advanced graduate students, research workers. Numerous diagrams. Bibliography. 440pp. 5⅜ × 8½.   64917-2 Pa. $8.95

UNBOUNDED LINEAR OPERATORS: Theory and Applications, Seymour Goldberg. Classic presents systematic treatment of the theory of unbounded linear operators in normed linear spaces with applications to differential equations. Bibliography. 199pp. 5⅜ × 8½.   64830-3 Pa. $7.00

LIGHT SCATTERING BY SMALL PARTICLES, H.C. van de Hulst. Comprehensive treatment including full range of useful approximation methods for researchers in chemistry, meteorology and astronomy. 44 illustrations. 470pp. 5⅜ × 8½.   64228-3 Pa. $9.95

CONFORMAL MAPPING ON RIEMANN SURFACES, Harvey Cohn. Lucid, insightful book presents ideal coverage of subject. 334 exercises make book perfect for self-study. 55 figures. 352pp. 5⅜ × 8¼.   64025-6 Pa. $8.95

OPTICKS, Sir Isaac Newton. Newton's own experiments with spectroscopy, colors, lenses, reflection, refraction, etc., in language the layman can follow. Foreword by Albert Einstein. 532pp. 5⅜ × 8½.   60205-2 Pa. $8.95

GENERALIZED INTEGRAL TRANSFORMATIONS, A.H. Zemanian. Graduate-level study of recent generalizations of the Laplace, Mellin, Hankel, K. Weierstrass, convolution and other simple transformations. Bibliography. 320pp. 5⅜ × 8½.   65375-7 Pa. $7.95

THE ELECTROMAGNETIC FIELD, Albert Shadowitz. Comprehensive undergraduate text covers basics of electric and magnetic fields, builds up to electromagnetic theory. Also related topics, including relativity. Over 900 problems. 768pp. 5⅜ × 8¼. 65660-8 Pa. $15.95

FOURIER SERIES, Georgi P. Tolstov. Translated by Richard A. Silverman. A valuable addition to the literature on the subject, moving clearly from subject to subject and theorem to theorem. 107 problems, answers. 336pp. 5⅜ × 8½. 63317-9 Pa. $7.95

THEORY OF ELECTROMAGNETIC WAVE PROPAGATION, Charles Herach Papas. Graduate-level study discusses the Maxwell field equations, radiation from wire antennas, the Doppler effect and more. xiii + 244pp. 5⅜ × 8½. 65678-0 Pa. $6.95

DISTRIBUTION THEORY AND TRANSFORM ANALYSIS: An Introduction to Generalized Functions, with Applications, A.H. Zemanian. Provides basics of distribution theory, describes generalized Fourier and Laplace transformations. Numerous problems. 384pp. 5⅜ × 8½. 65479-6 Pa. $8.95

THE PHYSICS OF WAVES, William C. Elmore and Mark A. Heald. Unique overview of classical wave theory. Acoustics, optics, electromagnetic radiation, more. Ideal as classroom text or for self-study. Problems. 477pp. 5⅜ × 8½. 64926-1 Pa. $10.95

CALCULUS OF VARIATIONS WITH APPLICATIONS, George M. Ewing. Applications-oriented introduction to variational theory develops insight and promotes understanding of specialized books, research papers. Suitable for advanced undergraduate/graduate students as primary, supplementary text. 352pp. 5⅜ × 8½. 64856-7 Pa. $8.50

A TREATISE ON ELECTRICITY AND MAGNETISM, James Clerk Maxwell. Important foundation work of modern physics. Brings to final form Maxwell's theory of electromagnetism and rigorously derives his general equations of field theory. 1,084pp. 5⅜ × 8½. 60636-8, 60637-6 Pa., Two-vol. set $19.00

AN INTRODUCTION TO THE CALCULUS OF VARIATIONS, Charles Fox. Graduate-level text covers variations of an integral, isoperimetrical problems, least action, special relativity, approximations, more. References. 279pp. 5⅜ × 8½. 65499-0 Pa. $6.95

HYDRODYNAMIC AND HYDROMAGNETIC STABILITY, S. Chandrasekhar. Lucid examination of the Rayleigh-Benard problem; clear coverage of the theory of instabilities causing convection. 704pp. 5⅜ × 8¼. 64071-X Pa. $12.95

CALCULUS OF VARIATIONS, Robert Weinstock. Basic introduction covering isoperimetric problems, theory of elasticity, quantum mechanics, electrostatics, etc. Exercises throughout. 326pp. 5⅜ × 8½. 63069-2 Pa. $7.95

DYNAMICS OF FLUIDS IN POROUS MEDIA, Jacob Bear. For advanced students of ground water hydrology, soil mechanics and physics, drainage and irrigation engineering and more. 335 illustrations. Exercises, with answers. 784pp. 6⅛ × 9¼. 65675-6 Pa. $19.95

NUMERICAL METHODS FOR SCIENTISTS AND ENGINEERS, Richard Hamming. Classic text stresses frequency approach in coverage of algorithms, polynomial approximation, Fourier approximation, exponential approximation, other topics. Revised and enlarged 2nd edition. 721pp. 5⅜ × 8½.
65241-6 Pa. $14.95

THEORETICAL SOLID STATE PHYSICS, Vol. I: Perfect Lattices in Equilibrium; Vol. II: Non-Equilibrium and Disorder, William Jones and Norman H. March. Monumental reference work covers fundamental theory of equilibrium properties of perfect crystalline solids, non-equilibrium properties, defects and disordered systems. Appendices. Problems. Preface. Diagrams. Index. Bibliography. Total of 1,301pp. 5⅜ × 8½. Two volumes. Vol. I 65015-4 Pa. $12.95
Vol. II 65016-2 Pa. $12.95

OPTIMIZATION THEORY WITH APPLICATIONS, Donald A. Pierre. Broad-spectrum approach to important topic. Classical theory of minima and maxima, calculus of variations, simplex technique and linear programming, more. Many problems, examples. 640pp. 5⅜ × 8½. 65205-X Pa. $12.95

THE MODERN THEORY OF SOLIDS, Frederick Seitz. First inexpensive edition of classic work on theory of ionic crystals, free-electron theory of metals and semiconductors, molecular binding, much more. 736pp. 5⅜ × 8½.
65482-6 Pa. $14.95

ESSAYS ON THE THEORY OF NUMBERS, Richard Dedekind. Two classic essays by great German mathematician: on the theory of irrational numbers; and on transfinite numbers and properties of natural numbers. 115pp. 5⅜ × 8½.
21010-3 Pa. $4.95

THE FUNCTIONS OF MATHEMATICAL PHYSICS, Harry Hochstadt. Comprehensive treatment of orthogonal polynomials, hypergeometric functions, Hill's equation, much more. Bibliography. Index. 322pp. 5⅜ × 8½. 65214-9 Pa. $8.95

NUMBER THEORY AND ITS HISTORY, Oystein Ore. Unusually clear, accessible introduction covers counting, properties of numbers, prime numbers, much more. Bibliography. 380pp. 5⅜ × 8½. 65620-9 Pa. $8.95

THE VARIATIONAL PRINCIPLES OF MECHANICS, Cornelius Lanczos. Graduate level coverage of calculus of variations, equations of motion, relativistic mechanics, more. First inexpensive paperbound edition of classic treatise. Index. Bibliography. 418pp. 5⅜ × 8½. 65067-7 Pa. $10.95

MATHEMATICAL TABLES AND FORMULAS, Robert D. Carmichael and Edwin R. Smith. Logarithms, sines, tangents, trig functions, powers, roots, reciprocals, exponential and hyperbolic functions, formulas and theorems. 269pp. 5⅜ × 8½. 60111-0 Pa. $5.95

THEORETICAL PHYSICS, Georg Joos, with Ira M. Freeman. Classic overview covers essential math, mechanics, electromagnetic theory, thermodynamics, quantum mechanics, nuclear physics, other topics. First paperback edition. xxiii + 885pp. 5⅜ × 8½. 65227-0 Pa. $17.95

HANDBOOK OF MATHEMATICAL FUNCTIONS WITH FORMULAS, GRAPHS, AND MATHEMATICAL TABLES, edited by Milton Abramowitz and Irene A. Stegun. Vast compendium: 29 sets of tables, some to as high as 20 places. 1,046pp. 8 × 10½.   61272-4 Pa. $21.95

MATHEMATICAL METHODS IN PHYSICS AND ENGINEERING, John W. Dettman. Algebraically based approach to vectors, mapping, diffraction, other topics in applied math. Also generalized functions, analytic function theory, more. Exercises. 448pp. 5⅜ × 8¼.   65649-7 Pa. $8.95

A SURVEY OF NUMERICAL MATHEMATICS, David M. Young and Robert Todd Gregory. Broad self-contained coverage of computer-oriented numerical algorithms for solving various types of mathematical problems in linear algebra, ordinary and partial, differential equations, much more. Exercises. Total of 1,248pp. 5⅜ × 8½. Two volumes.   Vol. I 65691-8 Pa. $13.95
                                                      Vol. II 65692-6 Pa. $13.95

TENSOR ANALYSIS FOR PHYSICISTS, J.A. Schouten. Concise exposition of the mathematical basis of tensor analysis, integrated with well-chosen physical examples of the theory. Exercises. Index. Bibliography. 289pp. 5⅜ × 8½.
                                                      65582-2 Pa. $7.95

INTRODUCTION TO NUMERICAL ANALYSIS (2nd Edition), F.B. Hildebrand. Classic, fundamental treatment covers computation, approximation, interpolation, numerical differentiation and integration, other topics. 150 new problems. 669pp. 5⅜ × 8½.   65363-3 Pa. $13.95

INVESTIGATIONS ON THE THEORY OF THE BROWNIAN MOVEMENT, Albert Einstein. Five papers (1905–8) investigating dynamics of Brownian motion and evolving elementary theory. Notes by R. Fürth. 122pp. 5⅜ × 8½.
                                                      60304-0 Pa. $3.95

NUMERICAL METHODS FOR SCIENTISTS AND ENGINEERS, Richard Hamming. Classic text stresses frequency approach in coverage of algorithms, polynomial approximation, Fourier approximation, exponential approximation, other topics. Revised and enlarged 2nd edition. 721pp. 5⅜ × 8½. 65241-6 Pa. $14.95

AN INTRODUCTION TO STATISTICAL THERMODYNAMICS, Terrell L. Hill. Excellent basic text offers wide-ranging coverage of quantum statistical mechanics, systems of interacting molecules, quantum statistics, more. 523pp. 5⅜ × 8½.   65242-4 Pa. $10.95

ELEMENTARY DIFFERENTIAL EQUATIONS, William Ted Martin and Eric Reissner. Exceptionally, clear comprehensive introduction at undergraduate level. Nature and origin of differential equations, differential equations of first, second and higher orders. Picard's Theorem, much more. Problems with solutions. 331pp. 5⅜ × 8½.   65024-3 Pa. $8.95

STATISTICAL PHYSICS, Gregory H. Wannier. Classic text combines thermodynamics, statistical mechanics and kinetic theory in one unified presentation of thermal physics. Problems with solutions. Bibliography. 532pp. 5⅜ × 8½.
                                                      65401-X Pa. $10.95

ORDINARY DIFFERENTIAL EQUATIONS, Morris Tenenbaum and Harry Pollard. Exhaustive survey of ordinary differential equations for undergraduates in mathematics, engineering, science. Thorough analysis of theorems. Diagrams. Bibliography. Index. 818pp. 5⅜ × 8½. 64940-7 Pa. $15.95

STATISTICAL MECHANICS: Principles and Applications, Terrell L. Hill. Standard text covers fundamentals of statistical mechanics, applications to fluctuation theory, imperfect gases, distribution functions, more. 448pp. 5⅜ × 8½. 65390-0 Pa. $9.95

ORDINARY DIFFERENTIAL EQUATIONS AND STABILITY THEORY: An Introduction, David A. Sánchez. Brief, modern treatment. Linear equation, stability theory for autonomous and nonautonomous systems, etc. 164pp. 5⅜ × 8¼. 63828-6 Pa. $4.95

THIRTY YEARS THAT SHOOK PHYSICS: The Story of Quantum Theory, George Gamow. Lucid, accessible introduction to influential theory of energy and matter. Careful explanations of Dirac's anti-particles, Bohr's model of the atom, much more. 12 plates. Numerous drawings. 240pp. 5⅜ × 8½. 24895-X Pa. $5.95

ORDINARY DIFFERENTIAL EQUATIONS, I.G. Petrovski. Covers basic concepts, some differential equations and such aspects of the general theory as Euler lines, Arzel's theorem, Peano's existence theorem, Osgood's uniqueness theorem, more. 45 figures. Problems. Bibliography. Index. xi + 232pp. 5⅜ × 8½. 64683-1 Pa. $6.00

GREAT EXPERIMENTS IN PHYSICS: Firsthand Accounts from Galileo to Einstein, edited by Morris H. Shamos. 25 crucial discoveries: Newton's laws of motion, Chadwick's study of the neutron, Hertz on electromagnetic waves, more. Original accounts clearly annotated. 370pp. 5⅜ × 8½. 25346-5 Pa. $8.95

INTRODUCTION TO PARTIAL DIFFERENTIAL EQUATIONS WITH AP- PLICATIONS, E.C. Zachmanoglou and Dale W. Thoe. Essentials of partial differential equations applied to common problems in engineering and the physical sciences. Problems and answers. 416pp. 5⅜ × 8½. 65251-3 Pa. $9.95

BURNHAM'S CELESTIAL HANDBOOK, Robert Burnham, Jr. Thorough guide to the stars beyond our solar system. Exhaustive treatment. Alphabetical by constellation: Andromeda to Cetus in Vol. 1; Chamaeleon to Orion in Vol. 2; and Pavo to Vulpecula in Vol. 3. Hundreds of illustrations. Index in Vol. 3. 2,000pp. 6⅛ × 9¼. 23567-X, 23568-8, 23673-0 Pa., Three-vol. set $38.85

ASYMPTOTIC EXPANSIONS FOR ORDINARY DIFFERENTIAL EQUA- TIONS, Wolfgang Wasow. Outstanding text covers asymptotic power series, Jordan's canonical form, turning point problems, singular perturbations, much more. Problems. 384pp. 5⅜ × 8½. 65456-7 Pa. $8.95

AMATEUR ASTRONOMER'S HANDBOOK, J.B. Sidgwick. Timeless, compre- hensive coverage of telescopes, mirrors, lenses, mountings, telescope drives, micrometers, spectroscopes, more. 189 illustrations. 576pp. 5⅜ × 8¼. 24034-7 Pa. $8.95

SPECIAL FUNCTIONS, N.N. Lebedev. Translated by Richard Silverman. Famous Russian work treating more important special functions, with applications to specific problems of physics and engineering. 38 figures. 308pp. 5⅜ × 8½.
60624-4 Pa. $6.95

OBSERVATIONAL ASTRONOMY FOR AMATEURS, J.B. Sidgwick. Mine of useful data for observation of sun, moon, planets, asteroids, aurorae, meteors, comets, variables, binaries, etc. 39 illustrations 384pp. 5⅜ × 8¼. (Available in U.S. only)
24033-9 Pa. $5.95

INTEGRAL EQUATIONS, F.G. Tricomi. Authoritative, well-written treatment of extremely useful mathematical tool with wide applications. Volterra Equations, Fredholm Equations, much more. Advanced undergraduate to graduate level. Exercises. Bibliography. 238pp. 5⅜ × 8½.
64828-1 Pa. $6.95

CELESTIAL OBJECTS FOR COMMON TELESCOPES, T.W. Webb. Inestimable aid for locating and identifying nearly 4,000 celestial objects. 77 illustrations. 645pp. 5⅜ × 8½.
20917-2, 20918-0 Pa., Two-vol. set $12.00

MODERN NONLINEAR EQUATIONS, Thomas L. Saaty. Emphasizes practical solution of problems; covers seven types of equations. ". . . a welcome contribution to the existing literature. . . ."—*Math Reviews.* 490pp. 5⅜ × 8½. 64232-1 Pa. $9.95

FUNDAMENTALS OF ASTRODYNAMICS, Roger Bate et al. Modern approach developed by U.S. Air Force Academy. Designed as a first course. Problems, exercises. Numerous illustrations. 455pp. 5⅜ × 8½.
60061-0 Pa. $8.95

INTRODUCTION TO LINEAR ALGEBRA AND DIFFERENTIAL EQUATIONS, John W. Dettman. Excellent text covers complex numbers, determinants, orthonormal bases, Laplace transforms, much more. Exercises with solutions. Undergraduate level. 416pp. 5⅜ × 8½.
65191-6 Pa. $8.95

INCOMPRESSIBLE AERODYNAMICS, edited by Bryan Thwaites. Covers theoretical and experimental treatment of the uniform flow of air and viscous fluids past two-dimensional aerofoils and three-dimensional wings; many other topics. 654pp. 5⅜ × 8½.
65465-6 Pa. $14.95

INTRODUCTION TO DIFFERENCE EQUATIONS, Samuel Goldberg. Exceptionally clear exposition of important discipline with applications to sociology, psychology, economics. Many illustrative examples; over 250 problems. 260pp. 5⅜ × 8½.
65084-7 Pa. $6.95

LAMINAR BOUNDARY LAYERS, edited by L. Rosenhead. Engineering classic covers steady boundary layers in two- and three-dimensional flow, unsteady boundary layers, stability, observational techniques, much more. 708pp. 5⅜ × 8½.
65646-2 Pa. $15.95

LECTURES ON CLASSICAL DIFFERENTIAL GEOMETRY, Second Edition, Dirk J. Struik. Excellent brief introduction covers curves, theory of surfaces, fundamental equations, geometry on a surface, conformal mapping, other topics. Problems. 240pp. 5⅜ × 8½.
65609-8 Pa. $6.95

ROTARY-WING AERODYNAMICS, W.Z. Stepniewski. Clear, concise text covers aerodynamic phenomena of the rotor and offers guidelines for helicopter performance evaluation. Originally prepared for NASA. 537 figures. 640pp. 6⅛ × 9¼.
64647-5 Pa. $14.95

DIFFERENTIAL GEOMETRY, Heinrich W. Guggenheimer. Local differential geometry as an application of advanced calculus and linear algebra. Curvature, transformation groups, surfaces, more. Exercises. 62 figures. 378pp. 5⅜ × 8½.
63433-7 Pa. $7.95

INTRODUCTION TO SPACE DYNAMICS, William Tyrrell Thomson. Comprehensive, classic introduction to space-flight engineering for advanced undergraduate and graduate students. Includes vector algebra, kinematics, transformation of coordinates. Bibliography. Index. 352pp. 5⅜ × 8½.   65113-4 Pa. $8.00

A SURVEY OF MINIMAL SURFACES, Robert Osserman. Up-to-date, in-depth discussion of the field for advanced students. Corrected and enlarged edition covers new developments. Includes numerous problems. 192pp. 5⅜ × 8½.
64998-9 Pa. $8.00

ANALYTICAL MECHANICS OF GEARS, Earle Buckingham. Indispensable reference for modern gear manufacture covers conjugate gear-tooth action, geartooth profiles of various gears, many other topics. 263 figures. 102 tables. 546pp. 5⅜ × 8½.   65712-4 Pa. $11.95

SET THEORY AND LOGIC, Robert R. Stoll. Lucid introduction to unified theory of mathematical concepts. Set theory and logic seen as tools for conceptual understanding of real number system. 496pp. 5⅜ × 8¼.   63829-4 Pa. $8.95

A HISTORY OF MECHANICS, René Dugas. Monumental study of mechanical principles from antiquity to quantum mechanics. Contributions of ancient Greeks, Galileo, Leonardo, Kepler, Lagrange, many others. 671pp. 5⅜ × 8½.
65632-2 Pa. $14.95

FAMOUS PROBLEMS OF GEOMETRY AND HOW TO SOLVE THEM, Benjamin Bold. Squaring the circle, trisecting the angle, duplicating the cube: learn their history, why they are impossible to solve, then solve them yourself. 128pp. 5⅜ × 8½.   24297-8 Pa. $3.95

MECHANICAL VIBRATIONS, J.P. Den Hartog. Classic textbook offers lucid explanations and illustrative models, applying theories of vibrations to a variety of practical industrial engineering problems. Numerous figures. 233 problems, solutions. Appendix. Index. Preface. 436pp. 5⅜ × 8½.   64785-4 Pa. $8.95

CURVATURE AND HOMOLOGY, Samuel I. Goldberg. Thorough treatment of specialized branch of differential geometry. Covers Riemannian manifolds, topology of differentiable manifolds, compact Lie groups, other topics. Exercises. 315pp. 5⅜ × 8½.   64314-X Pa. $6.95

HISTORY OF STRENGTH OF MATERIALS, Stephen P. Timoshenko. Excellent historical survey of the strength of materials with many references to the theories of elasticity and structure. 245 figures. 452pp. 5⅜ × 8½.   61187-6 Pa. $9.95

GEOMETRY OF COMPLEX NUMBERS, Hans Schwerdtfeger. Illuminating, widely praised book on analytic geometry of circles, the Moebius transformation, and two-dimensional non-Euclidean geometries. 200pp. 5⅜ × 8¼.
63830-8 Pa. $6.95

MECHANICS, J.P. Den Hartog. A classic introductory text or refresher. Hundreds of applications and design problems illuminate fundamentals of trusses, loaded beams and cables, etc. 334 answered problems. 462pp. 5⅜ × 8½.   60754-2 Pa. $8.95

TOPOLOGY, John G. Hocking and Gail S. Young. Superb one-year course in classical topology. Topological spaces and functions, point-set topology, much more. Examples and problems. Bibliography. Index. 384pp. 5⅜ × 8¼.
65676-4 Pa. $7.95

STRENGTH OF MATERIALS, J.P. Den Hartog. Full, clear treatment of basic material (tension, torsion, bending, etc.) plus advanced material on engineering methods, applications. 350 answered problems. 323pp. 5⅜ × 8½.   60755-0 Pa. $7.50

ELEMENTARY CONCEPTS OF TOPOLOGY, Paul Alexandroff. Elegant, intuitive approach to topology from set-theoretic topology to Betti groups; how concepts of topology are useful in math and physics. 25 figures. 57pp. 5⅜ × 8½.
60747-X Pa. $2.95

ADVANCED STRENGTH OF MATERIALS, J.P. Den Hartog. Superbly written advanced text covers torsion, rotating disks, membrane stresses in shells, much more. Many problems and answers. 388pp. 5⅜ × 8½.   65407-9 Pa. $8.95

COMPUTABILITY AND UNSOLVABILITY, Martin Davis. Classic graduate-level introduction to theory of computability, usually referred to as theory of recurrent functions. New preface and appendix. 288pp. 5⅜ × 8½. 61471-9 Pa. $6.95

GENERAL CHEMISTRY, Linus Pauling. Revised 3rd edition of classic first-year text by Nobel laureate. Atomic and molecular structure, quantum mechanics, statistical mechanics, thermodynamics correlated with descriptive chemistry. Problems. 992pp. 5⅜ × 8½.   65622-5 Pa. $18.95

AN INTRODUCTION TO MATRICES, SETS AND GROUPS FOR SCIENCE STUDENTS, G. Stephenson. Concise, readable text introduces sets, groups, and most importantly, matrices to undergraduate students of physics, chemistry, and engineering. Problems. 164pp. 5⅜ × 8½.   65077-4 Pa. $5.95

THE HISTORICAL BACKGROUND OF CHEMISTRY, Henry M. Leicester. Evolution of ideas, not individual biography. Concentrates on formulation of a coherent set of chemical laws. 260pp. 5⅜ × 8½.   61053-5 Pa. $6.00

THE PHILOSOPHY OF MATHEMATICS: An Introductory Essay, Stephan Körner. Surveys the views of Plato, Aristotle, Leibniz & Kant concerning propositions and theories of applied and pure mathematics. Introduction. Two appendices. Index. 198pp. 5⅜ × 8½.   25048-2 Pa. $5.95

THE DEVELOPMENT OF MODERN CHEMISTRY, Aaron J. Ihde. Authoritative history of chemistry from ancient Greek theory to 20th-century innovation. Covers major chemists and their discoveries. 209 illustrations. 14 tables. Bibliographies. Indices. Appendices. 851pp. 5⅜ × 8½.   64235-6 Pa. $15.95

THE FOUR-COLOR PROBLEM: Assaults and Conquest, Thomas L. Saaty and Paul G. Kainen. Engrossing, comprehensive account of the century-old combinatorial topological problem, its history and solution. Bibliographies. Index. 110 figures. 228pp. 5⅜ × 8½. 65092-8 Pa. $6.00

CATALYSIS IN CHEMISTRY AND ENZYMOLOGY, William P. Jencks. Exceptionally clear coverage of mechanisms for catalysis, forces in aqueous solution, carbonyl- and acyl-group reactions, practical kinetics, more. 864pp. 5⅜ × 8½. 65460-5 Pa. $18.95

PROBABILITY: An Introduction, Samuel Goldberg. Excellent basic text covers set theory, probability theory for finite sample spaces, binomial theorem, much more. 360 problems. Bibliographies. 322pp. 5⅜ × 8½. 65252-1 Pa. $7.95

LIGHTNING, Martin A. Uman. Revised, updated edition of classic work on the physics of lightning. Phenomena, terminology, measurement, photography, spectroscopy, thunder, more. Reviews recent research. Bibliography. Indices. 320pp. 5⅜ × 8¼. 64575-4 Pa. $7.95

PROBABILITY THEORY: A Concise Course, Y.A. Rozanov. Highly readable, self-contained introduction covers combination of events, dependent events, Bernoulli trials, etc. Translation by Richard Silverman. 148pp. 5⅜ × 8¼. 63544-9 Pa. $4.50

THE CEASELESS WIND: An Introduction to the Theory of Atmospheric Motion, John A. Dutton. Acclaimed text integrates disciplines of mathematics and physics for full understanding of dynamics of atmospheric motion. Over 400 problems. Index. 97 illustrations. 640pp. 6 × 9. 65096-0 Pa. $16.95

STATISTICS MANUAL, Edwin L. Crow, et al. Comprehensive, practical collection of classical and modern methods prepared by U.S. Naval Ordnance Test Station. Stress on use. Basics of statistics assumed. 288pp. 5⅜ × 8½. 60599-X Pa. $6.00

WIND WAVES: Their Generation and Propagation on the Ocean Surface, Blair Kinsman. Classic of oceanography offers detailed discussion of stochastic processes and power spectral analysis that revolutionized ocean wave theory. Rigorous, lucid. 676pp. 5⅜ × 8½. 64652-1 Pa. $14.95

STATISTICAL METHOD FROM THE VIEWPOINT OF QUALITY CONTROL, Walter A. Shewhart. Important text explains regulation of variables, uses of statistical control to achieve quality control in industry, agriculture, other areas. 192pp. 5⅜ × 8½. 65232-7 Pa. $6.00

THE INTERPRETATION OF GEOLOGICAL PHASE DIAGRAMS, Ernest G. Ehlers. Clear, concise text emphasizes diagrams of systems under fluid or containing pressure; also coverage of complex binary systems, hydrothermal melting, more. 288pp. 6½ × 9¼. 65389-7 Pa. $8.95

STATISTICAL ADJUSTMENT OF DATA, W. Edwards Deming. Introduction to basic concepts of statistics, curve fitting, least squares solution, conditions without parameter, conditions containing parameters. 26 exercises worked out. 271pp. 5⅜ × 8½. 64685-8 Pa. $7.95

CATALOG OF DOVER BOOKS

CHALLENGING MATHEMATICAL PROBLEMS WITH ELEMENTARY SOLUTIONS, A.M. Yaglom and I.M. Yaglom. Over 170 challenging problems on probability theory, combinatorial analysis, points and lines, topology, convex polygons, many other topics. Solutions. Total of 445pp. 5⅜ × 8½. Two-vol. set.
Vol. I 65536-9 Pa. $5.95
Vol. II 65537-7 Pa. $5.95

FIFTY CHALLENGING PROBLEMS IN PROBABILITY WITH SOLUTIONS, Frederick Mosteller. Remarkable puzzlers, graded in difficulty, illustrate elementary and advanced aspects of probability. Detailed solutions. 88pp. 5⅜ × 8½.
65355-2 Pa. $3.95

EXPERIMENTS IN TOPOLOGY, Stephen Barr. Classic, lively explanation of one of the byways of mathematics. Klein bottles, Moebius strips, projective planes, map coloring, problem of the Koenigsberg bridges, much more, described with clarity and wit. 43 figures. 210pp. 5⅜ × 8½. 25933-1 Pa. $4.95

RELATIVITY IN ILLUSTRATIONS, Jacob T. Schwartz. Clear non-technical treatment makes relativity more accessible than ever before. Over 60 drawings illustrate concepts more clearly than text alone. Only high school geometry needed. Bibliography. 128pp. 6⅛ × 9¼. 25965-X Pa. $5.95

AN INTRODUCTION TO ORDINARY DIFFERENTIAL EQUATIONS, Earl A. Coddington. A thorough and systematic first course in elementary differential equations for undergraduates in mathematics and science, with many exercises and problems (with answers). Index. 304pp. 5⅜ × 8¼. 65942-9 Pa. $7.95

FOURIER SERIES AND ORTHOGONAL FUNCTIONS, Harry F. Davis. An incisive text combining theory and practical example to introduce Fourier series, orthogonal functions and applications of the Fourier method to boundary-value problems. 570 exercises. Answers and notes. 416pp. 5⅜ × 8½. 65973-9 Pa. $8.95

THE THOERY OF BRANCHING PROCESSES, Theodore E. Harris. First systematic, comprehensive treatment of branching (i.e. multiplicative) processes and their applications. Galton-Watson model, Markov branching processes, electron-photon cascade, many other topics. Rigorous proofs. Bibliography. 240pp. 5⅜ × 8½. 65952-6 Pa. $6.95

AN INTRODUCTION TO ALGEBRAIC STRUCTURES, Joseph Landin. Superb self-contained text covers "abstract algebra": sets and numbers, theory of groups, theory of rings, much more. Numerous well-chosen examples, exercises. 247pp. 5⅜ × 8½. 65940-2 Pa. $6.95

GAMES AND DECISIONS: Introduction and Critical Survey, R. Duncan Luce and Howard Raiffa. Superb non-technical introduction to game theory, primarily applied to social sciences. Utility theory, zero-sum games, n-person games, decision-making, much more. Bibliography. 509pp. 5⅜ × 8½. 65943-7 Pa. $10.95

Prices subject to change without notice.

Available at your book dealer or write for free Mathematics and Science Catalog to Dept. GI, Dover Publications, Inc., 31 East 2nd St., Mineola, N.Y. 11501. Dover publishes more than 175 books each year on science, elementary and advanced mathematics, biology, music, art, literary history, social sciences and other areas.